A COURSE IN ELECTRICAL ENGINEERING • VOLUME 2 • CHESTER LAURENS DAWES

Publisher's Note

The book descriptions we ask book-sellers to display prominently warn that this is an historic book with numerous typos or missing text; it is not indexed or illustrated.

The book was created using optical character recognition software. The software is 99 percent accurate if the book is in good condition. However, we do understand that even one percent can be an annoying number of typos! And sometimes all or part of a page may be missing from our copy of the book. Or the paper may be so discolored from age that it is difficult to read. We apologize and gratefully acknowledge Google's assistance.

After we re-typeset and design a book, the page numbers change so the old index and table of contents no longer work. Therefore, we often re-move them; otherwise, please ignore them.

Our books sell so few copies that you would have to pay hundreds of dollars to cover the cost of our proof reading and fixing the typos, missing text and index. Instead we let most customers download a free copy of the original ty-po-free scanned book. Simply enter the barcode number from the back cover of the paperback in the Free Book form at www.RareBooksClub.com. You may also qualify for a free trial membership in our book club to download up to four books for free. Simply enter the barcode number from the back cover onto the membership form on our home page. The book club entitles you to select from more than a million books at no additional charge. Simply enter the title or subject onto the search form to find the books.

If you have any questions, could you please be so kind as to consult our Fre-quently Asked Questions page at www. RareBooksClub.com/faqs.cfm? You are also welcome to contact us there. General Books LLC™, Memphis, USA, 2012.

⊰⊱ ⊰⊱ ⊰⊱ ⊰⊱ ⊰⊱ ⊰⊱ ⊰⊱ ⊰⊱

A COURSE IN ELECTRICAL ENGINEERING VOLUME II ALTERNATING CURRENTS ELECTRICAL ENGINEERING TEXTS

A Series Of Textbooks Outlined By The *Following Committee.*

Harry E. Clifford, *Chairman and Consulting Editor,* Gordon McKay Professor of Electrical Engineering, Harvard University.

Murray C. Bbebe, Formerly Professor of Electrical Engineering, University of Wisconsin.

Ernst J. Berg, Professor of Electrical Engineering, Union College.

Paul M. Lincoln, Consulting Engineer, Professor of Electrical Engineering, University of Pittsburgh.

Henry H. Norris, Associate Editor, *Electric Railway Journal,* Formerly Professor of Electrical Engineering, Cornell University.

George W. Patterson, Professor of Electrical Engineering, University of Michigan.

Harris J. Ryan, Professor of Electrical Engineering, Leland Stanford Junior University.

Elihu Thomson, Consulting Engineer, General Electric Co.

A COURSE IN ELECTRICAL ENGINEERING VOLUME II ALTERNATING CURRENTS CHESTER L? GAWES, S. B.

Assistant Professor Of Electrical Engineering, The Harvard Engineering School; Member, American Institute Of Electrical Engineers, Etc. First Edition .third Impression

McGRAW-HILL BOOK COMPANY, Inc. NEW YORK: 370 SEVENTH AVENUE LONDON: 6 & 8 BOUVERIE ST., E. C. 4 Copyright, 1922, By The Mcgraw-hill Book Company, Inc. PHINTED IN THE UNITED STATES OP AMERICA THE MAPLE PRESS-YORK PA PREFACE

This volume is intended for those who have such a knowledge of direct currents as is given by Volume I. It pre-supposes no knowledge of alternating currents. The first two chapters arc de-voted to the development of the funda-mental laws of alternating currents and alternating-current circuits. Subsequent chapters consider the application of these fundamental laws to alternating-current measurements, to polyphase cir-cuits, to alternating-current machinery, and to power transmission. A chapter on illumination and photometry has been included, as a brief discussion of the un-derlying principles of light and of light measurements is important in a general course in electrical engineering.

The development of the various alter-nating-current formulas and of the op-eration of various types of machinery, transmission lines, etc., are based on the fundamental laws of electricity and magnetism as set forth in Volume I. Mathematical developments are occa-sionally introduced, as supplementary to the descriptive matter. As in Volume I, numerous illustrative problems and methods of making laboratory tests are given throughout the text.

This volume is intended to be ele-mentary in character and to act as a stepping stone to the more advanced texts of this series. In many cases rig-orous and detailed analysis is not given, particularly in the chapter on alternat-ing-current measurements and in the discussion of certain types of alternat-ingcurrent apparatus. A thorough analy-sis of these subjects is found in "Electrical Measurements" by F. A. Laws, and "Principles of Alternating Current Machinery" by R. R. Lawrence, both of which volumes-are included in this series of Electrical Engineering Texts.

The author is indebted to various manufacturing companies for their co-operation in supplying material and illustrations for the text; to Professor R. R. Lawrence of the Massachusetts Institute of Technology for his careful review of the manuscript and his many helpful suggestions given during its preparation; and particularly to Professor H. E. Clifford of The Harvard Engineering School, for his helpful advice during the preparation of the manuscript and for the thorough manner in which he has edited the material contained in this volume.

C. L. D.

Harvard University, Cambridge, Mass.

Jan., 1922.

CONTENTS

A COURSE IN ELECTRICAL ENGINEERING VOLUME II

ALTERNATING CURRENTS

CHAPTER I

ALTERNATING CURRENT AND VOLTAGE

1. General Field of Use of Alternating Current.—Over 90 per cent. of the electrical energy generated at the present time is generated as alternating current. This is not due primarily to any superiority of alternating over direct current in its applicability to industrial and domestic uses. In fact, there are many instances where direct current is absolutely necessary for industrial purposes, but even in these cases the energy is often generated as alternating current.

Some of the reasons for generating energy as alternating current are:

Alternating current can be generated at comparatively high voltages and these voltages can be readily raised and lowered by means of static transformers. This permits the economical transmission of alternating current over considerable distances by using high transmission voltages. These high voltages can be reduced efficiently at the receiving end of the transmission line. Direct-current voltages cannot be raised and lowered on an industrial scale without the use of rotating commutators, and the permissible voltage per commutator is low. Therefore, the voltage of direct-current circuits cannot be changed economically.

Alternating-current generators can be built in large units running at high speeds, are suited to turbine drive, and the cost per kilowatt of such alternators is low. The largest single unit today (1920) has a rating of about 50,000 kv-a. Owing to commutation difficulties, direct-current generators cannot be built in large units, particularly for high speeds. At 1,000 r. p.m., it is difficult to build a direct-current generator having a rating of even 1,000 kw. On the other hand, 5,000-kw. alternators, operating at speeds of 3,600 r.p.m., are not uncommon.

For constant-speed work, the alternating-current induction motor is cheaper in first cost and in maintenance than the directcurrent motor. This is due to the fact that the induction motor has no commutator. Therefore, it is occasionally desirable to generate power as alternating current in order to be able to use induction motors.

The high transmission efficiencies obtainable with alternating current make it possible to generate electrical energy in large quantities in a single station and to distribute it over a comparatively large territory. The large boilers, automatic stokers, superheaters, recording instruments, etc., which are possible in large stations, result in high boiler-room efficiency. Large turbines have an economy which may be three or four times as good as that of the steam units in a small plant. The generator has an efficiency of 95 to 96 per cent. in the larger sizes. Then again, as the boilers and large turbo-units require few attendants per kilowatt, the labor and superintendence charges per kilowatt are small.

For these reasons it is often more economical to generate power with large units, to transmit it long distances and even to convert it into direct current, than to generate the direct current at the place where it is to be utilized.

It must be remembered, however, that the reduced generating costs may be balanced by the distribution costs resulting from high investment charges in lines, cables, sub-stations, machinery, etc., in addition to the labor and maintenance costs of this distribution system.

Alternating current owes its importance to the fact that it can be generated economically with large units. Its voltage can be readily raised and lowered, so that energy can be transmitted economically for considerable distances. Alternating-current motors for constant-speed work are usually preferable to directcurrent motors.

2. Sine Waves.—It was shown in Vol. I, Chap. X, that when a single coil rotates at constant speed in a uniform field, Fig. 1, in alternating emf. is generated. This emf. is zero when the plane of the coil is perpendicular to the field, and reaches its maximum value when the plane of the coil is parallel to the field. (See Vol. I. page 219, Fig. 188.) The successive values of the emf. may be represented by a smooth curve called a *sine wave* since the values of the emf. are proportional to the sine of the angle x which the coil makes with the vertical, Fig. 1. This will be discussed more in detail later.

For various reasons, commercial generators do not give exact sine waves of emf. In fact, the emfs. of some generators differ materially from a sine wave.

Still, most commercial generators have waves of emf. which are sufficiently close to a sine wave to warrant their being treated as such.

If a wave is not a sine wave, it may be resolved into a series of sine waves of fundamental and higher frequencies. Each one of these components may then be dealt with as a sine wave.

The waves ordinarily encountered in practice are approximately sine waves and may be treated by simple methods of analysis. The formulas and equations which follow apply to sine waves of current and voltage, unless otherwise specified.

The sine wave may be produced graphically as follows: Draw a circle, Fig. 2, whose radius A is equal to the maximum value of

Fio. 2.—Graphical construction of a sine wave.

the sine wave. Divide the circumference of this circle into any number of equal parts, in this case 12, and number them 0, 1, 2... 12. Also draw a horizontal line *ab*, which, if extended, would pass through the center of the circle. Divide

Fig. 3.—Numerical values of the ordinates of sine waves for definite angles.

ab into the same number of equal parts as there are on the circumference of the circle, and give the points corresponding numbers. Erect an ordinate or perpendicular at each point. Project the points on the circle horizontally until they meet perpendiculars having corresponding numbers. A smooth curve drawn through the intersections will be a sine wave.

The sine wave may also be plotted from a table of sines. (Appendix, page 458.) Mark a horizontal axis, Fig. 3 (a), in degrees. At each point erect an ordinate equal to the sine of the corresponding angle. Thus at 30 the ordinate *ab* is 0. 5; at 60 the ordinate *cd* is 0.866; at 90 it is 1.0; etc. The wave passes through zero at 180, because the sine of 180 is zero. When the angle becomes greater than 180, the sine becomes negative and the wave falls below the line, as the sine is negative between 180 and 360. (See page 456.) The above is equivalent to plotting the sines of the angle x, Fig. 1,

x being the angle which the plane of the rotating coil makes with the vertical at any instant.

If the wave in question has a maximum value of B, Fig. 3 (6), instead of unity, the value of the ordinate at any point may be found by multiplying B into the sine of the corresponding angle. That is $y = B \sin x$ (1)

Where x is in degrees.

Example.—Find the ordinates of a sine wave at points corresponding to 65 and 210, the maximum ordinate being 40 units, Fig 3 (fc). From page 459 sin 65 = 0.906. 40 X 0.906 = 36.24. *Ans.* Sin 210 =-(sin 210-180) =-sin 30 =-0.5. (Page 456.) 40 X (-0.5) =-20. *Ans.* These values are shown in Fig. 3 (6). 3. Cycle; Frequency.—When the coil, Fig. 1, has completed one revolution, it has passed one pair of poles (a north and a south pole) and it has traversed 360 space-degrees. The voltage wave has gone through one complete *cycle* of values and the wave is now ready to 3co i repeat itself. This is illustrated Fl' "'—Alternation and cycle. in Fig. 4. Having gone through one complete cycle, the voltage has gone through 360 electrical time-degrees. Therefore, in a two-pole machine *one space-degree is the same as one electrical time-degree.*

When the voltage has completed only half a cycle or 180, it has gone through one *alternation.*

Assume that the coil in Fig. 1 is making 60 revolutions per second, or 3,600 r.p.m. Sixty complete cycles will be generated each second. Therefore, a two-pole, 60-cycle generator must have a speed of 3,600 r.p.m.

The abscissas may be graduated in *time* as well as in degrees. For example, Fig. 5, the time corresponding to 360 is o sec-and the tune corresponding to 180 is K20 sec.

Alternating-current waves may be plotted with either *time* or *degrees* as abscissas.

Fio. 5.—Sine wave as a function of time.

Fio. 6.—Two cycles per revolution in 4-pole alternator.

Figure 6 (a) shows a four-pole machine. A single conductor a of a coil is shown rotating, rather than the complete coil. As soon as this conductor has passed a north and a south pole, that is, after it has passed from 1 to 5, it has completed one electrical cycle or 360 electrical time-degrees, as is shown in Fig. 6 *(b)*. Mechanically, it has completed one-half a revolution, or 180 spacedegrees, so that in one revolution, or 360 space-degrees, the emf. in the conductor will have completed two cycles, and will have gone through 720 electrical time-degrees. Therefore, in this case one space-degree corresponds to two electrical time-degrees. That is, for every space-degree that the coil passes through, the voltage wave completes two electrical time-degrees. This conductor needs to make only 30 r.p.s. or 1,800 r.p.m. in order to generate a 60-cycle electromotive force. Likewise for a 25-cycle electromotive force, this conductor needs to revolve at only 12.5 r. p.s. or 750 r.p.m. For a given frequency, as the number of poles increases, the mechanical speed decreases proportionately. The relation between speed, poles and frequency may be written in the form of an equation: *PXS PXS _-, J a s en i on / 120* where / is the frequency in cycles per second, *P* is the number of poles, and S is the speed in revolutions per minute.

The table shows the relation of speed, frequency and number of poles for a few typical cases.

Example.—A 60-cycle, engine-driven alternator has a speed of 120 r.p.m. How many poles has it?

Using equation (2) and solving for *P* 120/ 120 X 60 S" 120 = 60 poles. *Ans.*

In practice, nearly all alternators have stationary armatures and rotating fields, and the above equations apply.

4. Commercial Frequencies.—In this country, frequencies are standardized at 60 cycles and at 25 cycles per second, although other frequencies are used. In California and in Mexico, for example, 50 cycles is used on some of the large transmission systems. In the early days of alternating-current development, 133 cycles was common, but few, if any, plants now generate at this frequency. The principal advantage of higher fre-

quencies is that transformers require less iron and copper, and so are lighter and cheaper. The flicker of lamps is not perceptible at 60 cycles, but at 25 cycles it is very evident. On the other hand, the voltage drop in transmission lines and in apparatus varies almost directly as the frequency, so that better voltage regulation throughout the system is obtained with low frequency. Power apparatus, such as induction motors, synchronous converters, alternating-current commutator motors, etc., operates better at low than at high frequencies. However, with one or two exceptions, the operation is satisfactory at 60 cycles per second. A power and lighting company would ordinarily operate at 60 cycles per second, because the flicker of lamps at 25 cycles per second is objectionable and the transformers at this lower frequency are heavier and more costly than they are at the higher frequency. On the other hand, an electric company generating strictly for power purposes would ordinarily use 25 cycles. This frequency is used by the New York, New Haven and Hartford R. R. for its electric locomotives; on the Norfolk and Western Ry. for operating electric locomotives; and by the Boston Elevated Ry. Co. for transmitting high-voltage power to its direct-current sub-stations. In Europe, frequencies as low as 15 and even 12.5 cycles per second are common. 6. The Alternating-current Ampere.—Figure 7 (a) shows an alternating-current sine wave, having a maximum value of 1. 414 amp. At first thought it might seem that the value in amperes of such a wave should be based on the *average* value. If the wave over one complete cycle is considered, the average value is zero, as there is just as much negative as positive current. A direct-current ammeter, if connected to measure this current, would indicate zero, as such an instrument reads *average* values.

The value of an alternating current is not based on its average value but on its *heating* effect and may be defined as follows: *An alternating-current ampere is that current which, flowing through a given ohmic resistance, will produce heat at the same rate as a direct-current*

ampere.

Assume that a resistance unit is immersed in a calorimeter and that when a direct-current ampere is sent through this resistance the temperature of the water is raised 20 in 10 min. An alternating-current ampere, if sent through this same resistance unit, will raise the temperature of the water by the same amount in the same time, other conditions such as radiation, etc., being the same. That is, both currents produce heat at the same rate.

Flo. 7.— Maximum and effective values of sine-wave alternating current.

The heating effect varies as the *square* of the current ($= i^2R$). Therefore, the value in amperes of the wave of current in Fig. 7 (a) must be based upon its *squared* values. Figure 7 (6) shows the current wave of Fig. 7 (a) plotted, together with its squared values. That is, each ordinate of the "i" wave is squared and these values plotted to give the i2 wave shown. The maximum value of this new wave will be 2.0 ($= 1.414)2$ since the maximum value of the original current wave is 1.414 or -/2. The squared wave also lies entirely above the zero axis, because the square of 'a negative value is positive.

This squared wave has a frequency twice that of the original wave and has its horizontal axis of symmetry at a distance of 1.0 unit above the zero axis, as shown in Fig. 7 (6). The average value of this squared wave is 1.0 amp., as shown by the dotted line, because the areas above the dotted line will just fit into the shaded valleys below the dotted line. Therefore, if an equivalent rectangle were made from this wave, its height would be 1.0 unit. This value, 1. 0, is the *average of the squares* of the current wave. Average heating varies as the average of the squares of the current, so this procedure for determining the ampere value of the wave of Fig. 7 (a) is correct.

To obtain the correct value of the current in amperes, the square root of the average square must be taken. That is, I (in amperes) = /1.0 = 1.0 amp. This value of the current is called the *root-mean-square* (r.m.s.) or *effective* value of the current.

Therefore, an alternating-current ampere, sine wave, which produces heat at the same rate as a direct-current ampere, has a *maximum* value of 1.414 (= /2) amp. In fact, for any sinewave current, the ratio of the *maximum* to the *effective* value is equal to the /2 or 1.414. The ratio of effective to maximum value is 1/1.414 = 0.707.

To obtain the *effective* value of *any* current wave, not necessarily a sine wave: (a) Plot a wave whose ordinates are equal to the squares of the ordinates of the given current wave. (6) Find the average value of this squared wave by obtaining the area of its loops with a planimeter and dividing this area by the base; or by averaging the ordinates. (c) Find the square root of this average.

The same result may be obtained by erecting equidistant ordinates on the original wave, averaging their squares and taking the square root of this average. This will give the *rootmean-square* value.

If a sine wave of current be *averaged* in the ordinary manner for *half a cycle,* it will be found that this average is equal to *2/w* or 0.637 times the maximum value. The ratio of *effective* to *average* value is then 0.707/0.637 = 1.11 and the ratio of average to effective value is 0.9. It is sometimes necessary to know the average value, and the ratio of effective to average value enters into computations of induced emfs. in alternators, transformers and other types of alternating-current machinery.

The ratio of effective to average value is called the *form factor* of the wave. The form factor of a sine wave is 1.11. The maximum, effective, and average values for a sine wave of voltage, whose r.m.s. value is 100 volts, are shown in Fig. 8.

Fio. 8.—Relation of maximum, effective, and average values of a sine wave.

6. Equation of Sine Wave of Current. —If *ut* is substituted for *x* in equation (1), Par. 2, the equation of a sine wave of alternating-current is given by $i = I$ *max sin ut* (3) where i is the value of the current at any time, *t, Imax* is the maximum value of the current, and co = 2ir/. The term *u* is equal to 2ir times the frequency /, and is the *angular velocity* in radians per second of the rotating vector which may be used to construct the wave. (Appendix, page 453.) Similarly, the equation of a sine wave of electromotive force will be given by e = *Emaz sin ut.* (4) *Example.*—What is the equation of a 25-cycle current, sine wave, having an effective value of 30 amp., and what is the value, i', of the current when the time is 0.005 sec.? The wave crosses the time axis in a positive direction when the time is equal to zero.

$I\ ma$ = 30/2 = 42.4 amp.

225 = 157 = a

i = 42.4 sin 157 t. Ana.

$i,$ = 42.4 sin 157 X 0.005 =

42.4 sin 0.785 radians

2ir = 6.28 radians = 360 (page 453)

0 785 eT28 X 360 " 45" the wave comPletes 360 in Ms or 0.04 sec. in 0. 005 sec. it will have completed fTnVx = H cycle. 360/8 = 45 i' = 42.4 sin 45 = 42.4 X 0.707 = 30 amp. *Ann.* 7. Scalars and Vectors.— Quantities in general are divided into two classes, scalars and vectors.

A scalar is a quantity which is completely determined by its magnitude alone. Examples of scalar quantities are dollars, energy, gallons, mass, temperature, etc. Such quantities are added algebraically. For example, two dollars plus five dollars equals seven dollars.

A vector has direction as well as magnitude. A common example of a vector is force. When a force is under consideration, not only its magnitude but its direction as well must be considered. When two or more forces are added, they are not necessarily added algebraically but must be combined in such a way as to take into consideration their directions as well as their magnitudes.

Figure 9 (a) shows two forces acting at the point *0* and represented by the vectors *Fl* and *F2.* The length of each of these vectors, to scale, is equal to the *magnitude* of the force which it represents. The direction of each of these vectors shows the *direction* in which the force acts. *ft* is the angle between Fi and Fj. Their sum, *F0,* or the single force which would have the same effect on their point of application, *0,* as *Fi* and F2 acting in conjunction, is called their *resultant.* Fo is one diagonal of the parallelogram having Fi and F2 as adjacent sides.

Figure 9 (6) shows a triangle having *F* i and *Fz* as two of its sides, Fi and *F2* being respectively parallel to, and acting in the same directions as, Fi and *F2* of Fig. 9 (a). The exterior angle between *F* and F2 is therefore equal to /3. The third side of the triangle *F0* is equal in magnitude and direction to *F0* of Fig. 9 (a). Therefore, the resultant of two vectors may be found by means of a triangle properly constructed, of which two sides are the two component vectors and the third side is their sum. Such a triangle is called a *triangle of forces.* It is usually simpler to use the triangle of forces than to use the parallelogram of forces.

To subtract one vector from another, reverse this vector and add it vectorially to the second vector. For example, in Fig. 9 (c) it is desired to subtract *F2* from *F. F2* is reversed giving — *F2. F'0,* the vector sum of *F* and —2, found by completingthe parallelogram, is equal to *Fi — F2.* Vectors may be subtracted by the triangle method as shown in Fig. 9 (d). The vector *F,0,* connecting the ends of the two vectors *Fi* and F2 whose difference is desired, is their vector *difference.* (a) Sum of two vectors by parallelogram (*I)* Sum of two vectors by method triangle method

If a parallelogram, Fig. 9 (e), having vectors *Fi* and *F2* as adjacent sides, be completed, one diagonal *F0* of the parallelogram is the vector *sum* of *Fi* and *F2.* The other diagonal *F'0* of the parallelogram, is the vector *difference* of Fi and *F2.*

A vector is often indicated by placing a dot under its symbol. For example, in Figs. 9 (a) and 9 (6) 'f 0 = f i + *F* 2 shows that F0 is the *vector* sum of *Fi* and F2 and not their algebraic sum.

When more than two vectors are added, the resultant of two is first found and this resultant is combined with a third vector, etc. This is illustrated in Fig. 10, in which three vectors F_i, F_z and F_3 are added.

F_i and F_z are first combined and the resultant F' is found. F' is then combined with F_3, giving F0 as the sum of all three vectors, F_i, F_2 and $F?$. That is,

F o = f, + F_z + F_3

Fio. 10.—Sum of three vectors.

F' is an intermediate vector and therefore does not appear in the ultimate result. 8. Ohm; Volt.—If a resistance of one ohm, as measured with direct current, has no inductance and is so designed that alternating current in flowing through it does not produce any secondary effects, such as eddy currents or skin effect, it offers a resistance of one *ohm* to alternating current.

When an alternating-current ampere flows through such a resistance, the drop across its terminals is equal to one alternatingcurrent *volt*. Hence, the relation between *maximum* and *effective* volts is the same as the relation between *maximum* and *effective* amperes. For a sine wave, the maximum voltage is /2, or 1.414, times the effective voltage.

9. Phase Relations.—The current and voltage in the ordinary alternating-current system have the same fundamental frequency under normal operating conditions, although they do not necessarily pass through their corresponding zero values at the same instant. Figure 11 (a) shows two sine-wave currents, one having an effective value of 8 and the other of 12 amp. Their respective maximum values are accordingly 8/2 or 11. 3 amp. and 12/2 or 17.0 amp. Both currents pass through zero, increasing positively, at the same instant and are therefore said to be *in phase* with each other. Figure 11 (6) shows two sine-wave currents of 8 and 12 amp. respectively, but not passing through zero at the same instant. The 8-amp. current passes through zero, increasing positively, later than does the 12-amp. current. It must be remembered that time is increasing from left to right. If the 12-amp. current is passing through its zero value at 2. 00 o,clock, the 8-amp. current is passing through its corresponding zero value

some time later, for any value of time to the right of 2.00 is later than 2.00 o,clock. Therefore, the 8-amp. current *lags* the 12-amp. current.

Fig. 11.—Phase relations of alternating currents.

The time of lag shown in Fig. 11 (6) corresponds to 60 and is represented by the angle 6. Therefore, the 8-amp. current lags the 12-amp. current by an angle 6 or by 60. Or the 12-amp. current may be said to *lead* the 8-amp. current by an angle 6 or by 60.

In Fig. 11 (a) the two currents are *in phase* with each other. In Fig. 11 (6) the two currents have a *phase difference* of 60.

These phase differences may exist between currents and voltages, between two or more voltages, or between two or more currents.

9A. Addition of Currents.—Figure 12 shows two currents, having effective values of 8 and 12 amp. respectively, uniting to flow in a common wire. If these two currents were direct currents, then by Kirchhoff's first law (see Vol. I, page 77), the current /3 could have only two possible numerical values, 12 + 8 = 20 amp. if the two currents flow in the same direction /2= 8 Amp /

Fio. 12.—Alternating currents meeting 'at a junction.

and 12 — 8 = 4 amp. if they flow in opposite directions.

If the two currents, Fig. 12, are alternating, their sum /3 may be equal numerically to *any* value *from 20 amp. to 4 amp.*, depending on the phase relation existing between /i and /2.

Fio. 13.—Addition of two currents in phase.

Figure 13 shows these two currents plotted *in phase* with each other. Their sum /3 is found by adding their ordinates at each instant. The resulting current obtained in this manner will be a sine wave and will have a *maximum* value of 28.3 amp. corresponding to an effective value of 28.3/V2 = 20 amp. That is, when two currents are in phase their sum is found arithmetically. Figure 16 corresponds to the condition of Fig. 11 (6), where the two currents differ in phase by 60. Their sum is found in the

same manner as in Fig. 13 by adding the two, point by point, and obtaining the resulting current /3. The resultant /3 will not have a maximum value of 28. 3 amp., as it did when the currents were in phase, but its maximum value will be less, actually being 24.7 amp. This corresponds to an effective value of 17.45

Fio. 14.—Instantaneous values of current from a rotating vector.

amp. for the sum of the two, rather than of 20 amp. as before. *Therefore, the sum of any number of alternating currents depends upon their phase relations as well as upon their magnitudes.*

If voltages rather than currents be added, it will be found that their sum depends upon their phase relations as well as upon their magnitudes.

10. Vector Representation of Alternating Quantities.—It was shown in Fig. 2 that a sine wave could be drawn by projecting a rotating radius, in its successive positions, to meet corresponding equally-spaced ordinates. The value of the current or voltage may be found at any instant by projecting a radius upon a vertical line.

This is illustrated in Fig. 14. A certain current has a maximum value /'. This value /' is laid off as a radius and this radius rotates at a speed in revolutions per second equal to the frequency of the current. For example, if the current /' has a frequency of 60 cycles, the radius /' must make 60 complete revolutions per second, in a counter-clockwise direction. Counter-clockwise rotation has been adopted internationally as the positive direction of rotation.

When the radius /'is at the right-hand horizontal position, the value of the current is zero. When /' has advanced 30, the point 6 on the current-wave has been reached. The value of the current at this instant is *ab*, or what is the same thing, the current value is given by the distance *a'b'*, the projection of /' upon the *vertical* axis. At this particular instant, the distance $ab = a'b' = I'/2$, since sin 30 = 0.5.

Fio. 15.—Current waves produced by two current vectors differing in phase by 60.

Consider two currents, /i and /2, Fig.

15, having effective values of 12.0 and 8.0 amp. respectively. The current $I2$, whose maximum value is 11.3 amp., lags current Ii, whose maximum value is 17.0 amp., by 60. When the radius $/i$ is in the horizontal position, the value of $/i$ is zero at this instant. At this same instant, the radius $/2$ will not have reached its horizontal position, the value of the current being represented by cd, Fig. 15. In fact, the radius $/2$ does not reach its horizontal or zero position until $/i$ has advanced 60 beyond the horizontal. Further, the horizontal distance ce is 60, the same as the phase angle between the two rotating vectors.

Therefore, these two current-waves can be constructed in their proper phase relation by means of two rotating vectors having lengths of 17.0 and 11.3 amp., having equal angular velocities, and differing in phase by 60, Fig. 15.

11. Vector Addition of Sine Waves.— Assume that it is desired to add the two currents of Fig. 15. This may be done by adding the ordinates of the two curves at each point, as in Fig. 16, and plotting a new curve, $/3$. This new curve is the sum of the two currents whose maximum values are 17.0 and 11.3 amp. and effective values 12 and 8 amp. respectively, and the maximum value of this resultant, if measured accurately, will be 24.7 amp. This corresponds to an effective value of 17.45 amp. Therefore, the sum of two sine-wave alternating currents, having effective values of 12 and 8 amp. respectively and differing in phase by 60, is 17.45 amp.

Fig. 16.—Relation of vector addition of vectors to scalar addition of ordinates.

If the rotating vectors, Fig. 16, be added vectorially by completing the parallelogram, a third vector $/3$ results. This vector 13 will be found to have a length of 24.7 amp., the exact value of the maximum of the resultant current wave as just found. If a wave be plotted using $/a$ as the rotating vector, projecting as before, it will coincide with $/3$ as obtained by the addition of the ordinates for the 12-and 8-amp. waves. The angle 6 by which the radius vector I leads $I3$ equals the angle 6 by which the current

wave $/i$ leads the current wave $/3$.

Hence, this problem can be solved without going through the somewhat lengthy process of plotting the waves and adding their ordinates. It is merely necessary to lay off the maximum values of the waves 60 apart and add them vectorially, just as forces are combined. The resulting vector will be the maximum value of the wave obtained by adding the waves of I and $/a$.

In practice, one generally has to do with effective rather than maximum values. If the effective values of the waves be added in this same manner, their vector sum is the sum of the two

FIO. 17.—Vector addition of currents, using effective values.
alternating currents in effective amperes. This is illustrated in Fig. 17, where the 12-and 8-amp. vectors are laid off 60 apart, the 12-amp. vector leading. By completing the parallelogram, the resultant current $0c$ is obtained. This has a value of 17.45 amp. Its value is readily found as follows: Project ac upon Ob, where $ac = 8$ $ab = ac \cos 60 = 4.00$ $be = ac \sin 60 = 7.45$ $Oc = \sqrt{(12 + 4.00)^2 + (7.45)^2} = 17.45$ amp. *Ans.*

FIO. 18. — Vector addition of two equal voltages having 90 phase difference.

The angle 6 can be readily determined.
$\tan 6 = 12 \div 4$ $B = 25$ *Example.*—Each of two alternator coils Oo and Ob, Fig. 18 (a), is generating an emf, of 160 volta. These voltages differ in phase by 90. Determine the voltage across their open ends if they are connected together at O as shown.
Let Eao and Eob, Fig. 18 (6), represent the respective voltages across coils aO and 06. Combining these two vectorially, the voltage Sat is obtained. As Eaa and E,j, are at right angles, their resultant ia readily found.
$+ E = \sqrt{160^2 + 160^2} = 226$ volts. *Ans. It must be kept constantly in mind that alternating voltages and currents must be combined vectorially.*
The only occasions when arithmetical addition is permissible are when the voltages or the currents are in phase.

CHAPTER II
ALTERNATING-CURRENT CIRCUITS
12. Alternating-current Power.—The power in a direct-current circuit under steady conditions is always given by the product of the volts across the circuit and the current in amperes flowing in the circuit. This same rule applies to alternating-current circuits, provided that only *instantaneous* values of amperes and volts are considered. The *average power,* however, is not necessarily the product of the effective volts and effective amperes, the values which are ordinarily measured with instruments.
Fio. 19.—Power curve; current and voltage in phase.

Figure 19 shows a voltage wave and a current wave in phase with each other. To obtain the power at any instant, the amperes and the volts at that instant are multiplied together and a new curve P may be plotted, the ordinates being the instantaneous products of E and I. The curve P then gives the power in the circuit at any instant. The ordinates of this power curve will *always* be positive when E and $/$ are in phase, because the voltage and the current are both positive together during the first half cycle and are both negative together during the second half-cycle, and the product of two negative quantities is positive. That is, the current and the voltage act in conjunction throughout the cycle and the ordinates of the power curve are always positive.

It will be noted that this power curve is a sine wave having double the frequency of either the voltage or the current. For every cycle of either voltage or current, the power wave touches the zero axis twice, so that in such a circuit the power is zero twice during each cycle. Since the peaks of the voltage and current waves occur at the same instant, the corresponding peak of the power curve is $(-/2-B)$ $(V2/) = 2EI$ where E and $/$ are the *effective* values of voltage and current.

This power curve has its horizontal axis of symmetry at a distance El above the zero axis. Consequently, El must be the *average* value of the power, since the upper half waves will just fill the shaded valleys below the axis of symmetry of the power curve. When-the current and the voltage are *ir, _ age* power is their product, as with direct currents *Example.*—An incandescenHamp load takes 30 amp. from 115-volt, 60-cycle mains. (In this type of load the current and voltage are substantially in phase.) How much power do the lamps consume? $P = EI = 115 \times 30 = 3,450$ watts. *Ans.*

Figure 20 shows the current and voltage 90 out of phase, or in quadrature, the voltage leading. Let it be required to determine the power curve for this condition. At points *a, b, c, d* and *e,* either the current or the voltage is zero, and the power must be *zero* at each of these points. Between *a* and 6 the voltage is positive and the current is negative, and they are therefore acting in opposition. The product of a positive and a negative quantity is negative. Hence the power between points a and *b* must be negative. This means that the circuit is *returning* power to the source of supply. Between points *b* and *c* both the current and the voltage are positive and therefore the power between these two points must be positive. Between c and *d* the current is positive, but the voltage is now negative. Therefore, the power is again negative between these two points. Between *d* and *e* both the current and the voltage are negative and the power now becomes positive.

Fio. 20.—Power curve; current and voltage in quadrature, current lagging.

This power curve is a sine wave having double the frequency of either the current or the voltage. Its axis of symmetry coincides with the axis of current and voltage. There must be as much of the power curve above the zero axis as there is below that axis, or the positive power above the axis must be equal to the negative power below the axis. That is, all the positive power received *from* the source is returned *to* the source of supply. Therefore, the net power is *zero.* When current and voltage differ in phase, by 90, or are in quadrature, the average power-is *zero.* If the current *leads* the voltage by 90, the average power is *zero,* as is shown later in Fig.

30, page 31.

If current and voltage are out of phase by an angle less than 90, but greater than 0, the resulting power curve *P* is that indicated in Fig. 21. At points *a, b, c, d* and *e,* either the voltage or the current is zero and the power is zero at each of these points. Between *a* and *b,* and between *c* and *d,* the current and voltage are in opposition, and the power is negative. Between *b* and c, and between *d* and *e,* they are in conjunction, and the power is positive. It will be noted that there is more.positive power than negative power. The average power is not zero, but is positive, and is less than the product of *E* and *I.* It will be shown later that this power $P = EI \cos 6$ (5) where 6 is the phase angle between voltage and current. Cos 6 is called the ower-acor of the.circuit. *P* is the *true watts* and *EI* the *apparent watter. volt-amperes.* The power-factor true watts *P* (6)

Fig. 21.—Power curve; current and voltage out of phase by angle *B.* apparent watts *EI* The power-factor can never be greater than unity.

13. Circuit Containing Resistance Only. —Figure 22 shows an alternating-current circuit containing resistance only. A potential difference of *E* volts is impressed across the resistance *R.* In virtue of this voltage a current having the equation $i = Imax \sin dit$ flows, where co is the angular velocity of the rotating vector in radians per second. (See page 11, par. 6, equation (3).) As one revolution of the rotating vector corresponds to 2i r radians, the vector must complete 2irf radians per second, where / is the frequency. Hence o = 2ir/. (For 60 cycles, ia = 377; for 25 cycles, o = 157.) From the definition of an alternatingcurrent volt (Par. 8), $e = Ri = RImax \sin co£.$

The current and the voltage are in phase. They have the same frequency and when $t = 0$, sin *ut* = 0, and both the current and voltage waves are crossing the zero axis simultaneously and increasing positively, as shown in Fig. 23 (a).

If effective values are used, $E = IR.$ Figure 23 (6) shows the vector diagram for this circuit, using effective values.

The *IR*

Fig. 22.—Circuit containing resistance only.

Flo. 23.—Current and voltage waves in phase, and vector diagram.

drop is in phase with the current *I* and is equal to the voltage *E,* since no other voltage exists in the circuit.

As the current and the voltage are in phase, the power $P = EI$ (7) as is shown in Fig. 19. Also $P = I^2R$

It will be observed that with resistance only, the alternatingcurrent circuit follows the same laws as the direct-current circuit, in regard to the relation existing among voltage, current, resistance and power.

14. Circuit Containing Inductance Only. —It was shown in Vol. I, Chap. VIII, that inductance always *opposes* any *change* in the current flowing in a circuit. For example, when the current starts to increase in an inductive circuit, the electromotive force of self-induction opposes this increase. This is illustrated in Fig. 24 (a), which shows the rise of current in a direct-current circuit containing resistance and inductance, when a steady voltage is impressed. The current rises *slowly* to its ultimate value. (a) (6)

Fio. *24.*—Increase and decrease of current in an inductive circuit.

On the other hand, when the current attempts to decrease in the circuit, the inductance tends to prevent this decrease, as is shown in Fig. 24 (6). In other words, if inductance is present in a circuit, it always opposes any change in the current. With a *steady* direct current, however, the inductance has no effect.

If in Fig. 24 (a) the voltage across the inductance be lowered when the current reaches point *a,* the current will not reach its Ohm's law value. This same effect occurs in alternating-current circuits. With inductance_n_the_circuit, the current does not havetimejtoafi-hjSrsTOhm's law value before the voltage beginsltodecrease either positively or negatively. The current change is opposed by the electromotive force of self-induction, which at *di* any instant is equal to — L r where *L* is the inductance in henrys *di* and -r. is the rate in amperes per second at which the current is changing at that instant. The minus sign signifies that this voltage is opposing the change in the current.

Figure 25 shows a current wave /. Starting at (a) the current is *changing* at its maximum rate in a positive direction. Therefore, at this instant the electromotive force of self-induction must be at its negative maximum value. At point 6, the top of the current wave is horizontal and, therefore, at this instant the current is not changing at all. Hence the electromotive force of self-induction is zero. At c the current is changing at its maximum rate negatively and the electromotive force of self-induct ion must be maximum positive, because of the negative sign in the

Voltage

Fig. 25.—Current and voltages existing in an alternating-current circuit containing inductance only.

formula. Continuing in this way the voltage curve *a'b'c'* is obtained. It will be observed that this wave is a sine wave and is lagging the current by 90. This is the only voltage in the circuit which opposes the change oTcuiTent. It corresponds to the back electromotive force of a motor. The line, in the case of the motor, must supply a voltage opposite and equal to the back electromotive force before any current can flow into the armature. This same condition exists in the alternating-current circuit. Before any current can flow into a circuit containing inductance, but no resistance, a voltage Oggosite and equal to the electromotive force of self-induction musVbe supplied by the line.,

Therefore, in Fig. 25 the voltage E, which is the line voltage, is opposite and equal to the electromotive force of self-induction.

It will be noted that the impressed voltage *leads* the current by 90, or the current *lags* this voltage by 90. With inductance only in the circuit, the current tegsjhe impressed voltage byJJQ. (In practice it is impossible to obtain a pure inductance, as inductance must necessarily be accompanied by a certain amount of resistance.)

The above may also be proved as fol-

lows: Let the current be given by $i = Imaz \sin ut$. The emf. of self-induction sin *(ut*-90) is a sine wave lagging 90 with respect to $Imax \sin cat$. The equation of the line voltage which balances this emf., $e = Lulmal \sin (o + 90)$ is a sine wave *leading* the current $Imaz \sin ut$ by 90.

The choking_£ffect of inductance is obviously proportional to the frequency and to the inductance. To express this choking effectnrTJhmB7the self-inductance in henries must be multiplied by w = 2rf = 6.28/, where / is the circuit frequency.

That is, 2ir/L is the resistance to the flow of current offered by inductance and is called the *inductive reactance* of the circuit. It is denoted by XL, and is expressed in *ohms*.

The current in a circuit having inductive reactance only is / = *EfafL* = E/XL (,448)

The impressed voltage is ' V'"' $E = 2ir/L/ = IXL$ (9) *Example.*—Figure 26 shows a pure inductance of 0.2 henry connected across 110-volt, 60-cycle mains. What current flows? $Xl = 2r \ 60 \times 0.2 = 377 \times 0.2 = 75.4$ ohms / = 110/75.4 = 1.46 amp. Ana.

Figure 27 shows a vector diagram for an inductive circuit in which the impressed voltage leads the current by 90,

15. Circuit Containing Capacitance Only.—When a directcurrent voltage is impressed across the plates of a condenser, (Vol. I, Chap. IX) there is an initial rush of current which charges the condenser to line potential. After this there is no further flow of current if the line voltage remains constant. If the condenser plates now are short-circuited, making the voltage across the plates zero, current flows out of the condenser.

Figure 28(a) shows an alternating voltage E impressed across the plates of a condenser C. When the voltage starts from its zero value at a, Fig. 28(6), and increases positively, current flows into the condenser. Therefore, this current is positive. As long as the voltage across the condenser plates continues to increase, current must flow into the condenser from the positive wire and this

current will be positive in sign. When point *b* is reached, the increase of voltage ceases and the current becomes zero. Between *b* and *c* the voltage is decreasing so that current is flowing *out* of the condenser into the positive line, and as the current flow has reversed, the sign of the current is now negative. After E passes through zero at c, the emf. is negative and charges the condenser in the opposite direction, so the current still remains negative. This continues until the voltage reaches its negative maximum. At this point the current reverses and again becomes positive.

An examination of Fig. 28 shows that when an alternating voltage is impressed across a condenser, the current into the condenser leads the voltage by 90. This is illustrated by Fig. 29, in which the relation is shown vectorially.

It will be seen from the foregoing that alternating current does not actually flow conductively through the insulation of the condenser. A perfect condenser offers an *infinite resistance* to alternating, as well as to direct current. However, with alternating current the condenser is alternately charged and discharged, so _,.j that a quantity of electricity flows into the positive plate, and then out again, etc. It is this quantity of electricity which flows to charge and to discharge the condenser which constitutes the alternating current. An ammeter placed in the line to such a condenser indicates a current.

E ?--T3T c, It is clear that this current is proportional to

Fig. 29.— Vector, ,,, diagram for circuit the frequency, for the more rapidly the voltage containing capaci-alternates, the greater the quantity of electricity tanceonly. ''& 'J ,, charged and discharged per second, and therefore the greater the flow of current. This current is also proportional to the capacitance, C, and to the voltage E. The actual value of the current in amperes is given by / = *2irfCE* (10) where C is in *farads*.

This equation may also be written

E J?

l J_X0 (11) 27T/C Xc is called the *condensive or capacitive reactance* of the circuit in ohms and is equal to l/(2ir/C).

Also *Example.* — What is the condensive reactance of a 10-microfarad condenser at 60 cycles per second and how much current will it take from 110-volt, 60-cycle mains? 10 mf. = 0.00001 farad.

,, 1 100,000...

Xc-2.60 X 0.00001 =-2.60-= 26S hmS'

$I = r$: = 0.415 amp. *Ans.*

zoo

The relations of current and voltage in a condenser circuit may also be proved as follows:

Let e be the instantaneous voltage across the condenser, C the capacitance in farads, and q the charge in coulombs at any instant.

Let $i = Im«x \sin ut$ be the equation of the current.

This equation shows that the sine wave of voltage lags the current wave by 90.

The average power in a circuit containing capacitance only is zero.

Fio. 30.—Voltage, current. and power curves; circuit containing capacitance only.

This may be shown by plotting the power curve from the current and voltage curves, as was done in Fig. 20. This is shown in Fig. 30, where P is the curve of power. There is as much of the power curve below as above the zero axis, so that the net power is *zero,* as in a circuit with pure inductance only. When the power curve is positive, energy is being delivered to the circuit and stored in the condenser; when the power curve is negative, this energy is being given back again to the source. which is the equation for the series alternating-current circuit in the steady state.

Although the net power is zero, there is a continual transfer of energy from the source to the condenser and back again to the source.

16. Circuit Containing Resistance and Inductance in Series.—

Figure 31 shows a circuit consisting of a resistance R and an inductive reactance XL connected in series across an alternating circuit whose frequency is / cycles per second. The voltage impressed across the circuit is E and a current / flows. Let it be required to determine the relations among I, E, R, and XL.

Figure 32 (a) shows a vector diagram for this circuit. As the current I is the same in both XL and R, it is laid off horizontally to scale. The position of the current vector / is arbitrary. (It1 is given the position shown merely for convenience.) From Fig. 23 (6), page 25, the voltage ER across the resistance $Rsirhase$ withjthe current. Therefore, it is laid off along the current vector. From Fig. 27, page 28, the voltage EL across the inductance *leads* the current I by 90 and is equal to IXL.

Fig. 31.—Circuit containing resistance and inductance in series.

Flo. 32.—Vector diagram for a series circuit containing resistance and inductance.

The line voltage E must be the vector sum of these two voltages, so the parallelogram is completed and the diagonal s the voltage E. The same result is obtained if IXL is laid off perpendicular to / at the end of the vector $IR,$ using a triangle rather than a parallelogram, as shown in Fig. 32 (6).

As a right triangle is formed by these three voltages, the hypotenuse $E = V(IRY + (iXrf = Vl2(R2+Xi) = IVR + XL'$ is the *impedance* of the circuit and is ex Z = pressed in ohms. It is orctinariry denoted by Z. Equation (13) corresponds to Ohm's law for the direct-current circuit. The current in an alternating-current circuit is *directly* proportional to the *voltage* across the circuit and *inversely* proportional to the *impedance* of the circuit. That is, if the voltage in volts be divided by the impedance in ohms, the value of the current in amperes is obtained.

Also the voltage $E = IZ$. (14)

An inspection of Fig. 32 shows that the angle 6 by which the current lags the voltage may be determined as follows: , $VR' + XL'$ *Example.* — A circuit containing 0.1 henry inductance and 20 ohms resistance in series is connected across 100-volt, 25-cycle mains. (a) What is the impedance of the circuit? (6) What current flows? (c) What is the voltage across the resistance? *(d)* What

is the voltage across the inductance? *(e)* Determine the angle by which the voltage leads the current. $Xl = 2ir25$ X 0.1 = 157 X 0.1 = 15.7 ohms.

(a) Z = V(20)2 + (15.7)2 = V646 = 25. 4 ohms. *Am.* (b) I = = -= 3.94 amp. Ans. (c) $EK = IR$ = 3.94 X 20 = 78.8 volts. *Ans. (d)* $EL = IXL$ = 3.94 X 15.7 = 61.8 volts. *Ans.*

As a check /(78.8)2 + (61.8)' = 100 volts.

(e) tan 6 = = = 0.785. From page 460, 9 = 38.1. *Ans.* 17. Power. — It has already been shown that a pure inductance consumes no power. Therefore, the inductance of Fig. 31 consumes no power. All the power expended in the circuit must be accounted for in the resistance. That is $P = PR = I(IR)$

IR is obviously equal to E cos 0 (Fig. 32).

Therefore, the power $P = I(IR) = IE$ cos 6 $= EI$ cos 6

As has already been shown, cos 6 is the *power-factor* of the circuit and is equal to the true power divided by the volt-amperes or apparent power.

P

It is usually

Obviously the power-factor can never exceed 1.0. less than 1.0.

Example.—How much power is consumed in the foregoing circuit and what is the power-factor? p = IR = (3.94)s X 20 =310 watts. *Ans. P*-310 EI 20 25.4 18. Circuit Containing Resistance and Capacitance in Series. Figure 33 shows a circuit containing a resistance R and a con densive reactance Xc in series. An alternating voltage $E,$ of frequency/ cycles per second, is impressed across this circuit and a current / flows. Let it be required to determine the relation existing among E, I, R and Xc.

The current / is the same in both R and Xc and is laid off horizontal in the vector diagram, Fig. 34. The voltage ER across the resistance is *in phase* with the current. The voltage Ec across the condensive reactance *lags* the current I by 90 (see Fig. 29), page 30. The line voltage E is obviously the vector sum of IR and $IX c$ and is therefore the hypotenuse of the right triangle having these two voltages as sides. Obviously E = v#2 +

(IX cz = / V#2 + Xc2 = IZ (17)

Solving the above for the current /, *I = E = E E*

The power taken by the circuit is obviously *P = PR = I(IR)* as the net power taken by the condenser is zero. *IR = E cos 6*

Therefore *P = El cos 6,* which is the same expression for power as with inductance and resistance in circuit.

The angle *6* may be determined as follows: *Example.* — A capacitance of 20 microfarads and a resistance of 100 ohms are connected in series across 120-volt, 60-cycle mains. Determine: *(a)* The impedance of the circuit. *(b)* The current flowing in the circuit. (c) The voltage across the resistance. (/) The voltage across the capacitance. *(e)* The angle between the voltage and the current. (/) The power, *(g)* The power-factor of the circuit.

20 mf. = 0.000020 farads. *Xc* = 260 X 0.000020 = 133 hms, (a) *Z* = V(100)2 + (133)2 = V27,700 = 166 ohms. Ana. (6) */* = = 0.723 amp. *An2.* (c) *EH* = *IR* = 0.723 X 100 = 72.3 volts. *Ans.* (d) *Ec* = *IXC* = 0.723 X 133 = 96.2 volts. *Ans.* V(72.3)2 + (96.2)2 = 120 volts *(check). 0* = 53.1. *Ans.* (/) *P = PR* = (0.723)2 X 100 = 52.2 watts. *Ans. (g)* cos *6* = = 55 = o.602. *An.*

P

Also P.P.--12' = 0.602 19. Circuit Containing Resistance, Inductance and Capacitance in Series.—Figure 35 shows a resistance *R,* an inductive reactance *Xl* and a condensive reactance *Xc,* all connected in series. The voltage across the circuit is *E* volts, the frequency is / cycles per second and the current is / amp.

As this is a series circuit, the current is the same in all parts of the circuit and for convenience the current vector / is laid off horizontal in the circuit vector diagram, Fig. 36. The voltage *Ek (= IR)* across the resistance is *in phase* with the current and is laid off to scale along the current vector. The voltage *EL (= IXL)* across the inductance is laid off at right angles to the current and *leading.* The voltage *Ec (= IXc)* across the condenser is laid off at right angles to the current and *lagging.*

An examination of Fig. 36 shows that the voltage across the inductance and that across the capacitance are in opposition, so that the resultant voltage of these two is their arithmetical difference. In this particular case, *IXL* is shown as being greater than *IXC.* Therefore, *IXC* is subtracted directly from *IXL.* The line voltage must be the vector sum of the three voltages and is the hypotenuse of a right triangle of which *IR* and *(IXL — IXC)* are the other sides. Therefore

Fio. 35.—Circuit containing resistance, inductance and capacitance in series.

Fig. 36.—Vector diagram for circuit containing resistance, inductance and capacitance, all in series.

The value of *Xl* and *Xc* may be substituted in equation (21). It then becomes / = *E* (22)

The phase angle *6* is found as follows: tan *0 = XLXc* (23)

If *XL* is greater than *Xc,* the tangent is positive and *6* is positive, as shown in Fig. 36. This indicates a *lagging* current. If *XC* is greater than *-XL,* the tangent becomes negative and the angle *6* becomes negative. This indicates a *leading* current.

The power-factor of the circuit .:. P. F. = cos *0* = —. — — *R* =- (24)

V#2 + (l-XcY Z' *Example.* — A series circuit consisting of a resistance of 50 ohms, a capacitance of 25 mf. and an inductance of 0.15 henry is connected across 120volt, 60-cycle mains.

Find: (a) The impedance of the circuit. (6) The current in the circuit. (c) The voltage across the resistance, *(d)* The voltage across the inductance (e) The voltage across the capacitance. (/) The power taken by the circuit. *(g)* The phase angle of the circuit. *(h)* The power-factor of the circuit.

Xl = 2ir60 X 0.15 = 377 X 0.15 = 56. 6 ohms. *Xc* =-1 "106 (a) *Z* =-/(50)2 + (56.6-106)2 = v/(50)2 + (-49.4'j' = 70.2 ohms. *Ans.* 120 (6) / = ijr-j. = 1.71 amp. *Ans. tj.2i* (c) *ER = IR* = 1.71 X 50 = 85. 5 volts. *Ans. (d) EL = IXL*= 1-71 X 56.6 = 96.8 volts. *Ans.* e) *Ec = IXc* = 1.71 X 106 = 181.1 volts. *Ans. P = PR* = (1.71)2 X 50 = 146 watts. *Ans.* f.nfl *xl Xc* 56.6-

106-49.4 tan *6* = — g— — —-0.988. *6* = —44.6. Therefore the current leads. *Ans. Tf Kf*

Figure 37 gives the vector diagram for the circuit conditions represented by this problem.

It will be observed that the voltage across the condenser exceeds the line voltage by a considerable amount. This would be impossible under like conditions in a direct-current circuit, for the voltage across any part of the circuit cannot exceed the line voltage. This condition can exist in an alternating-current circuiVbecause the condenser voltage and the inductance voltage are in direct opposition. Both may be large, provided their difference is less than the line voltage.

20. Resonance in a Series Circuit.—The general equation (22) for the current in a series circuit shows that for fixed values of resistance and

Fig. 37.—Vector diagram for series circuit, giving numerical values.

impressed voltage the current is a maximum when the expression in the parenthesis under the square root sign is equal to zero. That is, in the equation *T-*

That is, the voltage across the inductancejs. equal to. the voltage across the capacitance. __ As these two voltages are in exact opposition, they balance each other, so that the *IR* drop is equal to the line voltage. This is illustrated in Fig. 38.

WEeH the foregoing conditions exist, the circuit is said to be in *resonance,* The current is then in phase with the line voltage and the power *P = EI.* Solving equation (25) for the frequency

This is the frequency for which a circuit having fixed values of *L* and *C* will be in resonance. It is sometimes called the *natural frequency* of the circuit, because it is the frequency at which the current in the circuit will oscillate, if no external frequency is impressed on the circuit, provided the resistance *R* is less than */4L/C.* For example, in a radio circuit a condenser *C,* charged to a high voltage, is discharged into an inductance *L,* of negligible resistance. The frequency of the resulting oscillations as determined by the values of *L* and *C,* is given in equation (26).

As the voltage across the inductance equals the voltage across the capacitance, when the circuit is in resonance, and the two are in opposition, each may reach a high value, even with moderate line voltage. This is illustrated by the following example.

Example.—A circuit has a resistance of 20 ohms, an inductance of 0.3 henry, and a capacitance of 20 mf. (a) For what value of the frequency will the circuit be in resonance? (6) If the current is 5 amp. find the line voltage, (c) The voltage across the inductance, *(d)* The voltage across the capaci

Fio. 38.—Vector diagram for series circuit in resonance.

It will be observed that the voltage across the inductance and that across the capacitance are equal, each being 612 volts, or more than six times the line voltage.

It should be noted that the current is a *maximum* when a series circuit is in resonance.

21. Parallel Circuits.—In practice, parallel circuits are more common than series circuits, because of the extended use of the multiple system of transmission and distribution. The solution of problems with two or more loads in parallel involves the finding of the current in each branch of the circuit and the combining of these currents *vectorially* to give the resultant current. This is illustrated by the following example:

A resistance of 10 ohms, an inductive reactance of 8 ohms and a condensive reactance of 15 ohms are all connected in parallel across 120-volt, 60-cycle mains, as shown in Fig. 39 (a), (a) Find the total current. *(b)* Determine the circuit power-factor, (c) Determine the power.

The current taken by the resistance Ia = -tjt = 12 amp. in phase with E. 120 Il = -5-=15 amp. in quadrature with E and lagging.

o Ic = -r-r-= 8 amp. in quadrature with E and leading.

lo

These currents are shown vectorially in Fig. 39 (6).

The voltage is the same for all three

branches of the circuit and is laid off as the horizontal vector. The resistance current Ig is in phase with the voltage E. The inductive current *lags* the voltage by 90 and the condensive current *leads* the voltage by 90. As the inductive current and condensive current are in exact opposition) they subtract from each other, leaving 7 amp. lagging by 90. The resultant current /o is the vector sum of the 7 amp. and the 12 amp.

(o) /0 = /T22 + 7" = 13.9 amp. lagging. *Ans. (b)* Obviously the cosine of the angle *6* between the voltage and the current is cos 6 = = -,44 = 0.864 = P.F. *Ans.*

It lo.9

(c) $P = EIR$ = 120 X 12 = 1,440 watts. *Ans.*

Also $P = Ela$ cos e = 120 X 13.9 X 0. 864 = 1,440 watts. *Ans.*

For convenience the following equations are given for the parallel circuit. R and L in parallel $Z = 1$ RXL' = $E R$ and C in parallel Z, l RXc where /o is the total current and E is the circuit voltage.

22. Resonance in a Parallel Circuit.— Resonance in a parallel circuit exists when the resultant current and the line voltage are in phase with each other. Under these conditions the condensive current must be equal to the inductive current. These twobeing opposite and equal balance each other, leaving only the resistance current. This is illustrated in Fig. 40 (a). E is the voltage wave; IK is the current in the resistance; //, is the current in the inductance; /C is the current in the condenser and is equal to IL. As the inductive current *lags* the voltage by 90 and the condensive current *leads* the voltage by 90, they are in direct opposition and being equal, they balance. This leaves only /«.

Figure 40 (6) illustrates vectorially these circuit conditions. It will be observed that the total current is a *minimum* when the *parallel* circuit is in resonance, whereas in the *series* circuit, the current is a *maximum* at resonance. In the *parallel* circuit the inductive and condensive *currents* are opposite and equal; in the *series* circuit the inductive and condensive *voltages* are opposite and equal. If a pure capacitance and a

pure inductance were connected in parallel and adjusted for resonance, the line current would be zero, even though the inductance and condenser were each taking current.

Example. —-A resistance of 12 ohms, an inductance of 0.2 henry and a condenser are connected in parallel across 120-volt, 60-cycle mains. For what value of capacitance will the circuit be in resonance? Ic must be equal to Il 120 2. 60 X 0.2-

Ic = 120 X 2ir60 X C = 1.59 amp.

1 Q

c ' laoTcMo-aooo352 farads = 35.2 mf. *Ans.*

In the parallel circuit, as well as in the series circuit, LCw2 = 1.0 at resonance (co = 2ir/), *when the inductive and capacitive branches contain only pure inductance and pure capacitance.*

Also, under these conditions / =-;- *2T/LC*

When there is resistance in either the inductive or the condensive branch, L and C are the *equivalent* inductance and capacitance of the parallel circuit, and not the L and C, respectively, of the inductance and capacitance alone.

23, Polygon of Voltages; Three Voltages. — The inductances and condensers so far considered have been assumed as perfect, that is, as having no losses and with their currents exactly 90 from their respective voltages. In practice this is impossible. The wire of which the inductance is made has a certain resistance, and if an iron core is used, the core losses are equivalent to an added resistance, since they involve a power loss. Condensers are made having very small losses, and phase angles very nearly equal to 90, but even such condensers are not ideal.

When an inductance coil is being considered, its resistance must be added to the other resistances in the circuit, in order to find the total circuit resistance.

Figure 41 *(a)* shows a series circuit connected, across an alternating voltage E, having a frequency /. This circuit contains a resistance R and an impedance coil Z', having a resistance R' and an inductance L. The reactance X' of the impedance coil is equal to 2ir/L. Figure

41 (6) shows the vector diagram for this circuit. The voltage IR is in phase with the current I. The voltage Ez' across the impedance coil is not 90 ahead of the current, but leads the current by an angle d which is less than 90, due to the resistance of the impedance coil. The circuit voltage E is the vector sum of IR and Ez'. The impedance voltage Ez consists of two components, IR' in phase with the current and IX' in quadrature with the current. Therefore, the projection of Ez, the voltage across the impedance, on the current I is the voltage drop due to the resistance of the impedance. Divide this projected voltage by the current and the resistance of the impedance coil is obtained.

Fig. 41. — Circuit having resistance and impedance in series, and vector diagram,

Figure 42 (a) shows a series circuit, containing a non-inductive resistance R and an impedance coil Z' in series across the voltage E. Let it be required to construct the vector diagram of this circuit. A voltmeter across the resistance R measures the voltage ER; when across the impedance it measures the voltage Ee, and when across the line it measures the voltage E.

To construct the vector diagram of this circuit, the current vector I is laid off horizontally, as shown in Fig. 42 (6). The voltage EK is laid off to scale in phase with the current I; from the outer end of ER an arc is swung having Ez for its radius. Then from 0, the origin, another arc is swung having E for its radius. Lines drawn from the end of ER and from 0 to the intersection of the arcs complete the vector diagram. By trigonometry the angle 6, the circuit power-factor angle, and t, the impedance coil power-factor angle, can both be found. Knowing these, it is a simple matter to determine the power-factor and the power of the circuit.

Example. — A resistance and an impedance coil are connected in series across a 60-cycle alternating-current circuit, Fig. 42 *(a)*, and the current is 4.0 amp. The voltage across the resistance is found to be 60 volte, that across the impedance coil 80 volts, and the line volt-

age is 1 10 volts. Find: (a) The value of the resistance, (b) The circuit power-factor angle 0 and the power-factor, *(c)* The impedance-coil power-factor angle t and the corresponding power-factor, *(d)* The circuit power, *(e)* The impedancecoil power. (/) The impedance-coil resistance, *(g)* The impedance-coil reactance. (a) $R = Er/i = 60/4 = 15$ ohms. *Ans.* (b) Applying the law of cosines, page 457, to Fig. 42 (6), $80^2 = 110^2 + 60s$-$2 \times 110 \times 60 \cos 9$ Q UU $\cos 9 = ra = a704$-*Ana* $9 = 45.2$. Ans.

Fro. 42.—Triangle of voltages for circuit having resistance and inductive impedance in series.

(c) By the law of sines, page 457, sing 60 24. Polygon of Voltages; Four Voltages.—If three sides of a triangle are fixed, the triangle itself is fixed as regards both its area and its angles. If the four sides of a polygon are given, however, the polygon itself is not determined. In order to determine the polygon definitely, the angle included between two of its sides must be known. This is the condition which exists when there is resistance, inductance and capacitance in a series circuit. These three voltages and the line voltage give four voltages which in themselves make an indeterminate polygon. If the angle between two of these voltages is known, the polygon and its angles are completely determined.

Fio. 43.—Polygon of voltages for series circuit containing resistance, inductive impedance and capacitance.

This is illustrated in Fig. 43, in which resistance, impedance and capacitance are all connected in series and the current I flows in the circuit. Assume that the condenser power-factor angle is 90, which is practically the case in most commercial condensers. This constitutes the angle which determines the polygon of voltages. Along / lay off EK to scale, Fig. 43 (6). Ninety degrees behind / lay off Ec to scale. Add these two vectorially giving $E' = ER + Ec$. From the end of E' swing upward the vector Ez' and from 0 swing the line voltage E. Complete the polygon where these two arcs intersect. Then from 0 draw

Ez' parallel,to the Ez' swung from the end of E'.

It will now be seen that $\# + (\# + Ec) = E$

That is, the vector sum of the three voltages is equal to the line voltage, which condition exists in the circuit.

Example.—A resistance, an impedance coil and a condenser are all connected in series. The voltage across the resistance is 80 volts; that across the impedance coil is 70 volts; that across the condenser is 90 volts and the line voltage is 120 volts. A current of 5 amp. flows in the circuit and the condenser current leads its voltage by 90. Determine: (a) The circuit power-factor angle 6. (6) The resistance and reactance of the impedance coil.

The voltage polygon is shown in Fig. 44.

(a) $E' = /W + 80^2 = \sqrt{147500} = 120.5$ volts

Fio. 44.—Polygon of voltages for alternating-current series circuit.

Applying the law of cosines (page 457) to triangle *oab*, $70^2 = 120.5^2 + 120^2$-$2 \times 120.5 \times 120 \cos 0$. 26. Polygon of Currents.—Obviously, if the resistances, impedances, etc. are in parallel, the voltage is the same for each branch of the circuit, but the respective currents may differ. Therefore, the polygon is composed of currents rather than of voltages. Figure 45 (o) shows a circuit consisting of resistance, inductive impedance and capacitance all in parallel. Assume that the condenser current is in quadrature with its voltage. Figure 45 (6) then represents the polygon of currents. The voltage E, being common, is laid off horizontal. The current IB is laid off in phase with E and the current Ic leads E by 90.

Fig. 45.—Parallel circuit. consisting of resistance, inductive impedance and capacitance all in parallel, with vector diagram.

These two are combined to obtain I'. From the outer end of I', Iz is swung to meet / which is swung from 0. This completes the polygon, which is similar to those shown in Figs. 43 and 44, except that the vectors are currents instead of voltages.

26. Energy and Quadrature Currents.— Figure 46 shows the vector diagram for a load connected across alternating-current mains. This load is typical of most commercial loads, except incandescent lamps. It takes a current /, lagging the voltage E by 6 degrees. The current / may be resolved into two components, ii in phase with the voltage and iz in quadrature with the voltage. Obviously I is the vector sum of i and iz.

The power taken by the load is $P = EI$ cos 6 but / cos 0 = t'i

Therefore $P = Eii$ t'i is called the *energy component* of the current, because this component multiplied by the voltage gives the circuit power.

The component fj is in quadrature with the voltage and can contribute no power therefore. iz is called the *quadrature* or *wattless* component of the current.

Energy Current?i

Fig. 46.—Energy and quadrature currents.

If this load is being supplied over a transmission line, the line loss is proportional to $PR = (zi2 + ii)R = isR + i22R$ where R is the transmission line resistance.

It will be observed that the quadrature component produces line loss, yet contributes no power to the load. Therefore, it is ordinarily desirable to make i2 as small as possible or, in other words, have the system operate at a high power-factor. For example, when '6 = 45, P.F. = 0.707, the energy and quadrature currents are equal. Therefore, the quadrature current contributes as much to the line loss as Lhe energy current does, but it contributes nothing to the power supplied to the load.

Example.—A transmission line, Fig. 47 (a), supplies 50 kw. at 220 volts, single-phase, to a load having a power-factor of 0.60, lagging current. Each

Fig. 47.—Energy and quadrature currents in a transmission line.

wire has a resistance of 0.02 ohm. Find: (a) The energy current. (6) The quadrature current. (c) The line loss due to the energy current. *(d)* The line loss due to the quadrature current. *(e)* The total line loss. (/) The line loss which would exist

if the load power-factor were unity. The total current . 50,000

/

22006 (a) ii = 379 cos 6 = 379 X 0.6 = 227 amp. *Ans. (b)* iz = 379 sin 6 = 379 X 0.8 = 303 amp. *Ans. (c)* ii2 X 0. 04 = 2,070 watts. *Ans. (d)* ii' X 0.04 = 3,680 watts. *Ans. (e)* /' X 0.04 = 5,750 watts. *Ans.* (/) If the power-factor of the load were unity, the quadrature current iz would be zero and the line current / = i. Therefore the loss would be /2 X 0.04 = 2,070 watts. *Ans.*

In this particular case, the line loss due to the quadrature current is considerably in excess of that due to the energy current, yet the quadrature current contributes no power to the load.

From the foregoing it must not be inferred that the energy and quadrature currents exist separately. Only one current actually flows, but this current is resolved into two components, each of which produces different effects in the circuit. The effect of each component can then be studied, resulting in a much better understanding of the circuit relations than if an attempt were made to consider the current as a whole.

CHAPTER III ALTERNATING-CURRENT INSTRUMENTS AND

MEASUREMENTS

ELECTRO-DYNAMOMETER TYPE INSTRUMENTS 27. The Siemens Dynamometer. —Several types of alternating-current instruments operate on the electro-dynamometer principle. The Siemens dynamometer, Fig. 48, is an example of this type of instrument in simple form. It consists primarily of two sets of coils. The coil F is fixed and the coil M, whose axis is at right angles to the axis of F, is free to turn through a small angle. M is suspended by a silk thread and its turning moment is opposed by a helical spring. Current is led into the moving coil through two mercury cups. When used as an ammeter, the two coils are wound with a few turns of coarse wire and are connected in series. When current flows through these coils, there is a tendency for the moving coil to swing into the plane of the fixed coil. When the current reverses, it reverses in the two coils simultaneously so that

the torque is always in the same direction. The movable coil is not allowed to deflect, however, but is kept in its zero position by turning the knurled head at the top of the instrument which acts on the coil through the spring. The angle by which it is necessary to turn this head is proportional to the turning moment of the coil. The turning moment is proportional to the current *squared*, so that the deflection of the index $D = KP$ where K is a. constant.

Fig. 48.—Siemens dynamometer.

The current / = $K'vD$ (27) where K' = $1/VK$

As the deflections are proportional to the *square* of the current, the instrument gives effective values when equation (27) is applied. It therefore reads correctly for both alternating and direct currents. When direct current is used, it is advisable to reverse the direction of the current and average the set of readings. This eliminates the effect of stray fields.

This type of instrument is difficult to adjust and to manipulate, especially when the current fluctuates. It is not direct reading and because of its construction is adapted to laboratory work only.

If small wire be substituted for the coarse wire and an extension coil be connected in series, the instrument can be used as a voltmeter.

28. The Indicating Electro-dynamometer.— As it is neither portable nor direct reading, the Siemens dynamometer itself is not adapted to portable and to switchboard instruments. However, many types of portable and switchboard instruments operate on the Siemens dynamometer principle. The general construction of a portable type of electro-dynamometer instrument is shown in Fig. 49.

Two fixed coils, FF', are so connected that their magnetic fields act in conjunction. These coils may be considered as being two parts of a single coil, opened in the middle to allow the spindle of the moving coil to pass through. M is a movable coil mounted on a vertical spindle. There is a hardened steel pivot at each end of the spindle, which turns in jewelled bearings. Two spiral springs similar to those used in direct-current instruments (see Vol. I, page 129, Fig. 117) oppose the turning of coil M and at the same time carry the current into the coil. As the springs can carry but a very small current, the movable coil is wound with fine wire.

Fig. 49.—Principle of the electrodynamometer.

Assume that at some instant the direction of the magnetic field ti, which is due to the fixed coils, is from left to right. At the same instant the current in coil M produces a field fa whose direction is along the axis of M. Coils tend to align themselves so that the number of magnetic linkages in the system is a maximum. Therefore, the moving coil M tends to turn in a clockwise direction so that its field will act in conjunction with fa. The turning of M is opposed by the control springs.

Obviously the torque developed is proportional to fa, fa and sin $j3$, where ft is the angle between the axis of coil M and the axis of coils FF'. As fa and fa are proportional to the currents in the coils FF' and M reepectively, the torque is proportional to the product of the two currents and sin $j3$.

29. The Electro-dynamometer Voltmeter.—Some types of alternatingcurrent voltmeter operate on the electro-dynamometer principle. The fixed coils FF', Fig. 50, are wound with fine wire and are connected in series with the moving coil M. A high resistance R is connected in series with the dynamometer to limit the current when the instrument is connected across the line. The current passing through the dynamometer is therefore proportional to the line voltage. The current passing through the instrument causes coil M to turn and the pointer attached to it moves over a scale graduated in volts. The scale will not be divided uniformly like that of the direct-current voltmeter, for the deflections are very nearly proportional to the *square* of the voltage. The divisions at the lower part of the scale are so small that poor precision is obtained. The divisions at the middle

Fig. 50.—Diagram of a dynamometer-voltmeter.

and upper portions of the scale, however, are usually of such magnitude that they may be read with a high degree of precision.

This dynamometer-type of voltmeter takes about five times as much current as a direct-current voltmeter of the same rating and consumes an appreciable amount of power. As the moving coil operates in a comparatively weak field, this type of instrument is very susceptible to stray fields. Unless the instrument is shielded, wires carrying currents, inductive apparatus and even iron alone, if brought too near, may cause large errors in the indications of this type of voltmeter.

This instrument may be used for direct current as well as for alternating current. Reversed direct-current readings should be taken in order to eliminate the effect of the earth's field and of any other stray fields. As the deflections depend upon the *square* of the voltage, the instrument reads effective values.

30. Inclined-coil Voltmeters.—The inclined-coil type of voltmeter operates on the dynamometer principle. It differs from the

Fio. 51.—General Electric inclined-coil instrument.

previous types only in the geometrical relations of its fixed and moving coils. The axis of the fixed coil, Fig. 51, is set at a considerable angle with the vertical. The axis of the moving coil makes a considerable angle with the spindle. This moving coil is connected in series with the fixed coil, the current being carried to the moving coil through light springs. A resistance to limit the current is connected in series with the instrument.

When the pointer is at the zero position there is a considerable angle between the axes of the fixed and the moving coils. When current flows through the instrument, the moving coil tends to take such a position that its axis coincides with the axis of the fixed coil, so that their magnetic fields act in conjunction. In turning, the moving coil is opposed by flat spiral springs. The scale is calibrated in volts.

As this instrument is of the dy-

namometer type, its scale readings hold for both direct and alternating currents.

31. Dynamometer Ammeters.—Owing to the difficulty of leading a heavy current into the moving coil, dynamometer ammeters of the portable type and of the switchboard type are not common. It might appear that this type of instrument could be devised for use with a shunt, which would allow but a small portion of the total current to pass through the moving coil. This involves two difficulties. Alternating currents divide inversely as the circuit *impedances*. Impedances are determined by the frequency. Unless the ratio of inductance to resistance were the same in the shunt as in the moving coil, the instrument would be correct at only one frequency and might be in considerable error with an irregular wave shape owing to the presence of higher-frequency harmonics. There would also be a considerable voltage drop across such a shunt. Instruments of the dynamometer type, in which the above difficulties are in part overcome, are available, but the iron-vane type of instrument described in Par. 34 is so much simpler and less expensive that the shunted type is little used.

32. The Wattmeter.—Alternating-current power is equal to the product of the effective current and the effective voltage only when the power-factor is unity. Therefore, the ammeter and voltmeter method, as used with direct currents, can seldom be used to measure alternating-current power. Consequently, a *wattmeter* is necessary for measuring alternating-current power.

The wattmeter shown in Fig. 52 operates on the electrodynamometer principle. M is a moving coil wound with fine wire and is practically identical with the moving coil of the dyna

Load

Fio. 52.—Connections for a wattmeter.

mometer voltmeter, Fig. 50. It is connected across the line in series with a high resistance R. The current is led into this coil through springs. The two fixed coils FF are wound with a few turns of heavy wire, capable of carrying the load current. As there is no iron present, the field due to the current coils FF is proportional to the load current at every instant. The current in the moving coil M is proportional to the voltage at every instant. Therefore, for any given position of the moving coil, the torque is proportional at every instant to the product of the current and voltage or to the instantaneous power of the circuit. If the power-factor is other than unity, there is negative torque for part of the cycle. That is, during the periods when there are negative loops in the power curve, Fig. 21, page 24, the current in the fixed coil and the current in the moving coil reverse their directions with respect to each other, and so produce a negative torque. The moving coil takes a position corresponding to the *average* torque. The torque is also a function of the angle between the fixed and moving coil axes, but this factor is taken into account by the scale calibration.

As the torque acting on the moving coil varies from instant to instant, having a frequency twice that of either the current or the voltage, the coil tends to change its position to correspond with these variations of torque. If the moving system had little inertia, the needle would vibrate so that it would be impossible to obtain a reading. Because of the relatively large moment of. inertia of the moving system, the needle assumes a steady deflection for constant values of average power. The position taken by the coil corresponds to the *average* value of the power, which is the result desired.

It should be noted in Fig. 52 that the voltage terminal marked "0" is connected directly to one end of the moving coil. This terminal always should be connected directly to that side of the line to which the current-coil is connected. The fixed and moving coils are then at the same potential. If the moving coil is connected to the other side of the line, the potential difference between the fixed and moving coils is equal to the full line potential, as shown in Fig. 53. In this diagram, the fixed coils are considered as being at zero or ground potential. The moving coil is then at the

potential of the other side of the line, or 550 volts, and this is the difference of potential which exists between the fixed and moving coils. This is dangerous from the insulation standpoint, and electrostatic forces existing between the fixed and the moving coils may cause an error in the instrument reading. (The wattmeter is also briefly described in Vol. I, Chap. VII.)

Fig. 53.—Incorrect method for connecting a wattmeter.

33. Wattmeter Connections.—In Fig. 54 (a), wattmeter W is shown measuring the power taken by a certain load. In order to measure this power correctly, the wattmeter current-coil should carry the *load* current, and the wattmeter voltage-coil, in series with its resistance, should be connected directly across the *load*.

Fig. 54.—Methods for connecting a wattmeter.

The current in the wattmeter current-coil is the same as the load current, but the wattmeter potential-circuit is not connected directly across the load, but is measuring a potential in excess of the load potential by the amount of the impedance drop in the wattmeter current-coil. Therefore, the wattmeter reads too high by the amount of power consumed in its own current-coil. Under these conditions the true power $P = P'-PRC$ where P' is the power indicated by the wattmeter, $/$ is the current in the wattmeter current-coil, and Rc is the resistance of this coil. This loss is ordinarily of the magnitude of 1 or 2 watts at the rated current of the instrument, and may often be neglected.

If the wattmeter be connected as shown in Fig. 54 (6), the wattmeter potential-circuit is connected directly across the load, but the wattmeter current-coil carries the potential-coil current in addition to the load current. In fact, the wattmeter potentialcircuit may be considered as being a small load connected in parallel with the actual load whose power is to be measured. Therefore, the power consumed by this potential-circuit must be deducted from the wattmeter reading. The true power taken by the load, $P = P'-E2/RP$ where P' is the wattmeter reading, E the load volt-

age and Rp the resistance of the wattmeter potential-coil circuit.

An idea of the magnitude of this correction may be obtained from the following example.

Example.—A certain wattmeter indicates 157 watts when it is connected in the manner shown in Fig. 54 (6). The line voltage is 120 volts and the resistance of the wattmeter potential-circuit is 2,000 ohms. How much power is taken by the load? $P = 157-1202/2,000 = 157-7.2 = 149.8$ watts.

It will be observed that a considerable percentage error would result in this case if the wattmeter loss were neglected.

The Weston Electrical Instrument Co. manufactures an instrument which compensates for this loss. A small auxiliary coil, connected in series with the moving-coil system, is interwound with the fixed coils so that a small counter-torque is exerted, this counter-torque being proportional to the power consumed by the potential circuit.

The current-and potential-circuits of a wattmeter must each have a rating corresponding to the current and voltage of the circuit to which the wattmeter is connected. A wattmeter is rated in amperes and volts, rather than in watts, because the indicated watts show neither the amperes in the current-coil nor the voltage across the potential-circuit.

If the current in an ammeter or the voltage across a voltmeter exceed the rating of the instrument, the pointer goes off scale and so warns the user. A wattmeter may be considerably overloaded and yet the load power-factor be so low that the needle is well on the scale. For this reason a voltmeter and an ammeter should ordinarily be used in conjunction with a wattmeter so

Fig. 55.—Wattmeter, ammeter and voltmeter connections for measuring power.' that it is possible to determine whether either the voltage or the current exceeds the wattmeter rating.

Corrections for the power taken by ammeters and voltmeters are often necessary. For example, in Fig. 55 the PR loss of the ammeter and the $E2/R$ loss of the voltmeter must be deducted from the wattmeter reading, n in addition to the wattmeter potential loss. The ammeter reads too high by the current taken by the voltmeter. This voltmeter current must be subtracted *vedorially* from the ammeter reading in order to obtain the true load current.

Polyphase Wattmeter.— Ordinarily, it requires two or more wattmeters to measure the total power of a two-phase or a three-phase circuit. If the load fluctuates, it is diffi-Fig. 56.—Interior view, Weston poly cult to obtain accurate simul-phase wattmeter taneous readings of two wattmeters. At power-factors less than 0,5, in a three-phase circuit, one of the wattmeters reverses ita reading. (See page 91, par. 48.) This necessitates reversing the connections of one of the instruments, which is often inconvenient. If both wattmeters be combined in one, that is, if both moving coils be mounted on the same spindle, the turning moments for each element add or subtract automatically, and the total power is read on a single scale.

Fig. 57.—Connections for polyphase wattmeter on 3-phase circuit.

Figure 56 shows the construction of a Weston polyphase wattmeter in which the two elements are clearly shown. Figure 57 shows one method for connecting a polyphase wattmeter in a three-phase circuit.

Fig. 58.—Connections for calibrating a wattmeter.

Although it is often more convenient to use a polyphase wattmeter, two single instruments are better adapted to precision work, as there is no mutually inductive action between the elements of the instruments such as may occur between the elements of a polyphase instrument. With two individual instruments, it is a simple matter to apply scale corrections.

Wattmeter Calibration.—A dynamometer-wattmeter is ordinarily calibrated with direct current, the connections for calibration being shown in Fig. 58. The voltage across the potential-circuit is measured with a standard direct-current voltmeter. The current is accurately measured by means of a potentiometer, although a standardized direct-current

ammeter is often sufficiently accurate. Both the current and the potential are reversed at each reading so as to eliminate the effect of the earth's field or of any stray field. The true power in watts is given by the product of the current and the voltage, as direct current is used.

IRON-VANE INSTRUMENTS 34. Voltmeters.—In Vol. I, Chap. VII, it was pointed out that instruments depending upon the solenoid action of an iron plunger were not satisfactory as ammeters. By the use of light iron vanes, jewelled bearings, etc., satisfactory types of commercial alternating-current instruments, based on the principle of magnetized iron, have been developed.

One such type of instrument, manufactured by the Weston Electrical Instrument Co., is shown in Fig. 59.

A small strip of soft iron, *M,* bent into cylindrical form, is mounted axially on a spindle which is free to turn. Another similar strip, *F,* which is more or less wedge-shaped, and with a larger radius than *M,* is fixed inside IG' a cylindrical coil. The cylindrical coil is wound with fine wire and is connected in series with a high resistance. When connected across the line, the current through the instrument is substantially proportional to the circuit voltage. When current flows through this exciting coil, both iron vanes become magnetized. The upper edges of the two strips will always have the same magnetic polarity, and the lower edges will always have the same magnetic polarity, but when the upper edges are north poles, the lower edges are south poles. Therefore, there will always be a repulsion between the two upper edges, and also between the two lower edges of the iron strips. This repulsion tends to move the spindle against the action of two springs. A pointer mounted on the spindle moves over a graduated scale and indicates the voltage.

59. — Weston iron-vane type of instrument.

This type of instrument can be used for direct current with a precision of 1 or 2 per cent. Its obvious advantages are its simplicity, its cheapness, and the fact that there is no current carried to the

moving element. When carefully calibrated, a precision of 0.5 per cent., and better, can be obtained with alternating current. This type of instrument cannot be calibrated accurately with direct current on account of the effect of hysteresis on the vanes. It should be calibrated by comparison with an alternating-current standard. Air damping is obtained by the use of a light aluminum vane moving in a restricted space.

The iron-vane principle has been applied to the inclined-coil type of instrument. A small iron vane, mounted obliquely on the spindle, Fig. 60, replaces the inclined moving coil of Fig. 51, page 54. When the pointer is at zero, this vane lies at an angle to the coil axis, as at a, Fig. 60. When current flows in the coil, the vane attempts to take such a position that the direction of its axis shall coincide with that of the magnetic field, which acts along the coil axis. This position is shown at *b,* Fig. 60. The vane in seeking this position turns the spindle which carries the pointer. The turning moment is opposed by springs. In the later models, the coils of these instruments are surrounded by iron laminations which shield them from stray fields. In the cheaper models, air damping is used, being obtained by a light aluminum vane attached to the moving element. The more expensive models employ magnetic damping, such as is used with watthour meters, a light aluminum vane moving between the poles of permanent magnets.

Fig. 60.—Inclined coil. iron-vane type of instrument.

35. Ammeters.—Owing to the difficulty of carrying any except the smallest currents into the moving system of dynamometer instruments, iron-vane ammeters are practically the only type used for commercial instruments. The Weston iron-vane ammeter operates on the same principle as the iron-vane voltmeter, Par. 34. The magnetizing coil in the ammeter is wound with a few turns of heavy wire instead of with the large number of turns of fine wire used with the voltmeter.

The General Electric Co.'s inclined-coil ammeter is of the same construction as the voltmeter, except that the coil is wound with coarse, instead of with fine, wire. (See Fig. 60.) 36. Hot-wire Instruments.—This type of instrument, described in Vol. I, Chap. VII, page 136, reads equally well on both direct and alternating-current circuits. As its deflection depends upon the square of the current *(PR* loss) the hot-wire instrument can be used as a transfer from alternating to direct current and *vice versa.* This type of instrument lacks high precision. 37. Alternating-current Watthour Meter.—The direct-current watthour meter can be used with alternating current, as the reversal of line voltage reverses both its armature and its field current simultaneously and the direction of the torque remains unchanged. At low power-factors, however, considerable error may be introduced by the inductance of the armature circuit. This causes the armature current to lag the line voltage by a small angle and although this has negligible effect at or near unity power-factor, the error at low power-factor is quite pronounced. This error may be compensated by shunting the current coils of the meter with a low non-inductive resistance.

The induction watthour meter is so much cheaper and so superior to the direct-current type that there is little necessity for using the direct-current type on alternating-current circuits.

A rear view of one type of induction meter is shown in Fig. 61. P is a potential coil which is highly inductive and is placed on one lug of the laminated magnetic circuit, this lug being over the disc *D. CC* are two series or current coils placed on two projecting lugs beneath the disc. These coils are so wound that if one tends to send flux upward the other tends to send it downward. *cw* is a small auxiliary or compensating winding placed on the potential-lug and its ends are connected to the resistance *R.* In order that the meter may register correctly, the potential-coil flux must lag the line voltage by 90. As it is impossible to make the resistance of the potential-coil zero, its current will lag by an angle less than 90. At low power-factors this introduces considerable error in the

meter registration. However, by properly adjusting the resistance *R,* the potential-coil *flux* may be brought into the 90 relation and the meter will register substantially correctly at all power-factors. To adjust the compensation, the meter is made correct at unity powerfactor and then the powerfactor is dropped to some low value, as 0.5. If the registration is now in error, it is due to improper com - Diagram of induction watthour meter. pensation. The meter is again made to register correctly by changing the resistance *R,* the two small wires of this resistance being either twisted or untwisted and then soldered. If the meter under-registers when the load current lags, the resistance *R* should be *decreased;* if the meter over-registers with lagging current the resistance *R* should be *increased.* The reverse is true with leading current.

L is a small metallic stamping placed under the potential lug and can be moved laterally by means of the lever *K.* Its function is to provide the small torque just necessary to overcome the friction of the meter. The operation of this adjustment is as follows:

Figure 62 'shows the stamping under the lug, set off center. When the flux starts to pass down through the lug a current is immediately induced in the short-circuited stamping. This current, by Lenz's law, opposes the flux entering the stamping so that the flux is crowded to the left-hand side of the lug as shown. When the flux starts to decrease, the current in the short-circuited stamping tends to oppose the decrease in the flux. This retards the time-phase of the flux in the right-hand side of the lug with respect to that in the left-hand side of the lug. The result is a sweeping of the flux from left to right across the lug. This

Fig. 62.—Shaded-pole principle of the light-load adjustment.
N (O (6) (5)
Fig. 63.—Gliding field in air-gap of induction watthour meter.
sliding flux cuts the disc and sets up eddy currents in it. These currents, reacting with the flux, produce a torque tending to drive the disc in the direction in

which the stamping is displaced from its position of symmetry. This is the "shaded-pole" principle which is also used to start small single-phase induction motors. (See Par. 119, Page 298.)

The driving torque of the meter at unity power-factor is produced as follows: Figure 63 (a) shows the current and the voltage wave in phase. If the meter is properly lagged the potential flux *tpp* is 90 behind *E.* The current flux *p* is in phase with */.* Figure 63 *(b)* shows the magnetic polarities of the meter poles for the various times indicated in (a). At 1 the current is zero so that no flux is produced by the current-coils. The potentialcoil flux is a negative maximum so that the potential-pole is $. Therefore the two current lugs must be *N* poles. At 2 the potential-coil flux is zero, but the current is a maximum. Therefore, the lower poles will be *N* and *S as* shown and the potential-lug will have an *S* on one side and an *N* on the other. At 3 the upper lug is *N* and the two lower ones *S.* Times 4 and 5 are also shown, 5 corresponding to 1.

In (1), the entire upper lug is an S-pole. In (2), this S-pole has diminished in magnitude, has moved toward the left-hand side of the lug and an A-pole appears on the right-hand side of this lug. In (3), an N-pole occupies the entire upper lug, and in (4) this has diminished and moved toward the left side of the lug.

A similar cycle takes place on the two lower lugs. In (1), both lugs are AT-poles making one large N-pole. In (2), this large Af-pole has diminished and moved toward the left, being followed by an S-pole appearing on the right. In (3), the JV-pole has disappeared, both lugs becoming S-poles, etc. By following the cycle, it will be observed that an JV-pole moves from right to left on both the upper and lower lugs. Similarly an S-pole does likewise, following the JV-pole. Therefore, the field "glides" laterally through the gap. In so doing, it cuts the disc and induces eddy currents. These eddy currents induced in the disc react with this gliding field and by Lenz's law the disc tends to follow the field. (See Induction Motor, par. 98, page

225.)

If the power-factor be zero, *fa,* Fig. 63 (a), will be either in time-phase with *tp* if the current lags or will be 180 out of phase with *tp* if the current leads. In either case, if instantaneous values of flux be taken, as in Fig. 63 (6), it will be found that there is no lateral displacement of the field in the gap but merely a sinusoidal pulsation of flux up and down in the gap. Under these conditions the torque acting on the disc is zero.

The disc of the induction meter, like that of the direct-current meter, cuts a field of constant strength produced by *permanent* magnets. This causes a *retarding* torque which is proportional to the speed of the disc. Therefore, both the *driving* torque (motor action) and the *retarding* torque (generator action) are produced on the same disc.

Calibration and Adjustment of the Induction Watthour Meter.— The induction watthour meter is calibrated in much the same manner as the direct-current watthour meter. A standard indicating wattmeter is used to measure the average power over a stated interval and the revolutions of the disc of the watthour meter are counted with the aid of a stop watch. The average meter watts are calculated by means of the equation $K \times N \times 3,600\ t$

Fio. 64.—Connections for testing alternating-current watthour meter.

where K is the meter constant, N the revolutions of the disc and t the time in seconds.

As a rule, an ammeter and a voltmeter are used in connection with such a test, as shown in Fig. 64, in order to determine the power-factor. Instrument losses should be carefully investigated and corrections made if necessary.

After the meter is adjusted at full load and unity power-factor by means of the retarding magnets, it is adjusted at light load by means-of the light-load adjustment. The power-factor is then lowered. Any error occurring now must be due to improper lagging. The registration is then made correct by adjusting the resistance R, Fig. 61, which is in series with the lagging coil. If the meter registers low with lagging current, the resis-

tance R should be decreased; if it registers high the resistance R should be increased. With leading current these operations should be reversed.

The induction watthour meter has certain advantages over the direct-current meter. As there is no coil-wound armature in addition to the disc, the rotating element of the induction meter is much lighter than that of the direct-current meter. Moreover, it has no commutator or delicate brushes, which are frequent sources of trouble with the direct-current meter.

The induction meter is also made in the polyphase type. Two single-phase elements act on a common spindle. There are two sets of damping magnets. (For a more detailed analysis, see "Electrical Measurements" by F. A. Laws.) 38. Frequency Indicators.—Frequency indicators are based on two principles, that of electrical resonance and that of mechanical resonance. The latter type is the more common and is simpler in operation. A number of steel reeds, each having a white index on its end, are clamped between two metal strips. Each reed has its own mechanical frequency of vibration. Behind this bank of reeds there is an electromagnet, the coil of which is excited by the circuit whose frequency it is desired to measure. The reed whose frequency is that of the circuit will vibrate with the greatest amplitude, Fig. 65. With the exception of one or two reeds near this one, none of the others will be affected. Therefore the frequency is determined by noting the scale reading opposite this reed. Were the reeds unpolarized, they would be attracted equally well by either a north or a south pole. An adjacent permanent magnet keeps the reeds polarized, so that the reed of a particular mechanical frequency will respond to the same electrical frequency. The reeds are usually so arranged that there is a reed for every half cycle. Figure 65 shows the Frahm type of indicator, as manufactured by Hartmann and Braun.

39. Power-factor Indicators.—Power-factor indicators and synchroscopes are based on the principle of the Tuma phasemeter. In Fig. 66, F is a fixed coil

carrying the circuit current. MM' are two flat coils wound with fine wire. They are fastened rigidly together and mounted on a spindle free to rotate. There

Fio. 66.—Principle of Tuma phase-meter.

is no mechanical control whatever of this moving element, such as springs, for example. The angle between the coils is 90, or nearly so. The windings of the two coils MM' are connected together at the common point A, and A is connected to the same side of the circuit as F. A non-inductive resistance R is connected between M and the other side of the line. A high inductance L is connected between M' and the other side of the line. The currents in M and M' may be assumed to differ by 90 in time-phase. Assume that the power-factor of the load is unity. The current in coil M' lags the line voltage by 90, hence lags the flux due to coil F by 90 and therefore exerts no torque. The current in coil M is in time-phase with the line voltage and hence with the flux due to coil F and will therefore move into the plane of coil F as there is no restraining torque. Hence, at unity power-factor, the entire moving element takes such a position that the coil M is in the plane of coil F.

If the power-factor of the load is zero, the current and the voltage differ in phase by 90. Hence the current in coil M and the flux due to coil F have a time-phase difference of 90, and coil M exerts no turning moment. However, the current in coil M' is now in time-phase with the flux due to coil F, and therefore coil M' will move into the plane of coil F. The moving system will then have a position of 90 from its position at unity powerfactor. That is, when the current changes its time-phase by 90, the moving element of the indicator changes its space-position by 90. The direction in which the element turns depends on

Fio. 67.—Three-phase power-factor indicator.

whether the current lags or leads the voltage. For intermediate power-factors, it can be shown that the angle of the moving system corresponds to the circuit power-factor angle. If the scale is calibrated in degrees, the pointer can be made to indicate the power-factor *angle* of the circuit. To make the indicator read power-factor, it is necessary merely to make the scale divisions proportional to the cosine of the power-factor angle. In practice the current is led into the moving system through strips of annealed silver foil which exert no appreciable control on the moving system.

As it is impossible to obtain either a pure resistance or a pure inductance, the currents in coils *M* and *M'* will not differ exactly by 90 in time-phase. It can be shown that if the space angle between coils *M* and *M'* be made equal to the angle of phase difference of their currents, the instrument indicates correctly.

If the angle between the two coils *MM'* be made 120, as shown in Fig. 67, the instrument can be made to indicate threephase power-factor, if the system is balanced. Non-inductive resistances *R, R* are now connected in series with each of the moving coils. The fixed coil is connected in one line of the threephase system and the common terminal of the two moving coils connects to this same line. The other terminal of each of the moving coils connects to one of the other two lines of the threephase system, as shown in Fig. 67. This is the scheme of connections for the power-factor indicator of the General Electric Co., so often seen on switchboards. The instrument indicates the three-phase power-factor if the system is very nearly balanced. If the system is unbalanced, the reading has little significance.

40. Synchroscope.—Before connecting an alternator to the bus-bars and in parallel with other alternators, it is necessary not only that its voltage be the same as that of the bus-bars but that it be in phase opposition as well. This corresponds to having direct-current generators of the same polarity before connecting them in parallel.

A synchroscope is an instrument for indicating when machines are in the proper phase relation for connecting in parallel, and at the same time for showing whether the incoming machine is running fast or slow. This type of instrument is based on the principle of the power-factor indicator. A diagram of one type of synchroscope is shown in Fig. 68. A horse-shoe magnetic circuit is excited by a winding which connects to the incoming machine, usually through a potential transformer. The moving coils are the same as those of the Tuma phase-meter, except that the connections are made through sliprings. This allows the coils to revolve freely. The moving element is connected across the bus-bars, usually through potential transformers. If the incoming machine has the same frequency as the bus-bars, the pointer remains stationary. When the machines are in the proper phase relation for closing the switch, the pointer is over an index on the dial, this position being shown in Fig. 69. The direction of rotation of the pointer shows whether the incoming machine is fast or slow. The generator switch is usually thrown when the pointer is rotating slowly in the "fast" direction and is approaching the index. Figure 69 shows a General Electric synchroscope and its mounting.

41. The Oscillograph.—It is often desired to investigate transient conditions in electrical circuits, such, for example, as the current and the voltage relations during the blowing of a fuse, or during the short-circuit of an alternator, or in oscillations produced by switching, etc. Further, it is desirable to have apparatus which will show the current and the voltage waves in alternating-current circuits during steady conditions. The oscillograph is an instrument which is capable of meeting these requirements.

Its principle is quite simple, being that of a D'Arsonval galvanometer (Vol. I, Chap. VII, page 123), as shown in Fig. 70(a). A small phosphor-bronze strip or filament is stretched over two clefts, *CC,* around a small pulley *P* and back again. The spring S acting on the pulley keeps the two lengths of the strip in tension. This filament is placed between the poles of a strong electromagnet. When a current flows through the filament, one length of the filament moves outwards and the other inwards. A very small mirror *M* is cemented across the two lengths of the filament and is given a rocking motion by this movement of the filament. If a beam of light be reflected from this mirror, it will be drawn out into a straight line by the mirror vibration. If

Vibrating element of oscillograph. Method of drawing out vibrating beam into a wave. Fig. 70.

this beam of light be made to strike a rotating mirror, in the manner shown in Fig. 70 (6), the rotation of the mirror introduces a time element and the wave is drawn out so that its characteristics are shown.

The instrument is merely a galvanometer having a single turn and a very light moving element. This makes the moment of inertia very small. Also, the filament is under considerable tension, so that its natural frequency of vibration is very high, being from 3,000 to 10,000 cycles per second. These characteristics are necessary in order that the filament may respond accurately to the comparatively high frequency variations which it is called upon to follow. The moving element is usually immersed in oil so that its movement is properly damped and the filament is kept cool.

Figure 71 shows the general arrangement of a laboratory type of oscillograph.

The light from the arc lamp strikes the two total-reflecting prisms by means of which the beam is turned at right angles and

Fig. 71. — Typical oscillograph.

directed upon the vibrator mirrors at *V.* These mirrors reflect the light back through the cylindrical lens, which concentrates the beam. A plane mirror *M* reflects the light down to a rotating mirror which in turn reflects it, drawn out as a wave, on the

Fig. 72.—Method of connecting oscillograph vibrators in circuit.

viewing screen. It is often desired to obtain a photographic record of the phenomena which occur. For this purpose a sensitive photographic film is wound on the film drum, which is driven by a motor. The mirror *M* is then pulled up out of the way and a mechanism causes the shutter to open and close during one revolution of the drum. In this case the time axis is furnished by the movement of the film.

The oscillograph vibrators are connected into the circuit in the same manner as direct-current ammeter and voltmeter coils are connected, Fig. 72. As the current vibrator can carry but a small current, about 0.1 amp., it is connected in parallel with a non-inductive shunt which is in series with the line. The voltage vibrator is connected across the line in series with a high non-inductive resistance. The current vibrator will then vibrate with an amplitude proportional to the circuit current and in phase with it. The current through the voltage vibrator will be proportional to the circuit voltage and in phase with it.

CHAPTER IV

POLYPHASE SYSTEMS

42. Reasons for the Use of Polyphase Currents.—In many industrial applications of alternating current, there are objections to the use of single-phase power.

In a single-phase circuit, the power delivered is pulsating. Even when the current and voltage are in phase, the power is zero twice in each cycle, as shown in Fig. 19, page 22. When the power-factor is less than unity, the power is not only zero four times in each cycle, but it is also *negative* twice in each cycle. This means that the circuit returns power to the generator for a part of the time. This is analogous to a single-cylinder gasoline engine in which the fly-wheel returns energy to the cylinder during the compression part of the cycle. Over the complete cycle, both the single-phase circuit and the fly-wheel receive an excess of energy over that which they return to the source. The pulsating nature of the power in single-phase circuits makes such circuits objectionable in many instances.

A polyphase circuit is somewhat like a multi-cylinder gasoline engine. With the engine, the power delivered to the fly-wheel is practically steady, as one or more cylinders are firing when the others are compressing. This same condition exists in polyphase electrical sys-

tems. Although the power of any one phase may be negative at times, the *total power* is constant if the loads are balanced. This makes polyphase systems highly desirable for power purposes.

The rating of a given motor, or generator, increases with the number of phases, an important consideration. Below are the approximate capacities of a given machine for different numbers of phases, assuming the single-phase capacity as 100.

Single-phase 100
Two-phase 140
Three-phase 148
Six-phase 148
Direct-current 154

The same machine operating three-phase or six-phase has about 50 per cent. greater capacity than when operating single-phase. A machine has the same capacity whether connected three-phase or six-phase, because the same windings are used in the same manner for each. (The foregoing table does not apply to synchronous converters. The ratio of polyphase to single-phase capacity in converters is much greater than that shown in the above table. See page 355.)

A minor consideration in favor of three-phase power is the fact that with a fixed voltage between conductors, the three-phase system requires but three-fourths the weight of copper of a single-phase system, other conditions such as distance, power loss, etc., being fixed.

43. Symbolic Notation.—The solutions of problems involving circuits and systems containing a number of currents and volt *a* 00

Fig. 73.—Symbolic notation applied to voltage and current vectors.

ages are simplified and are less susceptible to error if the current and voltage vectors are designated by some systematic notation, of which the following is one type. If a voltage is acting to send current from point a to point b, Fig. 73 (a), it shall be denoted by *Eab*. On the other hand, if the voltage tends to send current from b to a it shall be denoted by *Eba*-Obviously, *Eab* = —*Eba*. It may seem as if alternating currents cannot be considered as having direction

since they are undergoing continual reversal in direction. The assumed direction of a current, however, is determined by the actual direction of the flow of *energy*. In an alternator the energy comes *out* of the armature and the current is considered as flowing *out* of the armature, even although it is actually flowing *into* the armature for half the Corresponding to the voltage *Eab*, Fig. 73 (a), the current *I&* flows from a to b in virtue of this voltage. The current flowing from 6 to a must be opposite in direction to that flowing from a to b. Therefore *ha* = — *lab*-This relation is illustrated in Fig. 73 *(b)*, in which *Iab* differs in phase from *Iba* by 180. *Eba* is 180 from *Eab*.

Figure 74 represents a circuit network *abcde*. The parts of the network, 06, *be*, etc., may be either resistances, inductances, capacitances, or sources of emf. such as alternator or transformer coils. It is obvious that the voltage from a to c is equal to the voltage from a to 6 plus the voltage from b to c. That is, *Eac* = *Eab* + *Ebc*. It is to be noted that when several voltages in series are being considered, the first letter of each subscript must be the same as the last letter of the preceding subscript. Figure 75 (a) shows vectorially the voltage *Eab* and the voltage *Ecb*. To obtain the voltage *Eac*, *Ebc* is necessary. Therefore *Ecb* is reversed giving *Ebc*. *Ebc* added vectorially to *Eab* gives *Eac*.

Fig. 74.—Circuit network.

Ibd --Ial=fbc+Ibd (a) 00

Fig. 75.—Examples of symbolic notation.

Currents may be treated in a similar manner, the principle involved being Kirchhoff's first law. For example, in Fig. 74, the current *Jab* = *Ibc* + *Im*-Figure 75 (6) shows currents *hc* and *Idb*. *la* is reversed giving *ha* and this is combined vectorially with *he* to obtain *Iab*.

This notation not only distinguishes the various currents and voltages but the directions in which they act as well, It is to be noted that the use of arrows is not necessary, the subscripts denoting the directions of the vectors.

44. Generation of a Three-phase Current.—As three-phase is now the most common of the polyphase systems, it will be (a)

Fig. 76. — Generation of 3-phase current.

considered first. Figure 76 (a) shows three simple coils, 120 apart and fastened rigidly together. These coils are mounted on an axis which can be rotated. The coils are shown rotating in a counter-clockwise direction in a uniform magnetic field. The current can be conducted from each of these three coils by

Fig. 77.—Three-phase voltage waves and vector diagram.

of slip-rings, as shown in Fig. 76 (b). The terminals of connected to rings a', those of b to rings b', etc., making six slip-rings in all.

Figur% 77 (a) shows as /?,,,, the voltage in coil a. E_{oa} is zero and is increasing in a positive direction when the time t is 0. Obviously the voltage induced in coil b will be 120 electrical timedegrees behind E_{oa} and that induced in coil c will be 240 electrical time-degrees behind E_{oa}, as shown in Figs. 77 (a) and 77 (b). These three voltages constitute the elementary voltages generated in a three-phase system.

An examination of Fig. 77 (a) shows that for any particular instant of time, the algebraic sum of these three voltages is zero. When one voltage is zero, the other two are 86.6 per cent. of their maximum values and have opposite signs. When any one voltage wave is at its maximum, each of the others has the opposite sign to this maximum and each is 50 per cent. of its maximum value.

Figure 77 (6) shows the vectors representing these three voltages, the vectors being 120 apart.

Each of the coils of Fig. 76 (a) can be connected through its two slip-rings to a single-phase circuit. This gives six slip-rings and three independent single-phase circuits. With a rotating field and stationary armature type of generator, which is the most common type met in practice, the six slip-ririjgs would not be necessary, but six leads would be taken dire#pfcom the armature. '

In practice, however, a machine seldom supplies tsree independent circuits

by the use of six wires.

46. Y-connection.—The three coils of Fig. 76 are sh'wvn simple diagrammatic form in Fig. 78. The three correspeWi ends, one for each coil, are tied together at the common 'pomt& This is called the Y-connection of the coils. Ordinarily o three wires, aa', bb' and cc', lead to the external circuit, althou the neutral wire oo' is sometimes carried along, making a th phase, four-wire system. Figure 79 (a) again shows the three coils.and Fig. 79 (b) three corresponding voltage vectors, E_{oa}, E, and E0ethree voltages are called the coil or,Y-voltages. Let required to find the three line voltages E_{ab}, $E_{i,c}$ and E_{ca}. T. voltage $E_{ab} = E_{ao} + E\&$ (Par. 43). $E_{ao}$is not on the diagram but is obtained by reversing E_{oa}. E_{ao} is the vectorially to E giving E_{ab}.

From geometry, E_{ai}, lags the coil voltage E by areaU equal and are 120 apart. _ Each line voltage. is 30 ouLo£

Fig. 78.—Y-oonnection of generator coils.

phase with one of its respective coil voUages. The three line voltages are each -y/3, or 1.73, times the coil voltage.

It is obvious from Fig. 79 (a) that the three coil currents I_{oa}, I and I_{oc} are respectively equal to the threp line currents I_{aa},, In and /.,», as the coil and line are in series

Fio. 79.—Y-connection and corresponding voltage vector diagram.

V _Therefore, in a Y-system the line currents and the respective coil currents are equal._ Moreover, as the three coils meet at a comlon p&nt, the vector sum of the three currents must be zero by irchhdff's first law, provided there is no neutral conductor and iyirrent. That is $I_{oa} + I_d, + I_{oc} = 0$

V' _Power in Y-system._—Figure 80 shows the three currents I_{oa}, I_{ob} and I_{oc} of coils oa, ob, and oc respectively. Unity power-factor is assumed and the three currents are-lhereforeJiTjThaat. with their respectige_-coil voltageg. A balanced system is assumed and_the _three currents are therefore equal in magnitude.

The coil current I_{oa} andtne line cur-

rent I_{aa} are the same current. Therefore, the line current /,,,,' is 30 out of phase with the line voltage E_{ca}, when the power-factor is unity. This is true for each phase.

The power delivered by each coil is $P' = E_{oa} I_{oa}$ (unity power-factor) and the total power delivered by the generator is three times this. $P = 3E_{coil}I_{coil}$ t-ft i N t" E Etc Fio. 80.—Relation of line to coil volt ages and currents in a Y-system, unity power-factor.

Fig. 81.—Relation of. line to coil voltages and currents in A Y-systent Powerfactor = cos 6. .

As the power in the line is the same as that delivered by the generator, substituting $E_{nre}/-/3$ for the value of E_{co}», $3 I_{coil} = \sqrt{3}E_{line} I$ line (29)

'vs1, the coil current and the line current being equal.

In a balanced three-phase system, the line power at unity powerfactor is equal to-/3 times the line voltage times the line current.

Figure 81 shows this same three-phase system when the powerfactor is no longer unity. Each coil current now lags its respective coil voltage by the angle 6.

The total coil power is now $P = SE_{coil} I cou \cos 6 Coii$ The system power is $P = /3E_uM Inne \cos 6coit$ (30) and the system kw. is equal to $E_{nne}I_{nne} \cos O_{cou}$ 1,000 Therefore, in a balanced three-phase system, the system powerfactor is the CQsine.of the argle between the coil current and the coil voltage.

The angles between theiine currents and the line voltages are not power-factor angles, for they involve the factors (6 — 30) (Fig. 81) and also (6 + 30), 6 being the coil power-factor angle.

Obviously the system power-factor, which is the coil powerfactor, is RF wjiere P is the total system power.

If the system is unbalanced, that is, if the currents or voltages are not equal or are not 120 apart, the question arises as to just what the system power-factor is under these conditions. Where such unbalancing is not very great, equation (31) is used, the line currents and voltages being averaged. The system power-factor has little significance when the unbalancing is considerable.

Example. — A three-phase alternator has three coils each rated at 1,330 volts and 150 amp. What is the voltage, kv-a. , and current rating of this generator if the three coils are connected in K? *Eiine* = V3 X 1330 = 2,300 volts. *Am.* Rating = /3 X 2300 X 150 = 600 kv-a. *Ans.*

Current rating = 150 amp. *Ans.*

46. Delta-connection. — The three coils of Fig. 76 can be connected as shown in Fig. 82 (a), the diagram being simplified in Fig. 82 (6). The end of each coil, which, in Fig. 78, was connected to the neutral, is now connected to the outer end of the next coil, as shown in Fig. 82 (a). As points o and a are now connected directly together, *Eca = Eca,* etc. The o's are now superfluous and are dropped. *Fig.* 83 (a) shows vectorially the three voltages *Eab, Ebc,* and *Eca,* acting from *a* to 6, 6 to *c,* and *c* to a, respectively.

At first sight Fig. 82 looks like a short-circuit, the three coils, each containing a source of voltage, being short-circuited on

Fio. 82.—The delta-connection of alternator coils themselves. The actual conditions existing in this closed circuit may be demonstrated by the use of the subscript notation. Assume that the coil *be* is broken at *c',* Fig. 84 (a). The voltage *Ebc = Eba + Eae.* The vector sum of these two voltages, shown in Fig. 84 (6), lies along voltage *Em,* and is equal to it. Therefore,

Fio. 83.—Relation of line to coil voltages and currents in a delta-system, unity power-factor.

the voltage *Ecc* = 0 and points c and *c'* can be connected without any resulting flow of current. This is the same condition which exists when two direct-current generators having equal vt»ltages are connected in parallel. No current flows between the two if the proper polarity is observed. .

The coil currents of Fig. 82 are shown in Fig. 83 in phase with their respective voltages, balanced conditions being assumed. The line current *4 oaf* /6a I *4 ca*

This addition is made vectorially in Fig. 83 (a), giving /,„,' 30 from *EM.* It will be observed that /,„,' is-/3 times the coil current. Line currents *lw* and *Icc* may be found in a similar manner, with the result shown in Fig. 83 (6). Therefore, in the delta-system there is a phase difference of 30 between the line currents and the line voltages at unity power-factor, just as in the Y-system.

It is obvious that the line voltage is equal to the coil voltage in a delta-system. Moreover, the sum of the three *voltages* acting around the delta must be zero by Kirchhoff's second law.

Fio. 84.—Showing that the sum of three delta voltages is zero.

In a balanced delta-system, the line voltage is equal to the coil voltage, but the line current is /3 times the coil current.

Figure 85 shows three lamp loads, each requiring 10 amp. at 115 volts. They are first connected in Y and then in delta. In order to supply the proper voltage in each case, there are 199 volts across lines in the Y-system and 115 volts in the deltasystem. There are 10 amp. per line in the Y-system and 17.3 amp. per line in the delta-system. The power supplied is the same in each system.

Power in Delta-system.—The total power in a delta-system is *P = 3 Ecou Icoil cos dcou*

This power is equal to that in the line, as there is no intervening loss. Also, the line current 115-Volt Lamp Banks Connected in Delta

Fig. 85. — Lamp loads in Y and in delta.

Hence, substituting in the above equation

This equation is the same as equation (30) (page 83) for the Y-system. This should be so, for the relations in a three-phase line are the same whether the power originates in a delta-or in a Y-connected generator.

The power-factor of the delta-system is the same as that for a Y-system.

(33) -cos where *P* is the total power of the system, and *E* and / are the line voltage and line current respectively.

The denominator, V3 *El,* equation (33), gives the *volt-amperes* of the three-

phase system. The *kilovolt-amperes* of a three-phase system are given by /3 *El*/1,000.

METHODS OF MEASURING POWER IN THREE-PHASE SYSTEM 47. Three-wattmeter Method.— Let (1), (2) and (3), Fig. 86 (a), be the three coils of either a Y-connected alternator or of a

Fio. 86. — The 3-wattmeter method of measuring 3-phase power.

Y-connected load. If the neutral of the Y is accessible, it is possible to measure the power of each phase by connecting the current-coil of a wattmeter in series with the phase and by connecting the wattmeter potential-coil across the phase, as shown in Fig. 86 (a). Therefore , *Wi, Wz* and *W3* measure the power in loads 1, 2, and 3 respectively, regardless of power-factor, degree of balance, etc. The total power

P = w i + W 2 + W3

If the loads are balanced,

If the potential circuits of the three wattmeters have equal resistances, these three potential circuits constitute a balanced Y-load, having a neutral *0'.* As coils 1, 2, and 3 and these three wattmeter potential-circuits are both symmetrical systems, *O'* must be at the same potential as *0.* Therefore, no current flows between *0* and *0'* and the line can be cut at *X* without changing existing conditions. Figure 86 (6) shows the three-wattmeter connection for a three-phase system. It can be shown that the total power is the sum of the wattmeter readings even though the wattmeter potential-circuits have different resistances. Under these conditions, however, the wattmeters may not all have the same reading, even with balanced loads.

The three-wattmeter method is well adapted to measuring power in a system where the power-factor is continually changing, as in obtaining the phase characteristics of a synchronous motor. If the three instruments have equal potential-circuit resistances, they read alike regardless of power-factor, if the loads are balanced. The three-wattmeter method is necessary in a threephase four-wire system, as a system of *n* wires ordinarily requires *n* — 1 wattmeters in order to measure the power correctly.

The Y-box.—The use of the Y-box is based on the principle that each of the three wattmeters of Fig. 86 reads the same, if the loads are balanced. Under these conditions the total power $P = 3W$. If two resistances, each equal to the resistance of the potential-coil of Wi, be used in conjunction with this potential-coil, the wattmeters Wz and $W3$ are not necessary. As a rule these two equal resistances are mounted in the same box and are connected as shown in Fig. 87. Accurate results can be obtained with this method only when the loads are balanced. 48. Two-wattmeter Method.—The power in a three-wire, three-phase system can be measured by two wattmeters connected as shown in Fig. 88. The current-coils of the two instruments are connected in two of the lines and the potential-coil of each instrument is connected from its respective line to the

Fio. 88.—Two-wattmeter method of measuring 3-phase power.

third line. Under these conditions the total power passing through the system $p = w$, $+ W2$ regardless of power-factor, balance, etc. The choice of the plus or the minus sign will be explained later.

One method of proving that these instruments give the correct power is as follows: Let *ei, ez, 63* and *ii, iz, ia* be the respective voltages and currents of the three loads at any particular *instant.* These being instantaneous values, the power at the instant under consideration is equal to their products regardless of power-factor. That is, the instantaneous power

Substituting

As the line voltages in a Y-system are the *differences* of the proper coil voltages (Page 80, Par. 45) Wi reads *(e — ez)i* and Wz reads *(63 — e2)t'3*

The same proof may be used for a delta-load, except that $ei + e2 + e3 = 0$

It is shown in Pars. 45 and 46, that a phase difference of 30 exists between the line voltage and line current at unity power

Fig. 89.—Vector diagram illustrating 2-wattmeter method for measuring 3-phase power, balanced load.

factor. For power-factors other than uni-

ty, this phase difference becomes (30 + 0), where *6* is the power-factor angle of the coil. Figure 89 (a) shows two wattmeters, Wi and Wz, measuring the power taken by a balanced three-phase, Y-connected load. The wattmeter Wi is so connected that the current */&,,* flows in its current-coil and the voltage *Eba* is across its potential-circuit. Therefore, the reading of Wi is equal to the product of */&,,, Eba* and the cosine of the angle between this current and this voltage. Figure 89 (6) gives the vector diagram of the load. The three coil voltages *Eao, Eb0,* and *Eco* are all equal and 120 apart. The coil currents */,,,,, I(,0,* and *Ico* are equal and lag their respective coil voltages by the angle *6.* The voltage *Eba* is found by reversing *Eao,* giving *Eoa,* and then adding *Ebo* and *Eoa* vectorially *(Eoa = Eba + Eoa).* The current */&,,* is given. The angle between *Eba* and *ho* is 30 — *6.* Therefore, the reading of this wattmeter is $Wi = Eba\ Ibo$ cos $(30-6) = Eliae\ Iline$ cos $(30-0)$

Likewise, the wattmeter Wz reads the product of *Eca, Ico* and the cosine of the angle between them. From the vector diagram, Fig. 89 (6), *Eca* is found by adding vectorially *Eco* and *Eo (Era = Eco + Eoa)*-The current */«,* is given. The angle between *Eca* and *Ico* is 30 + 0.

Therefore the reading of this wattmeter is

$Wz = ECJCO$ cos $(30 + 9)$
$= E!ine\ Iline$ cos $(30 + 0)$ Summarizing

$W\ i = EI$ cos $(30-0)$
$Wz = EI$ cos $(30 + 0)$

where E and $/$ are the line voltage and line current, respectively, the system being balanced.

Wi and $W2$ will read alike when 0 = 0 and 0 = 180. Both conditions correspond to unity power-factor. When 0 equals 180, however, the power has reversed. The two instruments also read alike at zero power-factor (0 = 90), although this condition is seldom realized.

When 0 = 60, corresponding to a power-factor of 0.5, Wz reads zero, as cos (30 + 60) = cos 90 = 0. In this case, the reading of Wi gives the total power. For angles greater than 60, corresponding to

power-factors less than 0.5, cos (30 + 0) becomes negative, Wz reads negative and the total power becomes $P = Wi-W2$

Therefore, discretion must be used when two single instruments are employed, as the total power may be either the *sum* or the *difference* of the readings.

It may also be shown that $W — W$ tan 0 = *VB* (34) where 0 is the coil power-factor angle. Therefore it is possible to obtain the power-factor in a balanced three-phase system by means of the wattmeter readings alone.

Another convenient method for determining the power-factor from the wattmeter readings is to divide the smaller wattmeter reading by the larger, and then to use the curve shown in Fig. 90. This curve is plotted with the ratio against power-factor.

When $Wz/Wi = 1.0$, the power-factor is 1.0; when $Wz/Wi = 0$, the power-factor is 0.5; when Wz/Wi is negative, that is, it becomes necessary to reverse T72, the power-factor is less than 0.5. By means of a curve like that of Fig. 90, the powerfactor may be read directly from the ratio of the two wattmeter readings.

— 1.0-9-.8-1-.6-5-.4-.3-.2-.1.0+.1+.2 +.3+.4 +.5+.6+.1+.8+.9+1.0 _ Smaller Keating _ Cot(J+30) ':'li" Larger Beading = Cos((-30)

Fig. 90.—Power-factor diagram, 2-wattmeter method.

Example.—In a test of a three-phase induction motor, two wattmeters are used to measure the input. Their readings are 1,900 and 800 watts respectively. Both instruments are known to be reading positive. What is the power-factor of the motor at this load? Using equation (34) ,-1,900-800 _-1,100 I,900 + 800 2,700 $6 = 35.3$ cos $6 = $ cos $35.3 = 0.815.$ *Ans.* This result may be checked by Fig. 90. If a polyphase wattmeter is used (page 60, Fig. 57), the adding or subtracting is done automatically, as both elements of the instrument act on the same spindle. Therefore, the polyphase instrument, if properly connected, reads the total power at all times.

The two-wattmeter method cannot be used to measure power in a three-phase, four-wire system unless the current in the neutral wire is zero. When the current in the neutral wire of Fig. 91 is zero, the power is correctly indicated by $Wi \pm Wz$. Now apply load $B'O$ between line B and the neutral. The current to'this load will complete its circuit from wire B through the neutral without going through the current-coil of either watt

Fio. 91.—Two-wattmeter method generally not applicable to a 4-wire system.

meter. As neither wattmeter can indicate this additional load, the two wattmeters are not sufficient to measure the power in such a four-wire system under all conditions of load. TWO-PHASE SYSTEMS 49. Quarter-phase or Two-phase Systems (Sometimes Called Four-phase).—Although three-phase systems are superseding other systems, there are still many quarter-phase or two-phase systems in existence. The two-phase system is rarely used for transmission, but is used for distribution and in some instances it is specially advantageous to use two-phase machines. Quarter-phase current is generated in the elementary generator, Fig. 92 (a), by two coils A and B, 90 apart. Figure 92 (6) shows the emf. waves generated by these coils. The voltage of A leads that of B by 90. When one voltage is a maximum the other is zero. Figure 92 (c) shows these two-phase voltages represented vectorially.

The two phases may be carried along, insulated from each other, to supply two separate single-phase circuits or they may Generation of 2-phase power

A

(c) Vector representation of 2-phuse emfs.

Two-phase emt. waves

Fig. 92.—Phase relations of 2-phase electromotive forces.

supply a common load such as an induction motor, Fig. 93. The two phases are *entirely insulated* from each other in Fig. 93

Fig. 93.—Two-phase circuit in which two phases are insulated.

and no single load can be applied between the two phases. Moreover, Qnly

one value of voltage is obtainable, as the voltages of the two phases are equal.

If, however, the generator coils be connected at their neutral points and a neutral conductor carried along with the other conductors, a quarter-phase, five-wire system results, as shown in Fig. 94 (o). Moreover, three different voltages are available.

Fig. 94.—Two-phase inter-connected system giving 4-phase, 5-wire system.

If the voltages between the outer wires of each phase be 200 volts, then 200,100 and 141 volts are available, as shown in Fig. 94 (6). Thja,system is more readily unbalanced than the three-phase system, which is an objection to itsuse. Another objection is the greater number of wires.

Fig. 95.—Two-phase, 3-wire system.

If one end of the coil A be connected to one end of the coil B, a three-wire, two-phase system results, as shown in Fig. 95. This gives two different values of voltage, 200 and 283 ($= 200 /2$) volts. This system is little used because of the considerable amount of voltage unbalancing which results, even when moderate loads are applied. It should be noted that the common wire N carries a current I-/2, where / is the current in each of the two outer wires.

141 a

Fig. 96.—Mesh-connected, 2-phase winding.

A two-phase alternator may have a winding which consists of four coils. These coils may be connected in mesh as shown in Fig. 96. This corresponds to the delta-connection in a three

Fig. 97.—Measurement of power in an insulated 2-phase, 4-wire system.

phase system. As in the case of the delta, if these coils are properly connected, the winding is not short-circuited on itself. The line voltages are each equal to the coil voltage. The diametrical voltages are equal to-/2 times the coil voltage. The line currents are equal to the /2 times the coil current, because the line currents are the resultant of two equal currents having 90 phase difference.

Fio. 98.—Measurement of power in a 2-phase inter-connected system.

Fio. 99.—Measurement of power in a 2-phase, 3-wire system.

In Fig. 96, the coil voltage is 200 volts, and the diametrical voltage is 200 /2 = 283 volts. The coil current is 100 amp. and the line current is 141 amp. The total kv-a. capacity of this system is 4 X 200 X 100 *Measurement of Power in Two-phase Systems.*—In a two-phase, four-wire system, connected as shown in Fig. 97, the total power may be measured by two wattmeters. If the system is inter-connected, the loads must be balanced or this method is incorrect. If, however, the two phases are insulated from each other, two wattmeters measure the power correctly regardless of unbalance, power-factor, etc.

If the loads of a four-wire inter-connected system are not balanced, three wattmeters must be used, as shown in Fig. 98. The power is the algebraic sum of their readings. The power in a two-phase, three-wire system may be measured by two wattmeters connected as shown in Fig. 99.

CHAPTER V

THE ALTERNATOR

60. Rotating-field Type.—The generation of an electromotive force in a conductor may take place with the magnetic field stationary and the conductor moving through this field, as in a direct-current generator, or with the conductor stationary and the field moving past the conductor. It is merely necessary that there be *relative* motion between the conductor and the field. In direct-current machines, the commutator makes it necessary that either the armature be the rotating member, or that the brushes revolve with the field.

As alternators have no commutator, it is not necessary that the armature be the rotating member. Most commercial alternators have stationary armatures, inside of which the field poles rotate, as shown in Figs. 102, 103, etc. This construction has two distinct advantages. A rotating armature requires two. or more slip-rings for carrying the current from the armature to the external circuit. Such rings must be more or less exposed, and are difficult to insulate, particularly for the higher voltages of 6,600 and 13,200 volts at which alternators are commonly operated. These rings may become a frequent source of trouble, due to arcovers, short-circuits, etc. A stationary armature requires no slip-rings, and the armature leads can be continuously insulated conductors from the armature coils to the bus-bars. It is more difficult to insulate the conductors in a rotating armature than in a stationary one, because of centrifugal force and the vibration resulting from rotation.

When the field is the rotating member, the field current must be conducted to the field winding through slip-rings. As the field voltage seldom exceeds 250 volts and the amount of power is small, no particular difficulty is encountered in the operation of such slip-rings.

Usually it is difficult to get sufficient copper on the surface of an armature. This is particularly true with high-speed, high. voltage machines having armatures of small diameter. Increased space for copper may be obtained by deepening the slots. If the armature be the rotating member, the deepening of the slots is limited by the contraction of the tooth necks, as shown in Fig. 100 (a). No such difficulty is encountered if the armature be stationary, since the tooth necks increase in width with the deepening of the slots, Fig. 100 (6).

Fio. 100.—Effect of slot depth on the width of tooth necks in a rotor and in a stator.

ALTERNATOR WINDINGS 61. General Principles.—The usual direct-current armature generates alternating current, and if provided with properly connected slip-rings, alternating current may be obtained from it. On the other hand, only certain types of alternator windings can be used for direct-current armatures. The ordinary directcurrent winding is a *closed winding* (see Vol. I, page 223), but alternator windings may be either open or closed.

The general principles which govern direct-current windings hold also for windings of alternators. The span of each coil must be approximately one pole pitch; that is, the two sides of any

coil must lie under adjacent poles. The coils must be so connected that their electromotive forces add.

Alternator windings are divided into several general classes. There are single-layer and two-layer windings, usually made up of former-wound coils. Windings may be either of the lapwinding type, shown in Figs. 101 (a), 104, 108, 112, and 114, or of the wave-winding type, shown in Fig. 101 (6). In the directcurrent machine the wave winding gives a higher voltage than the lap winding, if the number of series-connected conductors and other conditions are the same. In the alternator, the wave and lap windings give the *same* voltage, if the number of series-connected conductors and other conditions are the same. An inspection of Fig. 101 (a) and (6) shows that each winding has the same number of series conductors between terminals.

(6) Slngle-pha partially distributed wave

Singl«-pbate partially dtstributed lap winding-/ partially distributed wave winding'

Fig. 101. — Single-phase lap and wave windings.

The type of winding shown in Figs. 101, 104, 111, etc., is called the *barrel* winding. The type shown in Fig. 105 (a) is called the *spiral* winding. A development of the spiral into the *chain* winding is shown in Figs. 107 and 110.

52. Single-phase Windings. — Figure 102 shows a single-phase, single-layer, half-coil winding for a four-pole machine. This machine has four slots and four poles, making one slot per pole. This winding is called a *half-coil* winding, because there is but one-half coil or coil-group per pole. The two coils are shown connected in series, and *Ti* and *Ti* are the terminals of the winding. Figure 103 shows the same type of winding as Fig. 102, except that four coils are now used. Therefore, there are two coil sides per slot. This is called a single-phase, two-layer, *whole-coil* winding. It is called a whole-coil winding because there is one ri r,

Fio. 102.—Single-layer, half-coil winding, one slot per pole.

Fio. 103.—Two-layer, whole-coil winding, one slot per pole.

Fio. 104.—Whole-coil, 2-layer winding, two slots per pole.

coil or coil-group per pole. The winding of Fig. 102 may be obtained from Fig. 103 by swinging coil *B* into the plane of coil *A* and coil *D* into the plane of coil *C.*

One slot per pole is seldom found in practice, as the surface of the armature is not economically used, and in addition a poor voltage wave results. Figure 104 shows the winding of Fig. 103, except that there are now two slots per pole instead of one. This is also a two-layer winding, as there are two coil sides per slot, placed one above the other.

CO

Barrel-type Winding which can replace Spiral Winding of *(a)*
Fio. 105.

Instead of making the coils lap one another, as is done in Fig. 104, the winding may be placed on the armature in the manner shown in Fig. 105. This is called a *spiral* winding. It will be observed that in this particular winding the coils themselves have a pitch less than 180 electrical space-degrees. Notwithstanding this lesser pitch, the winding is not considered as having the properties peculiar to a fractional-pitch winding. The slot conductors may be re-connected by barrel-type end-connections, as shown in Fig. 105 (c), without changing the electrical characteristics of the winding. This gives a full-pitch, half-coil, *barrel* winding. Therefore, the spiral winding for which this can be substituted is considered as a full-pitch winding. The differential action of the coil sides of Fig. 105, due to their not having a full pitch, is taken into consideration by the beltfactor constant. (See page 121, Par. 58.) admirably adapted to high-voltage machines. Although coils of several different sizes must be kept in stock as spares, a coil may

The inside coil shown dotted at a may be added to the winding, but it contributes so little to the generated electromotive force, because of its small pitch, that to use it is wasteful. This winding has but one coil-side per slot, so that it is also a single-layer winding. As the ends of the coils may be bent so that they all lie in a single vertical plane, as shown at (6), Fig. 105, it is called a *single-range* winding. Two-and three-range windings are also used in practice.

At the present time, single-phase machines are somewhat limited in their field of application. They are used more or less extensively for single-phase railway electrification and for some electric furnace work. Instead of building a single-phase machine for these purposes, however, Y-connected, three-phase machines are commonly used, as such machines are standard. Two phases of the Y are used in series. A spare phase is also available.

63. Two-phase Windings.—Two-phase windings are merely two single-phase windings displaced 90 electrical space-degrees from each other on the armature. If another winding be add i to Fig. 103, the coil sides of this new winding being midway between those shown in Fig. 103, a two-phase winding results, as shown in Fig. 106. These two windings are 90 electrical spacedegrees apart, so that their voltages differ in time-phase by 90.

Figure 107 shows a two-phase spiral or chain winding. This is merely adapting the winding of Fig. 105 to two phases. There are now eight slots per pole rather than six. As the coil ends in this winding must necessarily lie in two different vertical planes, in order to pass one another, the winding is called a *tworange* winding.

The chief advantage of a chain winding is the considerable space between the coil ends, so that there is little opportunity for electrical breakdown at these points. Therefore, they are

Phase A, Phase B. (A) (B)

Fig. 106.—Two-phase, 2-layer winding, one slot per pole per phase.

Phase A Phase B

Fig. 107.—Two-phase chain winding, 2-range.

be replaced more easily than it can be in the lap winding, where a large number of coils must often be removed in order to replace a single coil. 54. Two-phase

Lap Winding.—The lap winding is the most common type of alternator winding. With it, there are very few limitations in the choice of number of slots, pitch, etc. The coils are all alike, requiring the minimum number of spares, and the winding is very flexible in the matter of connections. For example, with a lap winding it is a simple matter to change a 440-volt winding to one of 220 volts by paralleling.

To obtain a lap winding, more coils are added to the winding shown in Fig. 103. The connections of the coils of any one phase are almost identical with those in the direct-current windings described in Vol. I, Chap. X. Direct-current lap windings may be used for single-phase and for polyphase voltages by taps at suitable points, with connections to slip rings, as is done in the synchronous converter. (See page 342, Par. 138.)

Fio. 108.—Two-phase, full-pitch, lap winding, four slots per pole per phase.

Figure 108 shows a two-phase lap winding, in which there are eight slots per pole, making four slots per pole per phase. This is a full-pitch winding, the coil pitch being 8 slots, which is the number of armature slots per pole. The connections of phase B are omitted for the sake of clearness as they are identical with those of phase A. It will be observed that the coil sides in any one slot are both of the same phase. This is not the case with fractional-pitch windings.

56. Three-phase Windings.—The difference between twophase and three-phase windings is merely in the number of phase-belts per pole. Figure 109 shows the simple winding of Fig. 103 adapted to three-phase. For clearness the end connections of phase A alone are shown. It is necessary merely to add two more windings, equally spaced, between those of Fig. 103 in order to obtain a three-phase winding having one slot per pole per phase. The three phases of Fig. 109 may be connected either in Y or in delta.

FiO' 109.—Three-phase, 2-layer winding, one slot per pole per phase.

Figure 110 shows a three-phase chain winding, in which there are six slots per

pole, making two slots per pole per phase. This is a two-range winding, for the coil ends in order to pass one another must lie in two different planes perpendicular to the machine shaft. If the number of coil groups per phase is odd, which occurs if the number of poles is not a multiple of four, coils having one long side and one short side must be used to complete the winding. This occurs in the six-pole winding of Fig. 110, in which two. coils d and d' must be of trapezoidal shape in order to pass the coil ends of phase A and so complete the winding. A plan view of such coils is shown at (a), Fig. 110. Six different sizes of coils are required in this winding, making it necessary to carry a considerable variety of spares. The connections of the coils of phase B only are shown, the other phases being connected in a similar manner.

Fig. 110.—Three-phase chain winding, requiring special coils.

Figure 111 shows a three-phase, full-pitch, lap winding, in which there are 12 slots per pole. The coil pitch is therefore equal to 12. For clearness the connections of the A phase alone are shown, the connections of the B and C phases being similar. It will be observed that in this type of winding the two coil sides in any one slot belong to the same phase.

Figure 112 is similar to Fig. Ill, except that the winding is now ⅚ pitch. A coil, instead of having a pitch of 12 slots, now has a pitch of 10 slots, so its spread is no longer equal to a full pole-pitch. This is a *fractional-pitch winding,*

The advantages of this type of winding are that it improves the wave form, there is an appreciable saving of copper in the coil ii-pitch, 2-layer lap winding, four slots per pole per phase.

ends, and the inductance of the winding is reduced, because of the lesser mutual inductance between those conductors which lie in slots containing conductors of the other two phases. (See Fig. 112.) The coil-end inductance is also reduced because of the lesser length of free conductor. Such windings generate (a) Full pitch. (6) Fractional pitch.

Fig. 113.—Relation of coil-side voltages in full-pitch and in fractional-pitch windings.

slightly less emf. than full-pitch windings under the same conditions, since the two coil sides do not lie under the same parts of the pole at any given instant and therefore their emfs. are slightly less than 180 apart. This is illustrated in Fig. 113. E_i is the emf. induced in the conductors comprising one side of a coil and E_2 is the emf. induced in the conductors comprising the other side of the coil. E_i is equal to E_2 numerically as each is induced by the same number of conductors cutting the same flux. Figure 113(a) gives the relation of the induced emfs. E_i and E_z in the two coil sides respectively, when a full-pitch coil is used. When one side of a coil is under a north pole the other side is in a corresponding position under a south pole. Therefore the induced emfs. are 180 out of phase, but the coil connection is such that these emfs. are additive as shown in Fig. 113(a).

Fig. 114.—Showing winding and end connections of an alternator armature.

When a % pitch is used, the coil spread is equal to % X 180 or 150 electrical space-degrees. The emfs. E_i and E_z will therefore differ in phase by 150 electrical time-degrees, as shown by the angle a, Fig. 113(6). The total emf. E, which is their vector sum, is slightly less than when a full-pitch coil is used.

It will be observed that with a fractional-pitch winding only two of the slots of each phase under any pole contain coil sides of the same phase. In the other slots the two coil sides are of different phases. In Fig. 112, slots 1 and 2 contain both phase A and phase C conductors; slots 3 and 4 contain phase A conductors only; and slots 5 and 6 contain both phase A and phase B conductors. Of this group, slots 3 and 4 contain phase A conductors only. The fact that certain slots contain conductors of different phases reduces slightly the inductance of the winding, as has already been pointed out.

Figure 114 shows a portion of a finished armature winding. The end-connections, the binding down of the coil-ends, the wooden slot wedges, and the ventilating ducts, are clearly shown.

ALTERNATOR CONSTRUCTION 56. Stator or Armature.—The stator or stationary member of the alternator is almost always the armature, the field structure being the rotating member or rotor. When the machine is in operation the armature iron is continuously cut by the flux of the rotating field and must be laminated in order to reduce eddycurrent losses. In machines of small diameter, each lamination is a single circular punching.

High-speed turbo-alternators have armatures of small diameter and are usually built up of single circular stampings, as shown in Fig. 115. The perforations back of the slots are ventilating channels. Engine-driven alternators must rotate at comparatively low speeds and so must have a large number of poles, and armatures of comparatively large diameter. The pole pieces are made up of laminations riveted together and are dovetailed to the armature spider, as shown in Fig. 116. The armature is built up of small overlapping segments, dove-tailed to the frame of the machine in much the same manner as the armatures of engine-driven, direct-current generators are assembled (see Vol. I, page 249, Fig. 219), except that in the alternator the armature laminations are a part of the stationary member. Figure 116 shows the general construction of such an alternator. The frame itself is usually a hollow box casting. This gives the necessary mechanical stiffness, with the minimum weight, and the space within the frame, allows a free circulation of air for ventilating purposes. Figure 117 shows the complete armature of an engine-driven generator. The ventilating ducts and the bracing of the coil ends should be particularly noted.

Fig. 115.—Punching and frame of a turbo-driven alternator.

Large units must be so designed that they can withstand not only the stresses incident to normal operation, but also the enormous mechanical stresses which occur at short-circuit, due to the attraction and repulsion of the armature currents. The coil ends, unless well supported, are likely to be dragged out of

position by electromagnetic stresses produced by the short-circuit currents. This is particularly true of turbo-alternators, whose internal reactance is comparatively low, and whose short-circuit

Fig. 117.—Completely-wound stator of an engine-driven alternator.

currents, therefore, may be of considerable magnitude. Figure 118 illustrates the care taken in bracing the coil-ends in one of the largest types of turbo-alternator.

Alternator slots are divided into two general classes, the open slot and the semi-closed slot. The open slot, shown in Fig. 119 (a), is the more common because the coils can be form-wound and insulated prior to being placed in the slots, giving the least expensive and most satisfactory method of winding, s

Flo. 118.—Bracing of end connections of a turbo-alternator to withstand short circuit stresses.

The semi-closed or overhung type of slot, shown in Fig. 119 (6), is often necessary, especially in induction motors. The larger area of tooth face reduces the air-gap reluctance and also reduces the tufting of the flux which tends to produce ripples in the electromotive force wave. It is usually necessary to place the conductors in the slot one at a time, which is expensive and uneconomical of slot space. It is also difficult to apply insulation.

In both types of slot the conductors are usually held in the slot by wooden or fibre wedges, as shown in the figures. The effect of the semi-closed slot may unnnnf be obtained by the use of open slots and magnetic wedges. These wedges are only partly of iron so that the slot is not entirely closed.

The internal temperatures of modern alternators are so high that Fl0' 120.—Passage of ventilating air through the ducts of a turbo-alternator.

built-up mica is found to be the insulation best able to withstand the high-temperatures and high-voltage stresses simultaneously. Such mica is pressed around the active part of the conductor, forming a solid, homogeneous mass.

The problem of ventilating a large unit is not an easy one. A 20,000-kw.

unit, having an efficiency of 96.5 per cent., requires 700 kw. to be dissipated. Such a unit might require from 60,000 to 70,000 cu. ft. of air per minute. This air is usually supplied by separate blowers, and to remove the dirt and increase the cooling properties of the air, it is often passed through an air washer consisting of a curtain of water. Figure 120 shows the passage of the air through the axial ducts back of the laminations (see Fig. 115) and out through a center radial duct, in a turbo-alternator.

67. Rotating-field Structure.—From the standpoint of their field construction, alternators may be divided into three classes: the very slow-speed engine-driven alternator (75 to 150 r.p.m.); the medium-speed belt-driven and water-wheel-driven type (150 to 750 r.p.m.); and the high-speed turbo-alternator (750 to 3,600 r.p.m.).

Fio. 121.—Pole piece of a 75 kv-a., 60-cycle, 377 r.p.m. alternator.

The poles of practically all salient-pole generators have cores made up of laminations, Fig. 121, in order to reduce pole-face losses. (Vol. I, page 358.) In slow-speed machines, the poles are

Fig. 122.—36-pole rotor with strip-wound coils.

often bolted to a cast-iron spider, Fig. 122, or they may be dovetailed to the spider in the manner indicated in Figs. 116 and 123.

At higher speeds, centrifugal forces require that the poles be dove-tailed to the spider. In small machines, the spider may be

Fio. 123.—Revolving field spider with dove-tailed poles.

of solid steel, as shown in Fig. 123. The pole pieces dove-tail to this spider and are wedged in by keys driven one from each end.

Fio. 124.—Twelve-pole, 600 r.p.m. rotor.

In the larger types of generator, the spider is made of steel plates riveted together, as shown in Fig. 124. The poles are dove-tailed to the spider in the manner indicated. The slots in the pole faces of this rotor should be noted. Damping or *amortisseur* windings are placed in these slots as will be described later.

(See page 319, Fig. 295.)

Salient poles cannot be used for high-speed turbo-generators, owing to the large centrifugal forces developed and to the excessive windage. Therefore, a non-salient-pole rotor is used. There are two common types of such a rotor, the parallel-slot type shown in Fig. 125 and the radial-slot type shown in Fig. 126.

Fig. 125.—Parallel-slot. 2-pole rotor for a turbo-alternator.

The winding in the parallel-slot type is of strip copper, wound by hand in the slots. The wires are held in the slots by means of non-magnetic metallic wedges. There is not sufficient space to run the shaft through the center of the rotor, so it is bolted to the ends by phosphor-bronze flanges, Fig. 125. These flanges must be non-magnetic or they would short-circuit the magnetic poles. The fact that they are of phosphor-bronze makes them expensive. This construction gives a smooth rotor, with little windage loss and strong mechanically, especially as regards the support of the coil ends. Parallel-slot rotors are seldom used except for two-pole rotors in small machines. The metal back of the slots becomes too small in cross-section to withstand the centrifugal

Fio. 126.—Radial-slot type of rotor, having four poles.

Fio. 127.—150 kv-a., 900 r.p.m., 2400-volt. 60-cycle alternator, with directconnected exciter.

forces, when attempt is made to adapt this type of rotor to.taore than two poles. i

Figure 126 shows a four-pole, radial-slot rotor. Although the coil-ends are not held as strongly in this type of rotor as they are in the parallel-slot type, it is better adapted to rotors having more than two poles, because there is not the reduction of iron section back of the slots with increase in the number of poles, such as occurs in the parallel-slot type.

The winding on these types of rotor is called a *distributed* field winding. (See Fig. 131.)

The field connections are usually carried out through the center of the shaft

to slip-rings. Two or more carbon brushes resting on the slip-rings carry the current to the winding. The excitation voltage is usually 120 volts or 250 volts and in the larger stations is supplied by bus-bars devoted to excitation only. In smaller installations, the exciter is mounted directly on the alternator shaft, Fig. 127, or else is belt-driven from the alternator shaft. Large central stations usually have a storage battery floating on the exciter bus, and in addition, may have steamdriven exciters to be used in emergencies.

ALTERNATOR ELECTROMOTIVE FORCES AND OUTPUTS **68. Generated Electromotive Force.**—Figure 128 (a) shows the magnetic flux between the armature surface and a north and a south pole of an alternator. Assume that the flux distribution is sinusoidal, Fig. 128 (6), the flux density being a maximum under the center of the pole. Let B' be the average value of 2 the flux density. B' is equal to-times the maximum value B.

Let a be a conductor cutting this flux with a velocity of v cm. per second. This conductor a has a length of I cm. perpendicular to the plane of the paper.

The average voltage from equation (93), Vol. I, page 217, is $e' = B'MQ\text{-}2$ volts.

Let D be the pole pitch in centimeters and / the frequency in cycles per second.

The time in seconds necessary for the conductor a to move the distance D is sec. Therefore, $v \text{—.} = 2fD$ cm. per second.

£ i 2/ The total flux cut per pole 0 = $B'lD$

The effective voltage is 1.11 times the average, 1.11 being the *form factor* of a sine wave. (See Page 10.) The *effective* in

Fig. 128.—Generation of alternating elertromotive force.

duced volts per conductor, by substitution in the above equation for e'

If there are Z conductors in series per phase, the effective emf. per phase $E =$ $2.22ZtflQ\text{-}2$ volts. (36)

If the emf. wave is not a sine wave, the factor 1.11, which is the form-factor of a sine wave, should be correspondingly changed.

Owing to the fact that the electromotive forces in the different coils of a phase belt are not in time-phase with one another, Fig. 130, the conductor electromotive forces do not add algebraically. Therefore a factor kb, called the *breadth factor* or *belt factor*, must be introduced to correct for this relative phase displacement. This factor is unity for a concentrated winding and is less than unity for a distributed winding. The table gives values of fcj, for a few typical windings.

Values Op Breadth Factor *kt,*

If fractional pitch is used, the electromotive forces in the two coil sides are out of phase, as shown in Fig. 113 (6), page 109. This again reduces the voltage. Correction for this may be made by multiplying the voltage equation by another factor *kp,* called the *pitch factor.*

Values Of Pitch Factor *kf Example.*— A six-pole, three-phase, 60-cycle alternator has 12 slots per pole and four conductors per slot. The winding is % pitch. There are 2,500,000 lines entering the armature from each north pole and this flux is sinusoidally distributed along the air-gap. The armature coils are all connected in series. The machine is Y-connected. Determine the opencircuit voltage of the generator.

The total number of slots is equal to 72.

Therefore the series conductors per phase 4X72

Slots per pole per phase = 72/(6 X 3) = 4. fa (from table) = 0.958. *kr* = 0.961.

Therefore the total induced voltage per phase E = 2.22 X 0.958 X 0.961 X 96 X 2,500,000 X 60 X 10- = 294 volts.

As the generator is Y-connected, the terminal voltage is 294 V3 = 510 volts. *Ana.*

Ordinarily the flux distribution in a generator is not sinusoidal, especially with salient-pole machines, the wave being flat-topped, as shown in Fig. 129. The electromotive force wave *per conductor* has the same shape as the flux wave. If the coil is a fullpitch coil, the electromotive forces in the two sides of each coil will be 180 out of phase and of the same magnitude, as these coil sides

both lie at any instant in corresponding parts of opposite poles. Therefore, the electromotive force wave of each coil will have the same shape as the conductor electromotive force wave. If but one slot per pole per phase is used, the resulting electromotive force wave will have the same shape as the flux wave, which may be flat-topped as shown in Fig. 129.

Fio. 129.—Flux density in air-gap of salient-pole machine.

Figure 130 (a) shows a phase-belt, consisting of four coils, of a three-phase generator having 12 slots per pole, or four slots per pole per phase. The shape of the electromotive force wave for each of the four full-pitch coils forming one phase of the winding is the same as the shape of the flux wave, as is shown in Fig. 130 (6), at 1, 2, 3, and 4. As 12 slots represent 180 electrical spacedegrees, 180/12, or 15, is the interval in electrical space-degrees between successive slots. Therefore, the four electromotive forces are 15 electrical timedegrees apart, as shown in Fig. 130 (6). As the coils are connected in series, the resultant electromotive force is found by adding the ordinates of the four waves, as shown, The resultant wave, instead of being a flattopped wave like that of the individual coil, is well-rounded and is very nearly a sine wave. This is the reason that a distributed winding gives a better wave shape than a concentrated winding. With a non-salient pole rotor, having a distributed winding, Fig. 126, the field coils lie in the rotor slots in the manner indi

Fio. 130.—Resultant electromotive force wave in a 4-coil phase belt.

cated in Fig. 131 (a). The magnetomotive force of each coil is a rectangle, if the currents in the conductors be considered as concentrated at their centers. This is shown in Fig. 131 (6), which gives the mmf. of a single coil. The height of the rectangle is equal to the mmf. of this coil, which can be expressed in ampere-turns, gilberts, or in any convenient unit. The base of each rectangle is equal to the coil spread. In Fig. 131 (a) and (c) there are three coils per pole. Each coil produces a rectangle

of mmf. If there are the same number of turns in each of the three coils, the height of each mmf. rectangle is the same. The re

Fig. 131. — Distributed field winding and resulting magnetomotive force and flux waves.

sultant mmf. is found by superposing the three mmf. rectangles, as shown in Fig. 131 (c). The resulting magnetomotive force wave is "stepped," as shown. Due to fringing, the resultant flux wave is nearly sinusoidal, as indicated in the figure. Therefore, a

Fig. 132.—Connecting alternator coils in Y.

non-salient pole alternator, having a distributed winding, has usually a better wave shape than a salient-pole alternator. 59. Phasing Alternator Windings.— Three-phase alternator windings may be connected either in Y or in delta. Instances often occur in practice where six leads come from the machine, these leads being the three pairs of terminals from the three phases. The proper phase relations must be observed in making the connections, whether they are to be connected in Y or in delta.

Let aa', bb' and cc', Fig. 132, be the three coil windings of a three-phase machine.

Assume first that these three windings are to be connected in Y. First connect ends a and *b* together. Measure *Eav,* the voltage across their open ends. This should equal /3 times the coil voltage. It may be equal to the coil voltage, in which case one coil should be reversed. Next tie the end *c* of coil *cc'* to point *ab*. The voltages *Evj* and 2?,,v should each be /3 times the coil voltage. If not, the coil *cc'* should be reversed.

Voltage

AcIobs *c'a'* co *Ece,*

Fio. 133.—Connecting alternator coils in delta.

If it be desired to connect the coils in delta, the ends a and *b'*, Fig. 133 (a), should first be connected. The voltage *Ea,b,* across their open ends, should be equal to the coil voltage. If not, one of these two coils should be reversed. End c' of coil cc' should then be connected to *b.* The voltage *Eca,* across the open ends should be zero, as shown by the vector diagram in *(b)*. (See Par. 46, Page 83.) If this voltage is practically zero, the two ends c and a' may then be closed. The voltage *EM*-may be twice the coil voltage, as shown in (c). If this is found to be the case, coil cc' should be reversed.

60. Rating of Alternators.—The rating of electric machinery is determined in general by its temperature rise. This temperature rise is caused by the losses in the machine. The *PR* loss in the armature, due to the load current, limits the output of a machine. This loss depends upon the value of the armature current and is independent of power-factor. For example, 100 amp. in a single-phase, 200-volt generator will produce the same *PR* loss whether the load power-factor be unity, 0.4 or any other value. The output in *kilowatts,* however, is proportional to the power-factor. If the above generator is limited to 100 amp., its output will be 20 kw. at unity power-factor but only 8 kw. at 0.4 power-factor. The rating is 20 kv-a. (kilovolt-amperes) regardless of power-factor.

For the above reasons, alternators are ordinarily rated in kilovolt-amperes (kv-a.). If a machine is rated in kilowatts, unity power-factor is assumed, unless otherwise specified. In stating the output of a machine it is always well to state the power-factor.

The rating of the prime-mover driving an alternator is independent of the alternator power-factor. The same-turbine could be used to drive a 200-kv-a. machine operating at 0.5 powerfactor or a 100-kv-a. machine operating at unity power-factor, although the first alternator would have double the kv-a. rating of the second.

CHAPTER VI

ALTERNATOR REGULATION AND OPERATION

Alternator Regulation.—It is shown in Vol. I, Chap. XI, that the voltage of a shunt generator drops as load is applied. This is due to three causes: the *IaRa* drop in the armature, armature reaction, and the drop in field current which results from the decrease in terminal

volts. As commercial alternators are excited from a separate source, there is no decrease of field current due to the drop in the alternator terminal voltage. However, both the *IaRa* drop in the alternator armature and armature reaction ordinarily cause a drop of terminal voltage as load is applied. Another factor which causes the alternator voltage to drop with application of load is the *reactance* of the alternator armature. This will be discussed later.

The regulation of direct-current generators is inherently better than the regulation of alternators. For example, shunt generators of commercial size regulate very closely, and it is usually possible to so compound a shunt generator that its terminal voltage is practically constant at all loads. In the alternator, the armature reaction drop, which is not present in the directcurrent generator, and the greater effect of armature reaction, result in poorer regulation. In addition, alternators cannot be compounded readily.

The regulation of the alternator depends not only on the magnitude of the current, but on the power-factor as well. A knowledge of the regulation of an alternator at various powerfactors is usually essential, since the amount by which the voltage varies with the load has an important bearing on the operation of the system as a whole. If the machine supplies incandegcent lamps, it must regulate very closely or else special regulators are necessary on the lighting circuits. Moreover, alternators may regulate well at unity power-factor, while at low power-factors the regulation may be very poor, even if the *current* be the same in the two cases.

In the larger types of alternator, the large values of current which result from short-circuit may cause serious damage to the machine and to the system. The value of this short-circuit current is closely related to the regulation of the machine, so that a knowledge of the regultvtibn is helpful in designing the circuit breakers, switches, power-limiting reactances, etc.

It is very desirable, therefore, to understand the factors and the reactions

which affect the regulation and the operation of alternators. As it is usually impossible to obtain the requisite loads for testing an alternator under actual load conditions, it becomes necessary, in determining the regulation, to employ methods which do not require the actual loading of the machine. These methods will be described later.

61. Armature Reactance.—When a current flows in the conductors of an alternator armature, magnetic lines are set up around these conductors. Such lines are indicated around the conductors of one phase on a smooth-core armature, in Fig. 134. This magnetic leakage flux linking with the current gives *inductance* to the armature conductors. This inductance when multiplied by *2ir* times the frequency gives the *reactance* of the conductors. Therefore, alternating current flowing in these conductors will encounter not only resistance, but reactance as well.

Figure 135 (a) shows the conductors lying in a rather deep and narrow slot of an iron-clad armature. The current flowing in these conductors produces magnetic lines, whose path is across the slot and back through the armature iron, as shown in the figure. The reluctance of this local magnetic circuit lies almost entirely in the slot itself, as the reluctance of that part of the path which lies in the iron is practically negligible. A deep narrow slot will allow more lines per ampere-conductor to cross it than the shallow and wider slot shown in Fig. 135 (6). Hence an alternator with deep narrow slots will have a much

Armature

Fig. 134.—Flux linked with armature conductors on smooth-core armature.

higher armature *reactance* than one with wide shallow slots, other conditions being the same.

A semi-closed slot, like that shown in Fig. 135 (c), will have considerably more magnetic flux per ampere-conductor than either of the slots of (a) or (6), because the overhanging tooth-tips reduce the reluctance of the magnetic circuit. Thus, the reactance of a machine may be controlled hi part by the design of the slot. In a smooth-core armature, like that shown in Fig. 134, the armature reactance will be small as compared with that of the slotted type of armature,

A certain amount of reactance is due to the magnetic flux linking the coil ends. Although this is small compared to the reactance due to the slot linkage, it cannot be neglected as a rule.

(a) CO

Fio. 135.—Slot leakage flux which produces armature reactance.

It is pointed out in Vol. I, Chap. VIII, that the inductance varies as the *square* of the number of turns. This same rule applies to the conductors in alternator slots. If the number of series conductors in a slot is *doubled,* the reactance per slot is increased four times, other conditions remaining unchanged.

As the reactance is proportional to the frequency $(X = 2ir/L)$, the reactance of a 25-cycle alternator will be considerably less than that of a 60-cycle alternator, other conditions being the same.

62. Armature Resistance.—The armature iron forms a considerable portion of the path of the flux which links the armature conductors, Figs. 134 and 135. As this flux is alternating, it is accompanied by hysteresis and eddy-current losses, which occur in the iron immediately surrounding the slots. As this flux is produced by the armature current, the power represented by this loss must be supplied by the armature current. The eddycurrent loss varies as the square of the flux density and the hysteresis loss varies as the 1.6 power of the flux density. As the leakage flux is nearly proportional to the current, the eddycurrent loss varies as the square of the current and the hysteresis loss as the 1.6 power of the current. The combined loss varies nearly as the square of the current.

The effect of these local iron losses is to increase the total loss due to the flow of current through the armature. As these local losses vary nearly as the current squared, their effect is practically the same as if the resistance of the armature were increased.

Unless the armature conductors are small in cross-section, the effect of the slot leakage flux is to force the current towards the top of the slot, so that the

current density in the portions of a conductor near the top of the slot is greater than in those portions near the bottom of the slot. This also increases the effective resistance of the armature.

The effective resistance of an armature is therefore greater for aitgfn&trog; than for direct current, due to the alternating flux which accompanies the flow of the alternating current. The percentage increase depends to a large extent on the shape of the slots and the teeth and on the size of the conductors, and ranges from 20 to 60 per cent. As the armature resistance drop is very small as compared with the voltage drops due to armature reactance and armature reaction, considerable error in determining the resistance introduces little error in most computations. The effective armature resistance may be measured by running the machine as a generator, with weak field excitation. The input is measured with the armature open-circuited and then shortcircuited through ammeters. Neglecting the change in core loss caused by armature reaction, the difference of input divided by the number of phases is equal to the PR loss per phase. The effective resistance per phase is found by dividing this difference of input by the current per phase squared. A more common, though less accurate method, is to measure the ohmic resistance with direct current, and to increase this value by an estimated factor, such as 40 per cent., to cover the indeterminate losses.

63. Armature Reaction.—In direct-current machines, the armature ampereturns act on the magnetic circuit of the machine in such a way as to distort and to change the magnitude of the airgap flux. For a given armature current, the direction and magnitude of this armature reaction depend on the position of the brushes. In an alternator practically the same conditions exist. For a given armature current, the magnitude and direction of the armature reaction cannot depend upon brush position, but do

Fig. 136.—Distortion of alternator flux by an in-phase current.

depend on the phase relation existing between the current and the voltage and

hence on the power-factor of the load.

Figure 136 (a) shows the position of an armature coil whose sides are directly under the pole centers. At this instant the induced voltage in the coil is at its maximum value. If the current is in phase with this induced voltage, corresponding approximately to a load of unity power-factor, the current is at its maximum value at this same instant, and flows in the direction shown in (6) and in (c). The current flows in this coil in such a direction that its magnetomotive force acts downward, as shown in (6). The effect on the main magnetic circuit of the current flowing in this coil is shown in Fig. 136 (c). The flux is increased on the right-hand side of each pole and decreased on the left-hand side. Were there no effect of saturation, the total flux would not be changed, as the increase on one side of the pole would be balanced by the decrease on the other side.

(6) llnity Power-factor Load. Fig. 137. —Flux distribution in the air-gap of a salient-pole alternator.

Figure 137 (a) shows the no-load flux distribution of an alternator. The distribution is symmetrical about a vertical axis. Figure 137 (6) shows the flux as affected by the singln coil of Fig. 136 (c). Although the total area under the flux wave is practically unchanged, the curve is now peaked on one-side and depressed on the other. This occurs also in direct-current machines when the brushes are in the geometrical neutral (see Vol. I, page 268, Fig. 238) and cross-magnetization alone results.

It will be observed in Fig. 136 that the coil in question is acting principally on the interpolar space, whose reluctance is high. Therefore, in this position the effect of the coil ampereturns upon the magnetic flux of the generator is a minimum. This does not apply to a nonsalient pole machine, where the airgap is uniform.

Figure 138 is a vector diagram representing these conditions. *Fi* is the mmf. due to the field coils, *A* is the mmf. due to the armature coil and *F* is the resultant of the two. When the current is in phase with the induced emf., the space direction of the armature mmf., *A,* is 90 electrical space-degrees from the resultant mmf. *F.* It will be observed, Fig. 138, that the principal effect of *A* is to distort the alternator flux or to change it from its original position practically without altering its magnitude.

Figure 138 is a space-vector diagram of *mmfs.* The resultant *flux* is equal to the mmf. *F* divided by the reluctance of the magnetic circuit. In a non-salient pole machine, where the reluctance

Fio. 138. — Vector diagram showing effect of armature reaction when current is in phase with induced e.m.f.

Fig. 139.—Armature reaction due to current lagging 90.

of the air-gap is uniform, *t* may be found in terms of induced emf. from the saturation curve. In salient-pole machines, this method is only approximate, as the reluctance of the magnetic circuit varies. The reluctance is a minimum when the space direction of *F* is along the pole centers and is a maximum when the space direction of *F* is midway between pole centers. Figure 139 represents the conditions when the current lags 90

Fio. 140.—Vector diagram showing effect of armature reaction with current lagging 90.

with respect to the induced electromotive force. When the coil is in position (1), Fig. 139 (a), the electromotive force is a maximum, as in Fig. 136. The current is zero at this instant because it lags the induced voltage by 90. The current does not reach its maximum value until the coil has travelled 90 electrical spacedegrees further and has reached position (2). The coil then lies directly under a south pole. It will be noted that the magnetomotive force of this coil is *downward* and is therefore in direct opposition to the magnetic flux entering the south pole. *Therefore, when the current lags the induced electromotive force by 90, it acts in direct opposition to the main field.* As a result the field is materially weakened by a lagging current and this is accompanied by a reduction of the induced electromotive force. This result is similar to that of moving the brushes forward 90 in a direct-cur-

rent generator. All the armature ampere-turns are then demagnetizing, tending to weaken the field.

It will be observed that this coil is acting directly upon a part of the magnetic circuit where there is iron, rather than on an interpolar space. Therefore, for a given current in the armature, the coil in this position has a much greater effect upon the magnetic field of the machine than it had at unity power-factor, shown in Fig. 137. Figure 140 shows vectorially the conditions which exist at zero power-factor or 90 lagging current. The armature reaction *A* acts in direct opposition to the impressed field *Fi* so that the resultant field *F* is considerably smaller than *Fi.*

Figure 141 shows the conditions existing when the current leads the induced electromotive force by 90. As before, the electro 136 *ALTERNATING CURRENTS* motive force reaches its maximum value when the coil sides are directly under the pole centers, position (2), Fig. 141 (a). The current, however, reaches.its maximum value 90 electrical spacedegrees ahead of this position or at (1). It will be observed that the ampere-turns of the coil now assist or strengthen the main field, as they are acting directly in conjunction with it. The coil again is in the most favorable position so far as its effect upon the magnetic circuit of the machine is concerned. This condition may be represented by the vector dia.——. gram, Fig. 142. The magnetomotive "*A* 5 £ force of the field ampere-turns is *Fi,*

Fig. 142. —Vector dia-that of the armature is *A,* and the regram showing effect of suitant magnetomotive force is their sum armature reaction with current leading by 90. *F,* because the two are acting in the same direction.

It should be noted that for a given value of armature current, the effect of armature reaction varies with the power-factor in a salient-pole type of machine, due to the varying reluctance caused by the salient poles. In a non-salient-pole generator, the air-gap is uniform around the periphery of the armature, so that the armature magnetomotive force acts on a path of uniform

permeance.

If the power-factor has other than one of the three values just illustrated, the armature mmf. will add vectorially to the impressed mmf. according to the phase angle existing between the

Fio. 143.—Vector diagram of armature reaction with current lagging *0* degrees.

current and the induced voltage. This is illustrated in Fig. 143. The direction of the armature reaction is shown at a power-factor cos *6,* the current lagging, *0* being the angle between the current and the *terminal* voltage. *Fi* is the field mmf. When the current is in phase with the *induced emf.* the direction of the armature reaction is along *Ai,* 90 behind the resultant mmf., *F.* (See Fig. 138.) When the current is in phase with the *terminal* voltage, the armature mmf. acts in the direction *A«,* because the terminal voltage lags the induced emf. by an angle *a.* (See Fig. 145 (6).) If the current lags the terminal voltage by *9* degrees, its mmf. must act along *A, 6* degrees behind *Az.* Combining the mmf. , *A,* with the impressed mmf., *Fi,* gives the resultant mmf., *F.*

Under the usual conditions of operation, the power-factor is neither unity nor zero. Hence, the armature reaction may strengthen the field or may weaken it, according as the current leads or lags, and at the same time may distort it.

64. Armature Impedance Drop.—In a direct-current generator, the induced armature voltage is obtained by adding *numerically* the *IR* drop in the armature and the terminal voltage. In the alternator, the armature reactance drop as well as the armature (a) (6)

Fio. 144.—Alternator vector diagram for unity power-factor.

resistance drop must be added to the terminal voltage in order to obtain the induced armature voltage. These voltage drops must be added *vectorially* to the terminal voltage, in order to obtain the induced electromotive force. That is, the emf. induced in an alternator armature is the terminal voltage plus the armature *impedance drop,* this addition being performed vectorially. *Current in Phase with Terminal Voltage.*—Figure

144 (a) shows the conditions existing when the load power-factor is unity. V is the generator terminal voltage and I is the armature current in phase with V. The IR drop in the armature is in phase with the current I, R being the *effective* resistance of the armature. The IX drop leads the current by 90 and is laid off at the end of IR. The vector sum of these two gives the IZ drop in the armature. This impedance drop when added vectorially to the terminal voltage V gives the electromotive force E' induced in the alternator armature. The vector addition is performed by completing the parallelogram having V and IZ for its adjacent sides. The diagonal E' is the vector sum of IZ and V and represents the induced emf.

The same result is obtained by adding the IR drop directly to V, Fig. 144 (6), and then adding the IX drop, at right angles to I and leading, at the end of IR. The vector addition in this case is made by the use of the triangle of vectors described in Chap. I, page 12. The impedance drop IZ is shown dotted in Fig. 144 (6) as it is not used in obtaining E' by this particular method.

It is to be noted that with a load of unity power-factor the current is in phase with the *terminal* voltage, but lags the generator *induced* voltage by an angle a.

It is a simple matter to find E' if the other quantities are known. E' is the hypotenuse of a right triangle of which $(V + IR)$ is one side and IX the other.
(38) *Example.* — A 60-kv-a., 220-volt, 60-cycle alternator has an effective armature resistance of 0.016 ohm and an armature reactance of 0.070 ohm. What is its induced emf. when the machine is delivering its rated current at a load power-factor of unity? 60.000
The current I =-Kofi" = 273 amp.
IR = 273 X 0.016 = 4.37 volts. IX = 273 X 0.070 = 19.1 volts. E, = V(220 + 4.4)2 + (19.1)2 = 225 volts. *Ans.* *Lagging Current.* — When the current lags the terminal voltage by the angle 6, the same method is employed to calculate the induced emf. Figure 145 (a) shows the current I lagging the terminal voltage V by the angle 6. The IR drop is

along the current vector I, and the IX drop is in quadrature with I and leading, as before. The resulting impedance drop IZ is then found, being the resultant of IR and IX. This impedance drop is then added vectorially to V, giving the armature induced emf., E'. It will be noted, Figs. 144 and 145, that the position of the armature impedance triangle is determined by the current and not by the generator voltage. Therefore, when the current lags, this impedance triangle swings clockwise with the current.
As before, the impedance drop may be added at the end of V, if the proper phase relations are observed. The most direct method of finding the induced emf. E' is to use the method described under the triangle of vectors, page 12. IR, which is in phase with the current, is first added vectorially at the end of the ter . w 6 Fio. 145.—Alternator vector diagram for power-factor cos 6, current lagging. minal voltage V. Then the reactance drop IX, at right angles to the current and leading, is added at the end of IR. The resultant voltage found by completing the polygon is the induced emf. E'. This method is illustrated in Fig. 145 (6), where IR is parallel to I and IX is at right angles to I and leading. The geometrical solution of this diagram is quite simple. If IR is projected on the current vector I, a right triangle of voltages, *Obd*, is formed, of which E' is the hypotenuse. The values of the two legs of this right triangle may be found as follows: $0a = V \cos 6$
$ab = IR$
$aV = be = V \sin 6$
$cd = IX$
Example.—Determine E, for a load in which the power-factor is 0.7, current lagging, using the constants of the example on page 138.
The rating of an alternator, as has already been pointed out, depends on the current or kilovolt-amperes rather than the kilowatts. Therefore, the current rating of the generator will remain unchanged, although the kilowatts in this problem are reduced to 0.7 of their former value.
cos 0 = 0.70 IR = 4.37 volts as before. 6 = *45.6* sin 0 = 0.714 IX = 19.1 volts as

before. E, = /(220 X 0.70 + 4.4)2 + (220 X 0.714 + 19.1)2 = 237 volts. *Ans.*
It is to be noted that the induced emf. is now higher than before, although the value of the impedance drop itself is the same. Therefore, for a fixed value of induced emf., the terminal volts become less with increasing lag of the current, even though the value of the current remains unchanged. This is due to the angle at which the impedance drop subtracts from the induced emf. It would be expected, therefore, that the regulation of an alternator would be poorer for lagging current.

At unity power-factor, the armature resistance drop is the important factor in determinng the value of E'. With a lagging current, the resistance drop plays but a small part and the armature *reactance* drop becomes the important factor.

Leading Current.—Figure 146 shows the alternator vector diagram when the current *leads* the terminal voltage by an angle 6. As the current changes Fio. 146.—Alternator vec-its phase relation with respect to the tor diagram for power-factor voltage y thc impedance triangle cos 0, leading current., swings with the current in a counterclockwise direction about the end of V. E' is found in the same manner as in Fig. 145. The voltage drop IR, parallel to the current, is projected on the current vector. $0a = V \cos 6$
$ab = IR$
$aV = be = V \sin 6$
$cd = IX$
This differs from equation (39) only in the sign of IX, which is now negative.
Example.—Repeat the foregoing problem when the power-factor is 0.7, current leading. cos e = 0.70 IR = 4.37 volts sin e = 0.714 IX = 19.1 volts E, = V(220X 0.70 + 4.4) + (220 X 0.714-19.1)2 = 207 volts. *Ans.*
The induced emf. in the armature is now *less* numerically than the terminal voltage. This is a condition which cannot exist in a direct-current generator. It results from the phase position of the IZ drop with respect to V. 65. Alternator Regulation.—The voltage E', as determined in the preceding paragraph, is the

voltage *induced* in the alternator armature under load conditions. In practice it is a quantity difficult to measure and can be calculated only approximately. There is no simple method of making a direct measurement of the armature reactance X. As a matter of fact, it is seldom necessary to know either the value of E' or of the armature reactance X.

A knowledge of the voltage regulation is very important because it shows how closely a machine will maintain its voltage from no load to full load, under the various conditions of load.

If there were no armature *reaction, E'* would be the no-load voltage of the machine, just as in a separately excited directcurrent generator the induced voltage under load would be equal to the no-load voltage if there were no armature reaction. The effect of armature reaction is to change the value of the magnetic flux, and this is accompanied by a corresponding change in the value of the induced emf. The effect of armature reaction on the operation of the machine is analyzed in the methods for determining regulation.

It is usually impossible to find the regulation of an alternator by actual loading, particularly in the larger sizes, until after the machine has been put in service, and even then it may be difficult to secure the desired adjustment of the load. To set up a generator for a load test requires a machine for driving purposes, and considerable power may have to be supplied and absorbed. With polyphase generators there is the added difficulty of obtaining a balanced load.

The regulation of a machine, however, may be calculated with sufficient accuracy from data obtainable from open-circuit and short-circuit tests. These tests involve very little power supply and do not require any power-absorbing devices. There are three common methods for determining regulation, the *synchronous impedance or electromotive force method*, the *magnetomotive force method*, and the *A. I. E. E. method.* The application and limitations of each method will be discussed in some detail. 66. Synchronous Impedance Method, or Electromotive Force Method.—This

method is often called the pessimistic method, because it gives a value of the regulation poorer than that actually existing in practice.

The principle is as follows: The armature *reaction* is combined with the armature *reactance,* or what amounts to the same thing, the armature reactance is increased a sufficient amount to allow for the effect of armature reaction. That this may be done can be shown as follows:

In Fig. 147 a sine distribution of flux along the air-gap is assumed. The line ab is the coil axis. When the coil axis lies along the pole axis oo, as is shown in (a), the flux linking the coil is a maximum. When the coil axis ab reaches position $a'b'$, as shown in (6), the flux linking the coil is zero. Therefore, the flux linking the coil varies with the time and at a frequency equal to the frequency of the induced electromotive force. In position (a), the flux linking the coil is a maximum, and the induced emf. is zero. In position (6), the flux linking the coil is zero, and the induced emf. is a maximum. It is seen that the emf. induced in the coil reaches its maximum value 90 electrical space-degrees later than the flux linking the coil, and, therefore, later in time. The flux linking the coil may then be said to *lead* by 90 the emf. which it induces.

As the flux linking the coil and the emf. induced in the coil vary sinusoidally with the space position of the coil, their instantaneous values may be found by means of rotating vectors.

These space relations are shown graphically in Fig. 147 (c). When the coil axis ab lies along the pole axis oo the flux linking the coil is a maximum and the induced emf. E' is zero. As the coil axis ab moves to the right, the flux £ linking the coil decreases sinusoidally and the induced emf. E' increases sinusoidally. When the coil axis ab reaches $a'b'$, midway between pole centers, the flux linking the coil is zero and the induced emf. E' is a maximum. Under the conditions assumed, the flux wave leads the emf. wave by 90, as is shown in Fig. 147 (c).

These space relations may also be

shown by rotating vectors, Fig. 147 (d). The vector £ is equal to the maximum value of the flux linking the coil, and the vector E' is equal to the maximum value of the induced emf. Each position of these two

Fig. 147.—Relation of flux linking alternator coil to induced voltage in coil.

rotating vectors represents a different position of the armature coil relative to the field poles. The instantaneous value of either quantity, / or E', is found by projecting its vector on the vertical axis YY. It is seen that the flux t reaches its maximum value 90 space-degrees in advance of the emf. E'.

Figures 147 (c) and 147 *(d)* are *space-phase* diagrams. Figure 147 (c) shows the flux linking the coil and the induced emf. in the coil for different space positions of the coil as it moves relative to the field poles. Figure 147 *(d)* shows these same quantities as rotating vectors.

Ordinarily, the einf. and current vectors represent the phase relations of these quantities with respect to the *time.* (See Chap. I, Fig. 5.)

Although t, the flux linking the armature coil, and E', the induced emf. in the coil, vary with the *space* position of the coil, they vary also with the *time.* When the coil moves through 360 electrical degrees in *space* with respect to the poles, the emf. wave passes through 360 electrical degrees in *time.* The time of doing this is I//sec., where/is the frequency in cycles per second. Therefore, the time required for the coil to pass through a given number of electrical space-degrees is equal to the time required for the emf. to pass through an equal number of electrical *timedegrees.* For this reason a *space-phase* diagram and a *time-phase A*

Fio. 148.—Vector diagram of alternator mmfs. and emfs.

diagram may often be combined, just as the angular variation of emf., Chap. I, Fig. 3, page 4, was changed to the time variation of emf., Fig. 5, page 6. The space-phase diagrams of Figs. 147 (c) and 147 (d) may also be considered as time-phaso diagrams.

Figure 148 shows the vector diagram of an alternator, in which the current I is in phase with the induced emf. E'. As F is the resultant field, E' must lag F by 90. It was shown in Fig. 138, page 134, that under these conditions the armature reaction acts at right angles to the resultant field, F. Therefore, the armature mmf., or the armature reaction A, must have a space position of 90 behind the resultant field F. This brings it in phase with E', and therefore in phase with the current I, as it should be of course. A. S-F is the resultant field, it must be the vector sum of the impressed field Fi and the armature reaction field A, as shown in the vector diagram.

In a non-salient pole machine, the space-direction of the resultant flux will be the same as that of the resultant mmf. F. In a salient pole machine, the space-direction of the resultant flux usually is *not* the same as that of the resultant mmf. vector F, due to the fact that the flux tends to seek the paths of minimum reluctance. The flux therefore is distorted in the direction of the pole-pieces. In salient pole machines this introduces errors in the methods used for predetermining alternator regulation.

If the armature reaction were zero, due to there being no load on the machine, the resultant field would obviously be the impressed field F. The *no-load* induced voltage must be 90 behind Fi, as shown at E, Fig. 148, because the no-load induced emf. lags the no-load field by 90.

It will be recognized, Fig. 148, that Fi, F, A constitute a *space diagram of mmf. vectors* taken from Fig. 138. E' is also a space vector when considered as being combined with the mmf. diagram shown in Fig. 147 (c) and (d). As the linking of the resultant flux F with the armature coils also varies with time, as described on page 143, F may be considered as being a *time* vector. E' is also a *time* vector, just as I and E are time vectors, so that it may be combined with them also. Hence, E' and F are connecting links between the *space* diagram of mmfs. and the *time* diagram Of currents and "fгMtor field mmf.-current lagging induced

Fio. 149. — Relation of induced voltages to ternator f emf. by 90 voltages. Therefore, the space and the time diagrams may be combined into the one given in Fig. 148.

When the current lags the induced electromotive force by 90, the armature reaction is in exact opposition to the resultant field. (See Fig. 140, page 135.) Figure 149 shows the vector diagram for this condition. The current I lags the induced emf. E' by 90. The armature reaction A being in direct opposition to Fi is therefore in phase with the current. The resultant field F is found by subtracting the armature reaction A from the impressed field Fj. Eis the no-load voltage due to field F.

In either Fig. 148 or Fig. 149 the no-load voltage E is found by adding vectorially a voltage E,E to E', this voltage $E'E$ always being in quadrature with the current.

If the voltage $E'E$ adds in quadrature with the current, it must be in phase with the IX component of voltage already discussed. This is illustrated in Fig. 150. The current I is shown lagging the terminal voltage V by an angle 6. The inter

Fio. 150.—Complete vector diagram for the synchronous impedance method. nal voltage of the armature E' is found by adding IR and IX vectorially to V. The resultant field F is 90 ahead of E'. By adding voltage $E'E$ to E', and in quadrature with the current I, the no-load voltage E is found. The voltage $E'E$ does not actually exist under load, for E is the no-load induced emf. and E' the load induced emf. However, $E'E$ represents the drop in voltage due to the reduced flux caused by the *armature reaction A.* $E'E$ lags the armature mmf. vector —A by 90 and would be proportional to — A if there were no saturation of the iron. $E'E$ may then be considered as an emf. induced by the armature reaction, —A. *As a matter of fact, however, $E'E$ is a fictitious voltage which replaces the effect of change in flux due to armature reaction.*

It is also evident that if IX be increased in value to $IX,$, where $IX, = IX + E,E$, E may be computed without knowing

E'. This assumes that the voltage $E'E$ is always proportional to the armature current, which is not strictly true. The foregoing is the principle of the electromotive force or *synchronous impedance* method. The *rational* or *general* method is first to compute E'. Find from the saturation curve, the field current F corresponding to E'. Add —A to F vectorially to find Fi, and then from the saturation curve find E corresponding to the field current F. One serious objection to this method is the difficulty of determining the armature leakage reactance X. It cannot be readily measured and can only be roughly calculated. These calculations and the general solution of the diagram are both laborious. The determination of the regulation is very much simplified if X be increased to the value $X,$, so that E is found directly without knowing E'. X, is called the *synchronous reactance* of the alternator. The corresponding impedance Z, (=-/. R2 + X,2) is called the *synchronous impedance* of the alternator.

The synchronous reactance is determined experimentally as follows: The saturation curve of the alternator, E and $//$, is first determined in the usual manner and the curve plotted as shown in Fig. 151. The field is then made very weak and the alternator armature is

Fig. 151.—Open-circuit and short-circuit characteristics of an alternator. short-circuited through an ammeter. The field is then gradually strengthened and a new curve of armature current and $//$ is determined. The field is increased until the armature current is almost twice its rated value. These two curves are shown plotted in Fig. 151.

Consider some value of field current $/'$. On open-circuit this field current produces a voltage E. On short-circuit the terminal voltage of the machine is practically zero. The voltage E does not actually exist in the armature at short-circuit because of armature reaction. (The voltage actually induced is E', Fig. 152.) However, if the effect of the armature reaction is replaced by an armature reactance drop, the voltage Ei may be considered as being entirely used in sending the current $/i$ through the armature impedance. That is, $Ei = hZs$ where Z, is the *synchronous impedance* of the armature. This short-circuit condition is represented vectorially in Fig. 152, where $/i$ is the short-circuit current $/j$ and Ei the assumed internal emf. of the 1 armature. The synchronous impedance r drop is made up of two components, IlR, where R is the effective resistance of the armature, and IX,, where X, is the synchronous reactance of the armature.

Obviously, Z, = .(41) and X, =- v/ZTR1 (42)

In practice R is small compared with Z, and they combine almost in quadrature so that $Ei Xs = -j$-very nearly.
li

The value of the synchronous reactance depends to a large extent upon the degree of saturation of the iron. For example, at low saturation the armature mmf. will have a much greater effect on the magnetic circuit than if the iron were saturated. Therefore, under short-circuit conditions, where the iron is operating at low saturation, the synchronous reactance will be *too large*. The variation of synchronous impedance with field current is shown in Fig. 151. As the iron becomes more saturated, the synchronous impedance *decreases*. Under operating conditions, the iron is considerably more saturated than it is under short-circuit conditions. In order to approach as near as possible to operating conditions, it is desirable to obtain the synchronous impedance at the highest possible value of armature current, as at *1 i*, Fig. 151. Also, the synchronous impedance is determined at very low power-factor, corresponding to short-circuit condi tions, as shown by Fig. 152, where the angle *a'* between the current and the emf. *Ei* is nearly 90. Therefore, the armature current is a maximum when the axes of the armature coils are almost opposite the pole centers as shown in Fig. 139, page 134. As the armature magnetomotive force has its maximum effect under these conditions, the value of the synchronous impedance so determined is too large for other positions of the coil, as shown in Fig. 137, page 133.

It will be seen that the value of synchronous impedance determined at short-circuit is *too large* and will make the calculated value of regulation too high. Therefore, the synchronous impedance method is called the *pessimistic* method. It is a safe method to use when making a guarantee, because the machine always regulates better than the computed values indicate.

The following example will illustrate the use of this method: *Example. —* A 50-kv-a., 550-volt, single-phase alternator has an opencircuit electromotive force of 300 volts when the field current is 14 amp. When the machine is short-circuited through an ammeter, the armature current is 160 amp., the field current still being 14 amp. The ohmic resistance of the armature between terminals is 0.16 ohm. The ratio of effective to ohmic resistance may be taken as 1.2. (a) Determine the synchronous impedance of the machine. (6) The synchronous reactance, (c) The regulation at 0. 8 power-factor, current lagging.

The rated current of the machine $/ =$,' = 91 amp.
GOU 300
(a) The synchronous impedance Z, =. . ,,, = 1.87 ohms.
The effective resistance = 1.2 X 0.16 = 0.192 ohm.
(ft) X, = V(ilf)rW»2)i = 1.86 ohms. (c) cos 6 = 0.8 sin 6 = 0.6.
Applying equation (39), page 139.
E = V(550 X 0.8) + (91 X 0.192)2 + (550 X 0.6) + (91 X 1.86)I' = V209,000 +249,000 = 677 volts.
The definition of *regulation* for an alternator is the same as for a directcurrent generator (Vol. I, page 292, Par. 199), namely the percentage rise in voltage when rated load is taken off the machine. As the *synchronous* reactance was used in the foregoing problem, the armature reaction was taken into consideration, so that the no-load voltage of the machine is presumably 677 volts. Therefore, the regulation is cent-*Ans*

It is to be noted in the foregoing problem that the armature impedance Z, is practically equal to the synchronous reactance X,, and in most cases it may be assumed, without appreciable error, as

being equal to it.

67. Three-phase Application.—The preceding discussion and problems have all been applied to single-phase generators. This has been done merely to illustrate methods. Very poor results accompany the practical application of these methods to singlephase alternators. In a single-phase alternator the armature reaction is pulsating, even for a constant value of armature current. The flux in the poles pulsates, because of the variation of the current in the armature coils as they pass the poles. For this reason the synchronous reactance is an indefinite quantity and calculated results of regulation with single-phase machines are far from satisfactory.

In a polyphase machine, however, the armature reaction is substantially constant if the load be constant and balanced. When the current in one phase has decreased, the resultant current of the other two phases has increased, etc. Therefore, the magnetomotive force of the armature is practically constant in value and is stationary in space with respect to the field poles. That is, if the field poles rotate, the armature mmf. follows them at rotor speed and is practically constant in magnitude for a fixed value of armature current. This effect will be described more in detail under the induction motor.

Under these conditions the synchronous reactance becomes a more definite quantity and more satisfactory results are therefore obtainable with these various methods of testing. Figure 153 (a) shows the connections for making the open-circuit test of a three-phase alternator. This is substantially the same method as is used with direct-current generators. The field is excited from some direct-current source and the field current is measured with an ammeter. The armature is driven at the rated or synchronous speed and the open-circuit voltage measured for different values of field current. The voltage of one phase only need be measured as the phase voltages should all be equal. A frequency indicator, F, may be used for determining the speed of the machine. An additional resistance, Ri, in the field circuit is often necessary for obtaining the points on the lower part of the saturation curve.

In the short-circuit test all three phases must be short-circuited. There are two methods of connecting the ammeters in this test. They may be connected in Y, Fig. 153 (6), in which case the ammeters read the *line* current directly, or they may be connected in delta, Fig. 153 (c), in which case the line current is obtained by multiplying the ammeter readings by /3 or 1.73. With delta connection the ammeters need be only about half the

Short-Circuit Test-Ammeters in Delta Fio. 153.—Connections for making open-and short-circuit tests of an alternator.

range (1/1.73 or 0.58) necessary for the Y-connection. The average of the ammeter readings is usually taken, although there should be but little difference in the three readings.

In calculating the regulation of a three-phase alternator, only one of its three phases is considered when making computations. The regulation, efficiency, etc., of one phase is determined and the machine being symmetrical, the other phases will have similar characteristics. Therefore, only the single-phase calculations already described are necessary. Two conditions arise, one when the machine is considered as being Y-connected and the other when it is considered as being delta-connected. In each case only coil values of current and voltage are used.

68. Regulation of a Y-connected Generator.—If the machine is considered as being Y-connected, the coil voltage is equal to the 0 40 80 120 ICO 200 2-10 210 320 360 400 410 4bO

Field Current Fio. 154.—Open-and short-circuit characteristics of a 1500 kv-a. alternator.

line voltage divided by /3. The coil current and the line current are the same. The method of dealing with such a problem is illustrated by the following example: *Example.*—Figure 154 shows the open-and short-circuit characteristics of a 1,500-kv-a., 2,300-volt, 60-cycle alternator. Terminal volts and line

current are plotted as ordinates with values of field current as abscissas. Assume that the machine is Y-connected. The resistance between each pair of terminals as measured with direct current is 0.12 ohm. Assume that the effective resistance is 1.5 times the ohmic resistance. Determine the , (U? . t'

'v7'"" *ALTERNATOR REGULATION AND OPERATION* 153 synchronous reactance of the generator and its regulation at 0. 85 powerfactor, current lagging.

From Fig. 154, the maximum value of the short-circuit current is 1,400 amp. , which is equal to the coil current. This corresponds to 240 amp. in the field, and at 240 amp. field current the open-circuit terminal voltage is 2,180 volts. The corresponding coil voltage is --/-= 1,260 volts

Vo

i-?fio

Z. (per coil) = 'nn = 0.90 ohm = X, nearly.

1)4UU

If the resistance between terminals is 0.12 ohm, it includes two coils in series, as the Y-connection is assumed, so that the ohmic resistance per coil is 0.12/2 = 0.06 ohm. The effective resistance per coil is equal to 1.5 X 0.06 = 0.09 ohm.

1 500 000

The rated current of the machine = — - '— =-— 376 amp. per terminal.

2 300

The rated voltage per coil = '_ = 1,330 volts.

V3

The no-load volts per coil is found by applying equation (39), page 139.

cos B = 0.850 _ 6 = 31.8 _ sin 9 = 0.527 _ E = V(1,330X 0.850) + (376X 0.09)2 + (1,330 X 0.527) + (376 X 0.90)" = 1,560 volts. j 5gQ _ i 330 The percentage regulation per coil = — —, ,,,,,'-100 = 17.4 per cent.

IiOoU

Ans. The open-circuit terminal voltage = 1,560V3 = 2, 700 volts.

The percentage regulation using this value = — — '-= 17.4 per cent. Ans.

69. Regulation of a Delta-connected Generator. — It is im possible to determine whether a machine is Y-connected or deltaconnected unless the winding

itself be inspected. Fortunately it makes no difference, so far as calculation of the regulation is concerned, whether the machine be Y-connected or deltaconnected. It may be assumed to be either and the result is the same if the work is consistent. In the delta machine, the line *fc* voltage and jfte.coil. voltagejj£-gcu; but Hie coil current is t line current divided_by /3._ The ammeters connected in delta, aslshown i n Fig. 153)7 measure the coil current directly.

Let it be assumed in the problem of the preceding paragraph that the machine is delta-connected. Using 240 amp., the same value of field current as before, the coil voltage in the opencircuit test is now 2,180 volts, and the corresponding coil current 1,400 in the short-circuit test is *t* = 808 amp.

The synchronous impedance per coil --=2-70ohms or three times its previous value.

Figure 155 shows the circuits of the delta when the ohmic resistance is measured with direct current. Let the resistance per coil be *R* and the resistance measured between any two terminals be *R0*. The circuit consists of two parallel branches, one of *R* ohms and the other of *2R* ohms.

Fio. 155.—Measurement of delta coil resistance with direct current.

Therefore, the ohmic resistance per coil *R* = (3/2) X 0.12 = 0.18 ohm, or three times its previous value. This must be increased 50 per cent., in order to obtain the effective resistance.

1.5 X 0.18 = 0.27 ohm effective resistance. The rated *coil* current of the machine, 376

Applying equation (39), page 139.

E -

.85) + (217X0.27)2+(2,300X0. = 2,700 volts which checks the result obtained when assuming that the machine was Y-connected.

Therefore, a machine may be assumed to be either Y-or deltaconnected when it is desired to calculate the regulation.

70. Magnetomotive Force Method.—In the synchronous impedance method of determining regulation, a voltage was substituted for armature reaction or for a magnetomotive force. In the magneto-

motive force method, a magnetomotive force is substituted for a voltage, this voltage being the IX drop in the armature of the alternator. In other words," '" 'Ir

Fio. 156.—Vector diagram of magnetomotive force method at unity power-factor.

the armature *reactance* is considered as being zero, but the armature reaction is increased a sufficient amount to compensate for this.

The method involves a short-circuit and an open-circuit test and in this respect is similar to the synchronous impedance

Fio. 157.—Open-circuit and short-circuit tests, magnetomotive force method.

method. Figure 156 shows the principle of the method. This diagram is constructed for unity power-factor.. V is the terminal voltage. To this is added the IR drop, giving the voltage V, A certain field magnetomotive force Fi is required to produce this voltage V. The value of this magnetomotive force in terms of the field current is found on the saturation curve, Fig. 157. Corresponding to the value of Fi the field current Fi is found. *Fi* is laid off at right angles to Fi and leading it, as a mmf. leads by 90 the emf. which its flux induces. In the shortcircuit test the field current is adjusted until the rated current flows. The corresponding value of field current A (Fig. 157) is then read. The magnetomotive force represented by this field current is necessary to send rated current through the armature reactance and at the same time overcome the armature reaction, if the resistance be neglected. This magnetomotive force, A, replaces the combined effect of the armature *reactance* and the armature *reaction*. It is laid off 180 from the current as shown at —A in Fig. 156. (The total mmf. which is assumed to produce the total voltage drop is $+A$. The component which must balance this mmf. is —A.) The resultant magnetomotive force is F, which, at unity power-factor, is the square root of the sum of the squares of Fi and —A. F is the mmf. which exists at no load under the assumptions made. The no-load voltage E lags F by

90, Fig. 156, and is found on the saturation curve corresponding to field current F, Fig. 157.

To summarize the method at unity power-factor, the IR drop is added to the terminal voltage, and the field current corresponding to this sum is found on the saturation curve. The machine is then short-circuited and the field current necessary to send rated current through the armature is determined. The square root of the sum of the squares of these field currents is then found. The value of emf. on the saturation curve corresponding to this resultant field current is assumed to be the no-load voltage of the machine.

When the power-factor is less than unity, the diagram is similar to that shown in Fig. 158.

The voltage Fi is the vector sum of V and IR. Its value is readily found by projecting these voltages on the current vector.

Thus,

Fi = $V(V \cos 0 + IR)2$ + (Fsin 0)2

In most cases a numerical addition of F and IR is sufficiently accurate.

The value of the angle a may be found by first finding the angle $/3$.

F sin 6 sin /3 = — — a is usually so small that it may be neglected. The vector F leads V by (90 — a) degrees, but a is so small that it may be neglected. The armature reaction vector —A is 180 from the current vector. By geometry the angle between -A andFi

X = 180-(90 + /3) = 180-(90 + 0) nearly

= 90-0 nearly

Fio. 158.—Vector diagram of magnetomotive force method, lagging current

By the cosine law,

F2 = Fi2 + A2-$2FiA$ cos (90 + 6) (43)

The voltage E corresponding to F and found from the saturation curve, Fig. 157, is the no-load voltage of the generator.

Example.—Take the example of the preceding paragraph. The exact method will first be used. The machine will be considered as being Y-connected.

o Qnn

The coil voltage = =&£ = 1,330 volts.

V3

The IR drop is 376 X 0.09 = 33.8 volts.

cos 6 = 0.85 sin 6 = 0.527

Vi = V(l,330 X 0.85) + (34)" + 1,330 X 0.5272 = 1,359 volts.

Algebraic addition would have given 1,364 volts, , 1,330X0.527

--

1,359 0 = 31.1 6 = 31.8 a = 31.8-31.1 = 0.7 which is negligible.

From Fig. 154, the field current corresponding to 1,359 coil volts, or 2,350 volte on the saturation curve (2,350 = l,359/3) is F = 266 amp.

The rated current of the coils is 376 amp. Corresponding to this current (Fig. 154) the field current is 64 amp. from the short-circuit test. F z = 2662 + 642-2 X 266 X 64 cos (90 + 31.8) F2 = 92,840 F = 305 amp.

From the saturation curve, the terminal voltage corresponding to 305 amp. field current is 2,580 volts across the terminals or 1,490 coil volts.

The regulation = — — Toon'-= 12.0 per cent. *Ans.*

l,ooU

Because of the low saturation on short-circuit, a given mmf. will produce a greater increase of flux than an equal mmf. will produce under operating conditions, where the iron is saturated. Therefore, the emf. corresponding to a given mmf. at shortcircuit will be much greater than the emf. corresponding to an equal mmf. taken higher up on the saturation curve. This is illustrated in Fig. 157. On short-circuit, the voltage *ab* corresponds to the mmf. A. The additional voltage *de* corresponds to a mmf. *be* equal to A, but taken higher up on the saturation curve. The voltage *de* is obviously much less than the voltage *ab*. Hence, that part of the mmf. A which replaces a voltage is too small under load conditions. Therefore, the noload emf. E found on the saturation curve is too low, and the regulation as determined by this method is ordinarily less than the actual regulation. For this reason this method is often called the *optimistic method*. This is illustrated by the foregoing example, when the regulation as obtained by the synchronous impedance method is 17.4 per cent., whereas that obtained by the magnetomotive force method is 12.0 per cent.

That part of the mmf. *A* which actually is armature reaction is too high on short-circuit, due to the favorable position of the armature coils with respect to the field poles. As in the synchronous impedance method, this factor tends to give too high a value of regulation. These two sources of error tend to offset each other in the magnetomotive force method, whereas they both produce errors in the same direction in the synchronous impedance method. Therefore, the mmf. method usually gives results. closer to the actual regulation than the synchronous impedance method does. The actual value of the alternator regulation probably lies between the two values just determined. Were the saturation curve of the machine a straight line, both methods would give nearly the same result.

71. The A. I. E. E. Method.—This method, recommended by the American Institute of Electrical Engineers, has an advantage over the other two methods, in that the synchronous impedance is *(a)* Vector diagram for a load of very *(6)* Curves used for A. I. E. E. low'power-factor. method.

Fig. 159.—The A. I. E. E. method.

measured when the machine is operating at full voltage, and, therefore, at normal saturation. This is accomplished by applying a load of very low power-factor, usually an under-excited synchronous motor. The vector diagram for this condition is shown in Fig. 159 *(a)*. *V* is the terminal voltage under these conditions, *E* the open-circuit voltage, and *IXS* the synchronous reactance drop at rated current. As the *IR* drop is small and is nearly in quadrature with *V*, the open-circuit voltage is substantially equal to the numerical sum of the terminal voltage *V* and *IX*. Therefore, numerically, $IX. = E - V$

$$E - V$$

X, is ordinarily determined from a no-load saturation curve and a curve taken at low power-factor and rated current.

Thus, in Fig. 159 *(b)*, *0a* is the no-load saturation curve and *db* is a curve taken at low power-factor and rated current. This low powerfactor load is ordinarily obtained by using under-excited synchronous motors as the load. The rated terminal voltage of the machine is *cb*, and when the load is thrown off, the opencircuit voltage obviously becomes *ca*, as the field current remains constant. The synchronous reactance *X*,, which is practically equal to the synchronous impedance, is determined by dividing *db* by the

Fio. 160. — Characteristics of an alternator at different power-factors.

rated current of the machine, this being the current at which curve *db* was determined. $ac — be-ab$

$$I \, '' \, l$$

When the value of *X*, is determined, it may be utilized in finding the regulation in the same manner as described under the synchronous impedance method of Pars. 68 and 69.

If it is not possible to load the machine, the distance *0d* may be found from a short-circuit test and the curve *db* determined from a knowledge of machines having similar constants.

The A. I. E. method gives too large a value of regulation for machines having salient poles, as the armature reaction is too large at low power-factors since the coil is acting directly upon the magnetic circuit of the generator, as has been shown in Fig. 139, page 134.

From the foregoing it is obvious that for a given current the regulation depends on the *power-factor.* The regulation has the greatest values at low power-factors, lagging current. At unity power-factor, the regulation is usually some nominal value, that is, from 5 to 10 per cent. With leading current, the voltage tends to *rise* as load is applied and the regulation may be zero or even negative. Figure 160 shows three typical load curves of an alternator, one being taken at unity power-factor, the second at 0.8 power-factor, lagging current, and the third at 0.8 powerfactor, leading current. The regulation in each case is as follows: $ac — ab$ Regulation $= r—$

It should be kept in mind that for a fixed *kilowatt* output the regulation with lagging current is even poorer than the values obtained for fixed *current* output.

72. The Tirrill Regulator.—An automatic voltage regulator of the Tirrill type for direct-current machines is described in Vol. I, Chap. XI, page 306. An automatic voltage regulator is much more essential in the smaller alternating-current stations than in direct-current stations The voltage changes in the generator and throughout the system are greater with alternating current than with direct current because of the added reactance drop in the generator armature, transformers, feeders, etc. Alternating-current generators cannot readily be compounded to compensate for voltage drop as direct-current generators are.

The Tirrill regulator is also designed to be used with alternating-current generators. As with large direct-current machines, the regulator acts through the field of an exciter. The underlying principle of the regulator is the same whether used for alternating or for direct current, the voltage being controlled in each case by the rapid short-circuiting of the exciter field rheostat. Figure 161 shows the connections of the alternatingcurrent type as applied to a three-phase alternator.

There are two control magnets, an alternating-current control magnet and a direct-current control magnet.

The alternating-current control magnet is operated primarily by a potential winding connected across one phase of the generator, usually through a potential transformer. The plunger of this magnet acts upon one end of a pivoted lever. On the other end of the lever there is an adjustable counter weight and the lower contact of the main contacts.

The direct-current magnet is operated by a winding connected across the exciter terminals. The plunger of this magnet acts upon another pivoted lever. On the other end of this lever there is the upper contact of the main contacts, so that these main contacts are not fixed but are "floating." A spring on the contact end of the lever tends to keep the main contacts closed.

Fio. 161.—Elementary connections of alternating-voltage regulator.

There is a differential relay magnet just as in the direct-current regulator. One coil of this magnet always is connected across the exciter terminals and the other coil is connected across the exciter terminals when the main contacts are closed. A series resistance limits the current in each to its proper value. The armature of this relay magnet, when released, is pulled upward by a spring and closes the relay contacts. These relay contacts short-circuit the exciter field rheostat. A condenser is shunted across these contacts to minimize arcing.

The field of the exciter is first adjusted so that the alternator voltage is about 65 per cent. below normal. This weakens both control magnets so that the floating main contacts are closed. This closes the circuit of the second winding on the relay magnet which opposes the other winding. The relay armature is therefore released and the relay contacts closed. These contacts short-circuit the exciter field rheostat and the voltage for both the exciter and the alternator rises. When the voltage has reached the value for which the regulator has been adjusted, the control magnets open the main contacts, the relay contacts open and the voltage of the exciter and the alternator both drop. The cycle is then repeated. When in operation, the main contacts and the relay contacts vibrate continuously so that voltage fluctuations are scarcely noticeable.

A compensating winding, supplied by the secondary of a current transformer in series with the line, compensates for line drop. It increases the pull of the alternating-current control magnet so that the main contacts are drawn closer together and the duration of short-circuit of the field rheostat is increased.

73. Parallel Operation of Alternators.—The same reasons which make it necessary to operate direct-current generators in

Flo. 162.—Alternators in parallel and speed-load characteristics.

parallel (see Vol. I, page 372, Par. 235) apply to alternators. Alternators, however, are made in units of very much greater capacity than it is possible to make direct-current machines, since there are no commutation difficulties. The largest single alternating-current unit at the present time is of 50,000 kv-a. capacity.

In order to operate satisfactorily in parallel, direct-current generators must have drooping voltage characteristics. In order that alternators may operate satisfactorily in parallel, their *prime movers* must have *drooping* speed-load characteristics. Otherwise the operation will be unsatisfactory. The reason for this is as follows:

Two alternators 1 and 2 are operating in parallel as shown in Fig. 162 (a). If they are operating in parallel they must have the same frequency and the same terminal voltage. Figure 162 (6) shows the speed-load curve of each of the prime movers driving the alternators. (Instead of plotting speed in r.p.m., the frequency or electrical speed is plotted. For example, a six-pole alternator running at 1,200 r.p.m. would have the same electrical speed as an eight-pole alternator running at 900 r.p.m.) The speed-load curves of the prime movers are determined by their respective governors, if they are steam-, water-, or gasdriven units. If motor-driven, the speed-load characteristics depend upon the motor speed-load characteristics.

Let oc, Fig. 162 (6), be the frequency at which the system is operating. By projecting horizontally to intersect the spieedload curves, the load taken by each machine at this frequency is obtained. oa is the load on machine 1 and ob is the load on machine 2, as both machines are operating at system frequency. Let the field of 1 be strengthened by means of its field rheostat. At the same time weaken the field of 2 so that the line voltage does not change. If these were direct-current generators, machine 1 would immediately take more load. But 1 *cannot take more load* because its prime mover can deliver only the load oa at this frequency. Machine 2 cannot drop any load because its prime mover

can deliver only the load ob at this frequency. Both machines must always operate at the same frequency which is not true of direct-current machines. *Therefore, the kilowatt load delivered by alternators in parallel cannot be shifted appreciably by means of the generator fields.*

To change the kilowatt load of either machine, the speed-load characteristic of its prime mover must be changed. In enginedriven units this is done by changing the tension in the governor spring or altering in some manner the governing device. Assume, in Fig. 162 (6) that it is desired to make generator 1 take the same load as 2. The governor spring of 1 is so adjusted that the characteristic of 1 is raised, as shown in Fig. 163. Both machines now deliver the same load oa' at a frequency oc'. Under the conditions shown, Fig. 163, the frequency oc' is higher than the original frequency oc, Fig. 162. If the original frequency is to be maintained, the speed-load characteristic of 2 must be lowered at the same time that the characteristic of 1 is raised. Therefore, to adjust the load between alternators in parallel, the speed-load characteristics of the prime movers must be changed. If the alternators are driven by shunt motors, the speed-load characteristics of the motors may be changed by adjusting the motor field rheostats. It will be noted, in Fig. 163, that the loads of the two machines are equal at one frequency only.

If the prime movers had flat speed-load characteristics, the operation of the alternators would be unstable. That is, very small disturbances or changes of frequency would cause very large fluctuations in the kilowatt load delivered by each machine. This condition would result in serious operating difficulties.

It has been shown that directcurrent shunt generators operating Fig. 163.—Speed-load curves of in parallel are in *stable equilibrium* alternators in parallel—effect of ,,,. changing governor control.

(See Vol. 1, page 373, par. 235).

That is, any circumstance which tends to throw machines out of parallel is counteracted by reactions opposing this tendency. In the same way, any action tending to throw alternators out of parallel is opposed by reactions which tend to prevent the alternators pulling out. This is most clearly illustrated by the conditions existing when neither alternator is supplying external load. If the two alternators are considered as a local series circuit, their voltages are in *opposition*. These voltages are represented in Fig. 164 by Ei and $Z?2$ respectively. Ei and Ez are equal and opposite, so that the net

Load E,

Fig. 164.—Synchronizing current of alternators in parallel.

voltage acting in the local circuit of the two alternators is zero. Therefore, there is no current flowing between the alternators, just as there is no current circulating between two batteries having equal emfs. and connected with terminals of like polarity together.

Assume that the prime mover of generator 1 speeds up temporarily. The internal induced voltage of this generator will advance an angle a with respect to Ez. That is, Ei will advance to position $E'i$. The vector sum of the two alternator emfs. E and Ei will no longer be zero, but due to the change in their relative phase positions, the vector sum of $E'i$ and $E2$ will be $E0$.

J=2a) _ The result is the same as with the two + g=8v. batteries of Fig. 165. No. 1 has an emf.. — $Jr=.\&Q$ of iQ volts and 2 has an electromotive

No. i. No. 2. force of 8 volts. If the load current is

Fio. 165.— Batteries in zero, the current circulating between

"?"'''' these batteries is then found by dividing the sum of the two voltages, giving each the proper sign, by the sum of the resistances of the two batteries. That is, , 10 + (-8) ,, /0 = 0.5 + 0.5:= 2 amp'

In the same way, the current circulating between the two alternators is the *resultant* voltage divided by the sum of the impedances of the two machines.

Q

$?i + £2 V(Ri + R2 + (Zi + X2)2$

Where Zi, Zz, Ri, Rz, and Xi, $X2$, are the respective impedances, resistances and reactances of the two machines. As the resistance of an alternator armature is very small compared to its reactance, this circulatory current will lag by an angle /3, nearly 90, with respect to the voltage $E0$ producing it, as shown in Fig. 164.

It will be observed that /0 is nearly in phase with the voltage E'. Therefore, it puts a power load on generator 1, and this tends to slow down this generator. On the other hand /o is nearly 180 froml?2, that is, it is acting in opposition to Ez. Therefore, /o develops motor action in generator 2, as the induced electromotive force acts in opposition to the current. This motor action tends to speed up machine 2. *Therefore, if two alternators in parallel attempt to pull out of step, a current is developed which, circulates between the two machines. This current tends to accelerate the lagging machine and to retard the leading machine, and so acts to prevent the alternators from pulling out of synchronism.*

If the machines are operating under load, /0 merely puts more load on the machine which tends to lead and takes load off the machine which tends to lag. This last machine will not ordinarily operate as a motor, as it did under no-load conditions, but as its load is reduced its angular position will be advanced.

Because /o tends to hold the two machines in synchronism, it is called the *synchronizing current.*

It has already been stated that changing the field current does not vary the distribution of load between two alternators. However, it does affect the current delivered by the two machines. Figure 166 (a) shows the vector diagram for two similar alternators having a common terminal voltage V. Both machines are delivering equal currents I and / 2 respectively, which are in

FiG. 166.—Vector diagram of currents and voltages when alternators operate in parallel.

phase with the terminal voltage V. The resultant load current is their sum /', which is in phase with V. As both machines have equal resistances and reactances, their respective internal voltages Ei and Ez are the same. (In this diagram the machines are treated with reference to the *external* circuit in which case the voltages and currents are acting in *conjunction*.)

Let the field of generator 1 be weakened and that of 2 be strengthened. It has already been shown that this cannot affect the division of the kilowatt load between the machines. When the field of generator 1 is weakened, its internal voltage *decreases* and when the field of 2 is strengthened, its internal voltage *increases*. Now both machines must continue to have equal terminal voltage. It has already been shown that if a machine delivers a leading current, its internal voltage is less than when the machine delivers a lagging current. (See par. 64, page 137.) Moreover, a leading current in an alternator tends to strengthen the field and a lagging current tends to weaken the field, through armature reaction.

For generator 1 to operate with a reduced internal voltage it must deliver a leading current, making Ei, shown in Fig. 166 (b), less than its previous value shown in Fig. 166 (a). On the other hand Ez, Fig. 166 (b), is greater in magnitude than in Fig. 166 (a) because generator 2 now delivers a lagging current. Also the leading current in generator 1 tends to strengthen its field and the lagging current in generator 2 tends to weaken its field, through armature reaction. In both cases, the change of flux produced by change in field current is *opposed* by armature reaction. The load current /' cannot change in phase or in magnitude, as the phase and magnitude of /' is determined entirely by the character of the load which is connected to the system. Therefore, since /i and /2 are equal, they must make equal angles with V so that their resultant /' will still lie along V.

It will also be observed that each machine is carrying a larger current than it did before and yet the kilowatt output of each has not changed. This means that the heating *(PR)* loss in each machine has been increased without any compensating advantages. Therefore, this is not the best condition of operation. Figure 167 shows the diagram of Fig. 166 (6) with the voltage drops eliminated. *Eo* is now the *difference* of *E* and *Ez,* and /o, the circulating current, lags *Eo* by nearly 90 as in Fig. 165. It will be observed that /0 is nearly in quadrature with the terminal voltage *V* so that it transfers practically no power from one machine to the other. This substantiates what has already been demonstrated, that changing the field current cannot transfer appreFio. 167.—Vector diagram showing ciable load from one machine effect of excitation upon alternator circu-+,, 4-Ug other latory current.

74. Synchronizing.—Before direct-current generators can be safely thrown in parallel, two conditions must be fulfilled. The two terminal voltages must be equal, or substantially so, and the proper polaritymust be observed. These same two conditions must be fulfilled when alternators are connected in parallel. The equality of voltages can be readily determined by connecting a voltmeter first to one machine and then to the other. The voltmeter, when so connected, does not give any indication as to polarity, as the indications of an alternating-current voltmeter are independent of its polarity.

Lamps, however, can be used to determine the correct polarity. Figure 168 shows the connections for phasing a three-phase alternator with the bus-bars. A lamp is connected across each pole of the three-pole switch which connects the machine to the line. The voltage rating of the lamps should be 15 per cent. greater than that of the machineoFline. For example, if the system is 220 volts, two 115-volt lamps in series may be used across each pole, although these lamps will be subjected to overvoltage during a part of the synchronizing period. If the machines are

Fig. 168.—Connections for "3 dark" method of synchronizing with lamps.

properly connected, the three lamps should all become bright and dim together. If they brighten and grow dim in sequence, it means that the phase rotation of the two machines is opposite, so that one phase must be reversed.

The lamps nicker at a frequency equal to the *difference* in the frequencies of the two machines. As the machines approach synchronism the flicker becomes slower and slower. When the lamps are all dark the switch may be closed. The fact that the lamps are all dark indicates that the potential difference between each switch blade and its clip is nearly zero and the two alternators are in *opposition* so far as their local series circuits are concerned. Two points across which the potential difference is zero may be connected without any resulting disturbance, so that the switch may now be safely closed and the two alternators are in parallel.

The disadvantage of this method is that lamps are dark even although a very considerable voltage may exist across their terminals, and the machines may be connected in parallel therefore when considerable voltage difference exists between them. This may do no harm with slow-speed or small-capacity units, but with high-speed turbo-units, which have little armature reactance and are quite "sensitive," there may be considerable disturbance if there exists a substantial phase difference at the time of connecting in parallel. Another objection to this "three dark" method is that the lamps do not show whether the incoming machine is fast or slow.

The foregoing difficulties may be in part eliminated if the connections of two of the lamps, as 1 and 2, Fig. 169, be crossed.

Fio. 169.—Connections for "2 bright and 1 dark" method of synchronizing with lamps.

When the machines are in synchronism, 1 and 2 are bright and 3 is dark. As one of the bright lamps is increasing and one is decreasing in brilliancy near the point of synchronism, it is possible to determine very accurately the instant at which the switch should be closed. This is called the Siemens-Halske or "two bright and one dark" method. By noting the sequence of brightness of the lamps, it can be determined whether the incoming machine is fast or slow.

The best method is the use of the synchronism indicator or synchroscope described in Chap. Ill, page 71. Such an instrument shows very accurately the position of synchronism. The synchroscope is connected across but one phase. It is possible that one phase of each machine may be in synchronism, but the other two out of phase due to wrong phase rotation. The correct phase rotation must be determined by lamps or by other means before depending entirely upon the synchroscope. Synchronizing lamps are often used in conjunction with a synchroscope so that the operator has a check on the instrument.

75. Hunting of Alternators.—The driving torque of a reciprocating engine, or of a gas engine, is not uniform during a revolution of the fly-wheel, but varies from zero at the dead centers to a maximum at some intermediate position. Even with a heavy fly-wheel, this variation of torque may impart impulses to the induced emf., causing it to be ahead of its proper position at some instants and behind it at other instants. This causes heavy synchronizing currents to flow between machines in parallel and often causes their rotating members to "oscillate" as they are rotating. The angular effect of the crank position can be appreciated when it is realized that in a 60-pole alternator a displacement of one mechanical or space-degree in the rotating member makes a diffference of 30 electrical degrees in the phase angle of the electromotive force. The above impulses are often communicated to the system, causing synchronous motors and converters to oscillate. These oscillations are called "hunting." Hunting may become serious if the engine governors have a natural frequency of oscillation nearly the same as that of the machine rotors. The oscillations may then become cumulative and may even cause the machines to go out of synchronism.

Remedies for hunting are to use heavy fly-wheels, to put dashpots on the engine governors, and to use amortisseur or squirrelcage windings around the field, such as is shown in Fig. 295, page 319. Where several engine-driven units are used, they are often paralleled when their cranks occupy different angular positions. This minimizes the effect of the engine impulses on the system, although their effect is increased in the local interchange currents between generators.

CHAPTER VII

THE TRANSFORMER

The static transformer is a device for transferring electrical energy from one alternating-current circuit to another without a change in frequency. This transference is usually, but not always, accompanied by a change of voltage. A transformer may receive energy at one voltage and deliver it at a *higher* voltage, in which case it is called a *step-up* transformer. A transformer may receive energy at one voltage and deliver it at a *lower* voltage, in which case it is called a *step-down* transformer. A transformer may receive energy at one voltage and deliver it at the *same* voltage, in which case it is called a *one-to-one* transformer.

A static transformer has no rotating parts, and therefore it requires little attention and its maintenance is low. The cost per kilowatt of transformers is low as compared with other apparatus and the efficiency is much higher. As there are no teeth, slots, or rotating parts, and the windings can be immersed in oil, it is not difficult to insulate transformers for very high voltages.

Because of these many desirable characteristics, the transformer is a very useful piece of apparatus, and as it can transform from low to high voltage, and from high to low voltage, economically, it is largely responsible for the extensive use of alternating current.

76. The Transformer Principle.—The transformer is based on the principle that energy may be efficiently transferred by induction from one set of coils to another set by means of a varying magnetic flux, provided both sets of coils are on a common magnetic circuit.

Electromotive forces are induced by a change in flux linkages. In the generator, the flux is substantially constant in magnitude. The amount of flux linking the armature coils is changed by the relative *mechanical* motion of flux and coils. In the transformer, 172 the coils and magnetic circuit are all stationary with respect to one another. The electro-

motive forces are induced by the change in the *magnitude* of the flux with time. This is illustrated in Fig. 170.

A core is made up of rectangular stampings of sheet steel, clamped or bolted together.

A continuous winding P is placed on one side or leg of the iron core. Another continuous winding S, which may or may not have the same number of turns as P, is placed on the opposite side or leg. An alternator A supplies current to the primary winding P. As this winding is linked with an iron core, its magnetomotive force produces an alternating flux in the core.

Fig. 170.—Simple transformer, secondary open-circuited.

This alternating flux links the turns of the winding S. As this flux is alternating, it induces in the winding S an errif. of the same frequency as its own. Because of this induced emf., the secondary winding S is capable of *delivering* current and energy. Therefore, the energy is transferred from P, the primary, to S, the secondary, by means of the magnetic flux.

The winding P which *receives* the energy is called the *primary*. The winding S which *delivers* the energy is called the *secondary*. In a transformer, either winding may be the primary, the other being the secondary, depending upon which winding receives and which delivers energy.

77. Induced Electromotive Force.—The flux p, called the 'mutual flux, in passing through the magnetic circuit formed by the iron core, links not only the turns of the secondary winding S, but also the turns of the primary winding P. Therefore, an emf. must be induced in both the windings S and P. As this flux t is the same for each of the two windings it must induce the *same emf. per turn* in each winding. The *total induced emf.* in each winding must then be proportional to the number of turns in that winding. That is, where Ei and Ez are the primary and secondary *induced* emfs. and Ni and Nz are the number of turns in primary and secondary respectively. In the ordinary transformer, the terminal voltage differs from the induced emf. only

by a very small percentage, so that for most practical purposes it may be said that the primary and secondary terminal voltages are proportional to the respective number of turns.

The induced electromotive force in a transformer is proportional to three factors; the flux, the frequency, and the number of turns. The complete equation for the induced electromotive force, assuming a sine wave, is as follows: $E = 4A4fN max$ 10-8 volts (46)

Flo. 171.—Sinusoidal variation of flux with time.

where / is the frequency in cycles per second, N is the number of turns, and tmax is the maximum value of the flux in the core. The factor 4.44 is 4 times the form factor, which is 1.11 for a sine wave. (See Chap. 1, Par. 5, page 10.) This equation is derived as follows:

Figure 171 shows the mutual flux t varying sinusoidally with the time. Between points a and b the total change of flux is 2 tmaz lines or maxwells. This change of flux occurs in half a cycle or in a time T/2 sec. where T is the *period* or the time required for the wave to complete one cycle. The time T/2 is obviously equal to l/(2/) sec. From equation (74), Vol. I, page 185, the *average* induced emf. becomes

«, =-A1 $2-jf$ 10-' volts =-4fNtmax 10-" volts

Since with a sine wave the ratio of effective to average volts is 1.11 (see page 10, Par. 5), the *effective* induced emf. is $E = 4.44/Ar0moi$ 10-" volts

If the flux varies other than sinusoidally with the time, a factor kf called the *form factor* must be substituted for 1.11 in the above equation.

The maximum flux $4max = Bmal A$, where Bmax is the maximum flux *density* and A is the core cross-section Equation (46) may then be written: $E = 4A4fNBmaz$ A 10-o volts (47)

This equation is the more convenient to use, as will be shown later.

Example. — The core of a 60-cycle transformer has a cross-section of 20 sq. in. and the maximum flux density in the core is 60, 000 lines per square inch. There are 700 turns in the primary and 70 turns in the secondary. What is the rated voltage of the primary and of the secondary? Ei = 4.44 X 60 X 700 X 60,000 X 20 X 10-8 = 2,230 volts. *Ans.* Ez= 4.44 X 60 X 70 X 60,000 X 20 X 10-8 = 223 volts. *Ans.* Also #2 = 2,230/ 10 = 223 volts. *Ans.* **78. Ampere-turns.** — Figure 172 shows a transformer having a primary and a secondary winding. The directions of the flux, of the voltages and of the currents, as indicated on the figure, are those existing at the instant when the upper primary line is positive. Assume first that there is no load on the secondary. Under these conditions a very small current flows in the primary, usually from 3 to 8 per cent. of the rated current. This no-load current can be resolved into two components, one supplying the no-load losses, and the other in quadrature with the first and producing the flux *t.* (See Par. 80.) This quadrature current is called the *exciting* or *magnetizing current* of the transformer. As the energy current which is in phase with the back emf. is small, the quadrature current is very nearly equal, numerically, to the total no-load current. Therefore, the no-load current is often called the exciting current of the transformer. The back emf. is nearly constant for all loads, as it differs from the terminal voltage only by the primary impedance drop, which is small. Therefore the flux and hence the exciting current are practically independent of the load.

This exciting current produces a flux *t* in the core, the direction of the flux being as shown (corkscrew rule). The value of this flux must be such as to make the induced primary emf. practically

Fio. 172.—Simple transformer, load applied to secondary.

equal to the primary line voltage. This primary induced emf. is a *back* emf. and is therefore in opposition to the primary impressed voltage.

Now apply a load to the secondary. As a result a current *Iz* flows in the secondary. The direction of this current must be such as to *oppose* the flux *t.* This is in accordance with Lenz,s Law that an induced current always has such a direction as to oppose the cause which produces it. If the secondary current *h*

were producing the flux *t,* then by the corkscrew rule the current would flow *in* at the upper terminal, Fig. 172. Since /2 opposes the flux (£, it must actually flow *out* at the upper terminal. The secondary current /2 then tends to reduce the value of the flux in the transformer core. If the flux is reduced, the back electromotive force of the primary is also reduced, and hence more current will flow in the primary to supply the increase in power due to the load on the secondary. This is the sequence of reactions which follow the application of load to the secondary, enabling the primary to take from the line the increased power demanded by the secondary.

The change in the back electromotive force in the primary from no load to full load is ordinarily about 1 or 2 per cent. As the back emf. is proportional to the mutual flux *t, the value of* 0 *therefore does not change appreciably over the working range of the transformer.* If this flux does not change appreciably, the *net* ampere-turns acting on the core cannot change appreciably. Therefore, the increased ampere-turns due the secondary load must be just balanced by the additional ampere-turns due to the increased primary current.. Since the flux remains practically constant it follows that the exciting current must remain substantially constant.

The effect of any *increase of* primary ampere-turns, when not opposed by equal secondary ampere-turns, would be to increase the mutual flux. This would increase the back emf., and might cause the primary to deliver power back into the power source, which is in violation of the law of the conservation of energy. Therefore, any primary ampere-turns in excess of the exciting ampere-turns must be balanced by equal and opposing secondary ampere-turns.

The exciting current is of small magnitude and differs considerably in phase from the total primary current, as shown by /o in Fig. 174, page 180. Therefore, it is usually neglected in comparison with the total primary current. If it be neglected, the *primary* and *secondary* ampere-turns are *equal,* and tf 1/1 = Therefore,

That is, *the primary and secondary currents are inversely as the respective turns.*

The above relation also follows from the law of the conservation of energy. If the transformer losses be neglected and unity power-factor be assumed, vJl = vJz /i = Vj Nz /i Vi Nl tl **79. Leakage Reactance.**—In the preceding discussion it has been assumed that *all* the flux which links the primary also links the secondary. In practice it is impossible to realize this condition. All the flux produced by the primary does not link the secondary, but a part completes its magnetic circuit by passing through the air rather than around through the core, as shown by 0i, Fig. 173. That is, between planes a and *b,* Fig. 173, there is a mmf. due to the primary ampere-turns, plane *a* being at a higher magnetic potential than plane *b* at the instant shown. This mmf. is proportional to the primary current and tends to send flux from a to *b* both through the air and around through the core. That part of the flux which passes from *a* to *b* through the air follows a magnetic circuit which is acted upon by the primary

Fio. 173.—Mutual flux, primary leakage flux, and secondary leakage flux in a transformer.

ampere-turns only. This flux 0i is called the *primary leakage* flux. It is proportional to the total ampere-turns of the primary alone as the secondary turns do not link the magnetic circuit of 0i. Therefore 0i induces an emf. in the primary but not in the secondary. The flux 0i is in time-phase with the total primary current *I.* The emf. induced by 0i must lag 0i and /i by 90. (See page 27, Par. 14.) The emf. necessary to balance this counter emf. is opposite and equal to it, and therefore leads the current *h* by 90. As this counter emf. is proportional to the current and lags it by 90, it is nothing more than a reactance voltage, and is denoted by — IiX. The component of line voltage which balances this emf. is-r-/i-X'i. Therefore, a reactance drop exists in a transformer primary in precisely the same manner that a reactance drop exists in an alternator armature. The effect of the primary leakage flux, therefore, is to oppose the flow of current *into* the transformer.

The mmf. of the secondary coil, acting alone, is such that the top of the coil is at a higher magnetic potential than the bottom of the coil. That is, plane *c* is at a higher magnetic potential than plane *d*, and therefore a flux *fa* tends to pass from *c* to *d* through the air, as shown. Flux *fa* is called the *secondary leakage flux*. As its path is not linked by the primary, the secondary leakage flux is proportional to the secondary ampere-turns only. *fa* induces an emf. in the secondary, lagging the secondary current / 2 by 90. This is also a reactance voltage, and the component which balances it leads the secondary current by 90. This last voltage is denoted by /2 X 2. The secondary reactance opposes the current flowing *out* of the secondary just as the armature reactance of an alternator opposes the current flowing out of the armature. Both the primary and secondary reactances of the transformer have the same effect on the regulation of the transformer as the armature reactance of the alternator has on the regulation of the alternator.

In that part of the core which is surrounded by the secondary winding the mutual flux *t* and the secondary leakage flux *fa* are shown in opposition. As *t* is produced by the joint ampereturns of primary and secondary, and *fa* by the ampere-turns of the secondary alone, *t* and *fa* are almost never in phase with each other, but are usually out of phase by an angle greater than 90, as shown in Fig. 174 (a). Two separate fluxes in the core do not actually exist at the same instant, but merely the resultant flux, found by combining *t* and *fa*. The primary leakage flux *fa* and the secondary leakage flux *fa* have the same general direction in the space between the primary and secondary coils.

In the actual transformer, the leakage flux paths are not so simple as those indicated in Fig. 173. That is, part of *fa* links some of the secondary turns, but not all, etc. However, the equivalent effect of *fa* and *fa* is readily determined in the ordinary transformer by simple measurements, as is described later.

In practice, the primary and secondary windings are not placed on separate legs, as shown in Figs. 170, 172 and 173, for being widely separated, large primary and secondary leakage fluxes would result.

These large leakage fluxes would cause the transformer regulation to be too poor for commercial use. To reduce the leakage, the primary and secondary should be interleaved. Therefore, each is usually split into a number of coils, and alternate primary and secondary coils are placed together, as shown in Figs. 183 and 184, pages 196 and 197.

80. **Transformer Vector Diagram.**— Figure 174 (a) shows the relations existing among the currents and voltages in a transformer, when the secondary is delivering a current /2, at terminal voltage Vz and power-factor cos 02. A one-to-one ratio of transformation is assumed, in order that the lengths of all the vectors Fig. 174(6).

Fio. 174 (a).—Complete vector diagram for'a transformer. Fio. 174 (6).— Energy and magnetizing components of no-load or exciting current. in the diagram shall be of the same order of magnitude. This same diagram may be made applicable to any ratio of transformation, merely by multiplying the proper vectors by the ratio of transformation.

The secondary current /2 is laid off at phase angle 02 from the secondary terminal voltage F2. The secondary leakage flux #2 is in time-phase with /2 and induces the emf. which is balanced by IzXz, leading /2 by 90. The induced voltage /?2 of the secondary is determined by adding vectorially to F2 the secondary resistance drop /2fl2, in phase with /2, and the secondary reactance drop /2X2, due to *fa*, in quadrature with /2 and leading. As both the primary and secondary induced voltages are induced by the same flux, and both windings have the same number of turns, since the ratio is 1 to 1, the primary and secondary induced voltages will be equal in magnitude and will be in phase with each other. Therefore, $Ei = E2$. It has already been demonstrated that an

emf. induced by a flux varying sinusoidally with time is a sine wave and lags the flux 90 (page 27, Par. 14). Therefore, in Fig. 174 (a) the mutual flux *t leads* the induced emfs. by 90, as shown.

The line must first supply a voltage at least equal to the primary induced voltage, and in opposition thereto, before current can flow *into* the primary. This is analogous to the direct-current motor, where the line must first supply a voltage equal to the back electromotive force, and in opposition thereto, before any current can flow into the armature. Therefore, a voltage —*Ei* opposite and equal to *E* must first be supplied by the line. The primary must furnish at least a sufficient number of ampere-turns to balance the ampere-turns of the secondary. These primary ampere-turns and the secondary ampere-turns are equal and opposite. Therefore, if there are N2/2 ampere-turns in the secondary, there must be an equal number of ampere-turns in the primary to balance these. These primary ampere-turns NJ'i, Fig. 174 (a), are 180 from NJ2. It is not customary to show the ampere-turns on the diagram, however, but only the currents, as in Fig. 174. The ampere-turns may then be obtained by multiplying each current by its proper number of turns.

In addition to I'i, the no-load current /0 must exist to produce the mutual flux, /, and to supply the no-load losses. This current would be in phase with the flux *t* were it not for the core losses. These losses require that /o have an energy component shown by *Ie* in Fig. 174 *(b)*. That is, /0 is resolved into two components, a magnetizing component *Im* in phase with *tt*, and an energy component *Ie* in phase with the primary emf. —*Ei* and leading *Im* by 90.

The total primary current is *Ii*, the vector sum of /0 and *I'*.

The primary leakage flux *fa* is in phase with /i, and induces the emf. which is balanced by *IiX*.

The primary terminal voltage Fj may now be found by adding *IiRi* and *IiXi* vectorially to — 7?.

The transformer regulation is defined

as the rise in secondaryvoltage divided by the rated-load voltage, when rated-load is removed from the transformer. The primary voltage is assumed to be constant.

The regulation for a one-to-one transformer is given by

Ft-F2

F2

81. Simplified Diagram.—The diagram of Fig. 174 may be materially simplified if the magnetizing current /0 be neglected. As /0 is usually from 3 to 8 per cent. of /i and the two are con *Fio.* 175.—Transformer diagram with primary voltajsjs rotated to secondary side of diagram.

siderably out of phase, /o may ordinarily be neglected without serious error. Figure 175 shows the diagram of Fig. 174 with /o omitted. Note that — *Ei* is 180 from JB2; /i is 180 from /2; *IiRi* is 180 from /2fi2; and /iXi is 180 from *IzX2*. Therefore, if the entire left-hand side of the diagram be rotated through 180 with the origin as a center, as shown in Fig. 175, *Ei* and *Ez* coincide, *IiR* and *IKi* become parallel to *IJtz* and *jXz* respectively.

As /i equals /2, the two *IR* drops may be combined into a single drop equal to 12 *(Ri + Rz)* and the two *IX* drops may be combined into a single drop equal to / 2 *(Xi + Xz),* as shown in Fig. 176 (o). Let fli + fl2 = R0 and Xi + Xz = X0. It is to bo noted that the vector diagram of Fig. 176 (a) is similar to that of the alternator. The voltage *Vi* is given by *Vi* = V(F2 cos 6 + /2K0)2 + *(Vz* sin 6 + /2Xg)2 (49) *Rg* is the *equivalent resistance* of the transformer and *Xo* is the *equivalent reactance.* Obviously /i could be substituted for /« in the foregoing equation.

Example.—A 40-kv-a., 220-volt, one-to-one transformer has a primary resistance and a secondary resistance each equal to 0.009 ohm. The leakage reactance of the primary is 0.037 ohm and that of the secondary is 0.043

This regulation is not strictly true because F2 is assumed to be fixed and *Vi* to vary as the load changes. In a transformer, the *primary* voltage is ordinarily *fixed* and the secondary voltage drops

as load is applied. As the regulation is small, little or no error is introduced by using the secondary voltage as a basis for calculating the primary voltage.

82. Equivalent Resistance and Reactance.—The preceding discussion refers specifically to transformers having one-to-one ratios. There is little difference, however, when other than one-to-one transformers are considered. For example, in Fig. 174, *Ei* and J52 are considered as being equal. If there were *Ni* primary turns and *Nz* secondary turns, the true primary voltage would be v *F _ # i*

v

El-NlE2

Likewise the primary current would be *T-N2T*

Zl"FII'

In testing transformers and in computing their performance, it is more convenient to work with one side of the transformer only. The method of treating such a problem is as follows:

First, consider the resistance of the primary and of the secondary. The total copper loss in the transformer

Pt = /i2fli + /22fl2 (I) where /i and *Ri* are the primary current and resistance respectively and /2 and #2 are the secondary current and resistance respectively.

If the exciting current is neglected, /2 _ *Ni, , N, Ii. Ni '2 ll N2.* Substituting in (I) Pc = /12fl1 + /

This means that the *total* copper loss can be found by multiplying the primary current *squared* into the expression

This expression is equal to the primary resistance added to the secondary resistance when multiplied by the ratio of primary to secondary turns *squared.* This quantity is called the *equivalent resistance of the transformer referred to the primary* and is denoted by Bo. 2?2 (50)

The total copper loss may then be found by using the primary current alone. That is Pc = /i'fioi

The equivalent resistance of the transformer referred to the secondary may be found in a similar manner and is equal to: #02 = #2 + (j) 'fli (51)

The total copper loss may be found by using the secondary current alone.

That is *Pc* = /22«0=

These values of equivalent resistance may also be used in determining the regulation of the transformer, as will be shown later.

In well-designed apparatus and for the most economical use of the materials, the various parts should all come to their limiting temperatures at the same time. As primary and secondary windings ordinarily occupy approximately the same volume, their losses should be the same for equal temperature rise. That is, /1'72, = /«'TZg

Since *!2 Hi /i N2*

Ri _ W W

#2 /,2 *NJ*

That is, under these assumptions the ratio of primary and secondary resistance is equal to the square of the ratio of the primary and secondary turns.

The primary leakage reactance drop is *IX* and the secondary leakage reactance drop is *IzX2*. *IiXi* is a voltage and must be to the same scale as the primary induced emf. *Ei*. *IXz* is likewise a voltage and is to the same scale as the secondary induced emf. *E2*. It is convenient to use the same length of vector to represent *Ei* and *E2* and later correct these values by multiplying by the ratio of transformation. (See Fig. 174.) Assume that *Ez* is ten times *Ei* and that everything is to be drawn on the basis of *E2* being equal in length to *Ei*. In order that *Ez* may be represented by the same length of vector as *Ei*, *Ez* must be multiplied by-tv-or by Tt;' *IXi* is to the scale of *Ei.*

As *IzXz* is to the same scale as *Ez,* it likewise must be multiplied by -That is, JY2 -gives the secondary reactance drop one-tenth its actual value as it is reduced to the same basis as the primary reactance drop and the emfs. *Ei, Ez,* etc.

Substituting for /2, / r the above expression becomes *N-N*

That is, the secondary reactance drop may be referred to the primary side by multiplying the primary current into the secondary reactance-X'2, when multiplied by the ratio of primary to secondary turns *squared.*

The total reactance drop in the transformer, to primary scale, becomes, *Xqi* is called the *equivalent reactance of the transformer referred to the primary side.* Its use in determining the characteristics of the transformer will be considered later.

Likewise, the equivalent reactance referred to the secondary side

X1 (53)

If the permeance of the leakage flux paths is the same for both primary and secondary, the leakage reactances of the primary and secondary are to each other as the *square* of the number of turns. This follows from the fact that inductance varies as the *square* of the number of turns, as was demonstrated in Vol. I, Chap. VIII. In the actual transformer it is practically impossible to separate-$X'i$ and Xz, because the paths of the leakage flux are complicated, some of the flux linking only a part of the turns, etc. However, it is not necessary to know Xi and Xz separately, but rather their combined effect. This effect may be found by multiplying $X0i$ by the primary current $/i$ and adding this voltage in its proper phase, or it may be found by using $X02$ and the secondary current $/a$ and adding this voltage in its proper phase.

The relations which follow from the preceding equations are:

« »

The equivalent impedance referred to the primary

The equivalent impedance referred to the secondary

Z02 Also

That is, the equivalent resistance, reactance, and impedance referred to the primary are to the equivalent resistance, reactance, and impedance referred to the secondary as the ratio of primary to secondary turns *squared. Example. —* A 50-kv-a. 4,400 to 220-volt transformer has a primary resistance and reactance of 3.45 and 5.40 ohms, respectively. The secondary resistance and reactance are 0.0085 ohm and 0.014 ohm, respectively. Find (a) the equivalent resistance referred to the primary; (6) the equivalent resistance referred to the secondary; (c) the equivalent reactance referred to both primary and secondary; (d) the equivalent impedance referred to

both primary and secondary; (e) the total copper loss using the individual resistances of the two windings and using the equivalent resistance referred to each side.

(e) $Pc = (11.36)\ 3.45 + (227)2\ 0.0085 = 883$ watts. *Am. Pe =* 7,tf,, = (11.36) 6. 85 = 883 watts. *Ans.*

P. -Iz2Rn -(227)» 0.0171 = 883 watts. *Ans.*

The equivalent resistance, reactance, and impedance referred to either side may be used in determining the transformer characteristics, such as regulation, efficiency, etc. That is, the transformer may be treated as a simple impedance in series with a load which is connected across the line,. Fig. 176 (c). If the primary current and voltage are to be used, the primary equivalent constants Z0i, K0i, and-X'01, must be used, as is shown in Fig. 176 (c). The secondary terminal voltage F2 must be multi plied by the ratio of transformation f'j as shown in Fig. 176 (c) in order to refer it to the primary side. The secondary current must be multiplied by (j in order to refer it to the primary side. The problem is then merely one of a simple series circuit

If the secondary current and voltage are to be used, the secondary equivalent constants, Z02, Rw, and Xn, must also be used. This will be demonstrated in the methods of transformer testing which follow.

83. Open-circuit Test.—Figure 177 shows a transformer having the low side connected to an alternating source of supply and the high side open-circuited. Either an auto-transformer or a drop wire is shown as a means of varying the voltage supplied to the low side of the transformer. A voltmeter, an ammeter, and a wattmeter are connected in the primary circuit. The voltmeter reads the voltage across the primary terminals, the ammeter reads the no-load current, and the wattmeter reads the power taken by the transformer under these conditions.

Low High

Fig. 177.—Connections for open-circuit test.

This power goes to supply the prima-

ry PR loss and the core loss of the transformer. As the exciting current is very small, the primary PR loss due to it may be neglected. Therefore, the wattmeter reads the transformer core loss. If the primary voltage be varied and the core loss be determined for different values of voltage, a curve is obtained showing the relation of core loss to voltage. At no load the flux is practically proportional to the terminal voltage, as the primary impedance drop due to the no-load current is negligible. (See equation 46, page 174.) The eddy-current loss varies as the square of the voltage and the hysteresis loss as the 1.6 power of the voltage.. The core loss will increase, therefore, nearly as the square of the voltage, as shown in Fig. 178 (a).

Transformers are usually so designed that the most economical use of materials is obtained. Therefore, the core is operated at as high a flux density as the allowable core loss will permit. A study of Fig. 178 (a) shows that a slight increase of voltage, above rated voltage, produces a very large percentage increase in core loss. As transformers are rated by their maximum safe operating temperatures, this increased core loss may cause overheating of the transformer. Therefore, the effect of operating transformers at over-voltage is to produce a large increase in temperature.

If the magnetizing current be plotted as abscissas, and the voltage as ordinates, a saturation curve similar to that of Fig. 178 (6) is obtained. The point marked "rated voltage" is the point on the saturation curve at which transformers are generally operated, and is well beyond the knee of the curve. Outside the question of increased core loss, the usual transformer cannot

Voltage Im (Magnetizing Current; (o) (6)

Fio. 178 (a).—Relation of core loss to voltage in a transformer.

Fio. 178 (6).—Relation of magnetizing current to voltage in a transformer.

be operated at a voltage very much in excess of its rated voltage, for the exciting current increases very rapidly with small increase in voltage, as indicated in Fig. 178 (fe).

The flux density in the core is determined primarily by the permissible core loss. Open-hearth annealed sheet steel, such as is used in dynamos, can be used for transformer cores. For a given flux density and frequency, however, silicon steel has much less core loss per unit volume than open-hearth steel, the effect of the silicon being to increase the electrical resistance, and hence reduce the eddy-current loss. Because of its small core loss, silicon steel may be operated safely at very high flux densities. The greater cost of silicon steel is more than offset by the saving in iron and in copper, and in the general reduction of the transformer dimensions.

To obtain the true value of the exciting current, the current I_0 measured by the ammeter, in Fig. 177, should be resolved into two components, one of which lies along the voltage — E_i, or V, and is shown as I_e in Fig. 179 (— E_i and V are practically equal at no load). This current $I_e = I_0 \cos 6$ is the *energy* component of the current and supplies the core losses. The quadrature component $I_m = I_0 \sin 6$ is the true magnetizing current, shown plotted in Fig. 178 (6). In most commercial transformers $I_0 = I_m$, very nearly.

84. Short-circuit Test.—Figure 180 shows the transformer of Fig. 177 reversed and the low side short-circuited. The reversal is made in order that the line current may not be excessive, and also in order that a reasonable voltage drop may be obtained. In a transformer, the impedance drop seldom exceeds 5 per cent. of the rated voltage. If the 2,200-volt side of a transformer, Fig. 180, be used as the primary, the voltage necessary to send rated current through the windings on short-circuit is about 5 per cent. of 2,200, or 110 volts, which is a standard voltage for instrument coils. If the secondary of the transformer were rated at 220 volts, the voltage at short-circuit would be only 11 volts and the current would also be high. At this low voltage, high precision could not be obtained with ordinary instruments.

When a primary current I_i flows, Fig. 180, the secondary current I_2 is equal to I_i (j-j. There is, therefore, no need

of using an ammeter for measuring I_2. The power delivered to the transformer, Fig. 180, goes to supply three losses; the primary copper loss, I_iR_i, the secondary copper loss, IR_z, and the core loss at short-circuit. The core loss is negligible, as 5 per cent. primary voltage means only about 2 per cent. of the rated value of flux, since half the impressed voltage on short-circuit is consumed in the primary impedance drop. The core loss at 2 or 3 per cent. of the rated flux *is* so small as to be negligible, for the core loss varies nearly as the square of the flux. Therefore, the power at shortcircuit $P = I_{11}RI + I2^2\#2 = Ji'floi = I22fl(2$ where R_{oi} and R_{02} are the transformer equivalent resistances referred to the primary and secondary, respectively.

The value of equivalent resistance as found in this manner may be checked with the value determined by measuring the resistance of each winding with direct current. The ratio of effective to ohmic resistance is only a few per cent. greater than unity in most transformers.

Figure 181 shows the equivalent-circuit vector diagram for the short-cirf cuit test. This diagram is merely that of Fig. 176,

Fio. 181.—Vector diagram for short-circuited except that V_z HOW equals transformer. zero, and all quantities are now referred to the primary side. It will be recognized in Fig. 181 that the entire voltage V_i is consumed in the impedance drops of the two windings. From this it is obvious that if Z_{0i} be the equivalent impedance of the transformer, referred to the primary side, $Z_n = T_1$ (58)

Knowing the equivalent impedance and equivalent resistance, the equivalent reactance is readily found. for either primary or secondary side.

In making the short-circuit and the open-circuit tests, the question of instrument losses should be investigated and correction made if this be found necessary. As the losses in a transformer are very small, the power taken by the instruments may be a considerable percentage of the power being measured.

86. Regulation and Efficiency.—The data obtained from the short-circuit and open-circuit tests are sufficient to com-

pute the regulation and the efficiency of the transformer at any load.

As the equivalent resistance and reactance referred to either side are known, it is merely necessary to proceed by the method of Par. 81, page 182, to determine the regulation. The procedure will be demonstrated by an example which follows.

It has been pointed out that with constant voltage the mutual flux of the transformer is practically constant from no load to full load. It usually does not vary more than from 1 to 3 per cent. Therefore, the core loss is practically constant at all loads and may be determined by the open-circuit test, Fig. 177. For most purposes it is necessary merely to measure the loss at the rated voltage of the transformer.

The only other losses are the primary and secondary copper losses. These can be calculated readily, knowing the resistances of primary and secondary, or they may be computed from the equivalent resistance determined at short-circuit. The efficiency of the transformer may then be computed, since the losses are known. That is, the efficiency

Fff _, F2/2(P.F.)

$$F2/2(P.F.) + \text{Core Loss} + IR + I22fl2$$

$$(' = F2/2(P.F.) + \text{Core Loss} + I22\#02$$

(62) *Example.*—A 20-kv-a., 2,200 to 220-volt, 60-cycle distributing transformer is tested for efficiency and regulation as follows: A wattmeter, an ammeter and a voltmeter are used to measure the input to the low side, the high side being open-circuited as shown in Fig. 177. The wattmeter reads 148 watts, the ammeter 4.2 amp. and the voltmeter 220 volts. The transformer is then reversed, the low side being short-circuited, and 220 volts applied to the high side. Instruments having the proper ranges are connected in circuit as shown in Fig. 180. The ammeter now reads 10.5 amp., the wattmeter 410 watts and the voltmeter 220 volts.

Find: (a) Transformer core loss. (6) Equivalent resistance referred to high side, (c) Equivalent resistance referred to low side, *(d)* Equivalent reactance referred to high side, *(e)* Equivalent reactance referred to low side. (/) Regulation of transformer at 0.8 power-factor, lagging current. *(g)* Efficiency of transformer at full and at half load, load being at 0.8 powerfactor, lagging current.

(a) Core loss is indicated directly by the wattmeter and is equal to 148 watts. Ans. (b) Bo, = jry-2 = 3.72 ohms. Ans. (OOft

2jJ 2 = 0.0372 ohm. *Ana.*

(d) Z0i = *Tfi-£* = 21.0 ohms. Xo = V(21. 0)2-(3.72)2 =-v/427 = 20.7 ohms. Ans. (990 2

jJ = 0.207 ohm. Ans.

(/) Work on high side.

The rated high-side current is 20,000/ 2,200 = 9.1 amps. Using equation (49), page 183.

Vi = V (2,200 X 0.8 + 9.1 X 3.72)2 + (2,200 X 0.6 + 9.1 X 20.7)8

V5,492,000 = 2,340 volts.

2 340 2 200

Regulation = ——_ „' = 6.36 per cent. Ans.

The same result is obtained using the low-side constants.

V, = V(220 X 0.8 + 91 X 0.0372)2 + (220 X 0.6 + 91 X 0.207)2

= V54,920 = 234 volts 234 220

Regulation = on = 6.36 per cent. Ans.

(g) Full-load eff. (using high-side constants) 20,000 X 0.80 = 16,000 =

"20,000 X 0.80 + 148 + (9.1)2 X 3. 72 16,460 97.2 per cent. An«. Half-load eff.

10,000 X 0.80 = 8,000 = 10,000 X 0.80 + 148 + (4.55)2 X 3.72 8,225 97.3 per cent. Ans.

The same values of efficiency are obtained if the low-side current and resistance are used.

This method of determining the efficiency is much more accurate than an actual measurement of the output and input, because the losses are so small a part of the output that a large percentage error in their measurement will make only a small error in the efficiency. As the output and the input are so nearly equal, it is difficult to determine the efficiency accurately by direct measurement.

Figure 182 shows the voltage characteristic and the efficiency of a 100-kv-a. , 60-cycle, 2,200/122 to 144-volt transformer plotted against load. It will be noted that the efficiency is high and is practically constant from load to 25 per cent. overload.

120

110

100

I" tso 8 TO 10 20 SO 40 60 CO 70 60 90 100 110 120

Kilowatt Output; P. F.=1.0

Fio. 182.—Characteristics of a 100 kv-a., 60-cycle transformer.

86. Core-and Shell-type Transformers. —Transformers are divided into two general types, the core type and the shell type. These two types differ in the arrangement of the iron and copper with respect to each other.

In the core type of transformer the winding or the copper surrounds the iron core. Figures 170, 172, and 173 are diagrammatic merely, but they represent core-type transformers. Figure 183 (a) shows the general arrangement of the core-type transformer. The core is in the form of a hollow square made up of sheet-steel laminations about 14 mils thick. These laminations are usually built up with rectangular strips, the joints of which butt, in the individual layers. However, the joints lap in alternate layers, as indicated by Fig. 183 *(b)*, which shows the arrangement of joints in two adjacent layers. When a large number of transformers of a single type are being manufactured, the laminations are often made of L-shaped stampings stacked so

Fig. 183 (c).—Coil and core assembly of Wagner core-type distribution transformer.

that the joints alternate. Figure 183 (c) shows a core-type transformer assembled, with leads, etc., but without the case.

If a transformer were made with the primary and secondary coils on separate legs, as indicated in Figs. 170, 172, and 173, an unsatisfactory transformer would result, as the large leakage flux for both primary and secondary would give very poor regulation. By having both a primary and a secondary on each leg, as shown in Fig. 183 (a) and Fig. 183 (c), the leakage flux is reduced to a very small value. If the high-voltage winding were placed next the core, it would be necessary to insulate it, both from the core and from the low-voltage winding. Thus two layers of high-voltage insulation would be necessary. By placing the high-voltage winding outside, and around the low-voltage winding, only the one layer of high-voltage insulation, that between the high-and low-voltage windings, is necessary.

In the core type of transformer the mean length of turn is less, but the mean length of magnetic path is greater than it is in the shell type. The core type of transformer is well adapted to high voltages, especially in the smaller capacities, because the insulation problem is not difficult.

(a).—Arrangement of coils and core in shell-type transformer. (6).—Coil and core assembly of Wagner shell-type distribution transformer. Fio. 184.—Shell-type transformer.

In the shell type of transformer, the iron surrounds the copper, as shown in Fig. 184. The core has the form of a figure 8. The entire flux passes through the central part of the core, but outside this central core it divides, half going in each direction as shown in Fig. 184 (a). This results in a much shorter effective length of magnetic path, but a greater mean length of turn. The coils are made in the shape of pancakes, usually wound with strip copper. These coils are taped, and the primary and secondary are usually stacked so that each primary is adjacent to a secondary. In this manner the leakage flux of both primary and secondary is reduced to a very small value. In Fig. 184 (a) the primaries are the high side and the secondaries are the low side

The secondaries, or low-side coils, are placed adjacent to the iron in order to minimize the amount of high-voltage insulation required. 87. *Type H Transformer.*—In designing a transformer it is desirable that the mean length of turn be as short as possible. This reduces both the weight of copper and the resistance and reactance of the winding. This is accomplished in the Type H transformer of the General Electric Co. by making a shell-type transformer in which the core is cru faccondary ciform in shape, as shown in Fig. 185. The central core around which the coils are wound is operated at much higher flux density than the four wings. Although the reluctance and on channel« iosses i n this core are high, they are not Flo. 185.— Core and windings excessive when the entire magnetic of type H transformer. circuit is considered. These transformers are used mostly as distribution transformers for stepping down from 2,200 and 1,100 volts to 220 and 110 volts, so that the primary is the high side. It will be observed that the low side, the secondary, is next the iron. That is, one of the two low-side coils is next the central core and the other is next the iron of the four wings. The two high-side coils lie between the two low-side coils, and are not adjacent to the iron. The two high-side coils are insulated from the low-side coils by the mica shields represented by heavy lines. The advantage of this design is that only moderate insulation is required between the low-side coils and the core. As high-voltage insulation need be used only between the high-and the low-voltage coils, a minimum amount of high-voltage insulation is required.

In designing a transformer, provision should be made for keeping it cool. Spaces or ducts should be left between coils and between coils and core. Such ducts, or channels, are shown in Fig. 185. The oil in these ducts becomes heated, its specific gravity decreases, and the oil rises. When it comes in contact with the transformer case it cools, which increases its specific gravity, and it therefore flows downwards outside the transformer coils, and is subjected to further cooling. There is a continuous circulation of oil up through the coils and the core which carries away the heat.

Figure 186 shows a Type H transformer assembled, but removed from its case.

88. Cooling of Transformers.—All the energy lost in a transformer must be dissipated as heat. Although this energy is but a small proportion of the total energy undergoing transformation, it becomes quite large in amount in the larg-

er capacity transformers. The larger the transformer the more difficult it becomes to dissipate the heat, for the kilowatt capacity of the transformer increases much faster than the radiating surface.

Transformers are divided into two classes, self-cooled types and artificially-cooled types. The self-cooled types are usually immersed in oil. The oil within the windings and core becomes heated and, because of the lesser density of the heated oil, it rises to the top of the case, where it becomes cooled. The cooled oil has a greater density than the warm oil and so flows downwards, in close contact with the case, where it is further cooled. After it reaches the bottom of the transformer it again rises, passing up through the windings. In addition to carrying heat away from the windings and core, the oil is an excellent insulator and dielectric.

In the moderate sizes of transformers, the radiating surface i increased by corrugating the case (Fig. 187). As transformer increase in capacity, it becomes difficult to dissipate the heat means of the surface of the case alone. One method of in the radiating surface is to use exterior tubes running from the tc

Fio. 188.—Westinghouse 1000-kv-a. , 3-phase, 60-cycle, 46,000-23,000-volt tubular-tank transformer.

of the case to the bottom (Fig. 188). The hot oil passes out from the tank into the top of the tubes, is gradually cooled and descends through the tubes to the bottom of the tank where it again passes up through the windings and core.

Transformers having this tubular construction are limited in size by the side and overhead clearances of the railroads. The tubular principle can be utilized, however, by bolting radiators to the casing to take the place of the tubes, as is shown in Figs. 189 and 190. As these radiators are held by bolts, they may be removed during shipment and bolted in place when the transformer

Fig. 189.—General Electric 1,500-kv-a.. 60-cycle, 22,000/44,000-2,200/4,400volt transformer; round corrugated tank equipped with three radiators. is installed. Figure 191 shows the core

and windings of the radiator type of transformer of Fig. 190. It will be observed that this is a shell type of transformer. When the leads are carried out through bushings in the transformer cover it is necessary to support the core and windings entirely from the cover. The cover with the core and windings can then be lifted as a unit.

There are two different types of artificially-cooled transformers, air-cooled and oil-cooled transformers. The air-cooled or airblast type is ordinarily mounted on a platform under which air pressure is maintained by means of blowers (Fig. 192). The air is forced up through the transformer windings, keeping them at the proper temperature. The advantage of air-cooled transformers is that fire risk is reduced because the danger of flooding the station with burning oil is eliminated. They are

Fig. 190.—General Electric 8,000 kv-a., 25-cycle, 4,400-6,600/6,390/6,270-volt, outdoor transformer. (Radiator tank.) therefore used extensively in sub-stations which are located in congested districts. On the other hand, this type of transformer does not have the dielectric strength which the oil adds to the insulation. For this reason air-cooled transformers are seldom manufactured for potentials in excess of 30,000 volts.

The most common method of artificially cooling the oil type of transformer is to place a copper coil in the top of the transformer tank and circulate cooling water through the coil. This coil is located where it is in contact with the hot oil (Fig. 193). Careful tests should be made at regular intervals for leaks in the cooling coils, as the presence of a very slight amount of water in the oil greatly impairs its insulating and dielectric properties.

-Core and windings of the radiator type of transformer shown in Fig. 190. 89. Three-phase Transformers.—Three-phase transformers have considerably less weight and occupy much less floor space than three single-phase transformers of equal capacity. For this reason they are commonly used in practice. The principle of the three-phase core-type transformer is illustrated by Fig.

194 (a). Three single-phase transformers (secondaries not shown) have each a primary winding upon one leg. These transformers are symmetrically wound and each winding is connected to one wire of a three-phase system. These cores are placed 120 apart

Fio. 193.—Water-cooled, shell-type transformer removed from its case.

so that the empty legs of the three are in contact. The center leg formed by these three carries the sum of the three fluxes

Fig. 194 (6).—Practical arrangement of windings on 3-phase, core-type transformer, connected Y-Y.

produced by the three-phase currents /i, /2 and /3. As the sum of the three currents at any instant is zero, the sum of the three fluxes must also be zero. No appreciable flux exists in the common leg and this leg may be eliminated, therefore, without disturbing existing conditions. A more practical arrangement, from the construction standpoint, is shown in Fig. 194 (6). The reluctance of the magnetic circuit for the center coil is less than it is for the two outer coils. This makes the magnetizing current of the middle phase slightly less than that of the two outer phases, but the magnetizing currents are so small that this has no noticeable effect on the operation of the transformer.

Figure 195 shows a threephase, shell-type transformer. It does not differ from three single-phase, shell-type transformers laid side by side. Owing to the joint use of the magnetic paths between the coils, there is less iron in this type of transformer than in three equivalent single-phase units As each phase has a magnetic circuit independent of the others, the three phases .' are more independent of one another than they are in the complete three-phase trans ,

Fig. 195. — Arrangement of coils and laminations in 3-phase, shell-type transformer, core type. Figure 188 shows a former with a tubular tank.

The lower cost of three-phase transformers and the smaller space occupied by them is often balanced by the fact that if any one phase becomes disabled, the whole transformer must ordinarily be removed from service. (The shell type may be operated open delta at 58 per cent. of its rating, but this is not always feasible.) If one transformer of a three-phase bank of singlephase transformers becomes disabled, the system may run open delta at reduced capacity or the transformer may be replaced by a single spare which can be readily substituted.

90. Auto-transformers. — Figure 196 shows a drop wire, having a resistance of 10 ohms, connected across a 100-volt supply. The supply may be either direct or alternating current. A load having a resistance of 5 ohms is tapped to the middle of the drop wire and to one side of the line. The parallel resistance of the two 5-ohm coils is 2.5 ohms, which being in series with 5 ohms makes a total resistance of 7.5 ohms across the 100-volt circuit. The total current flowing is 100/7.5 = 13.3 amp. This current of 13.3 amps. divides equally between the two parallel resistances, making 6.67 amp. in each. The voltage across the single 5-ohm resistance is 66.7 volts and that across the 5-ohm resistance which constitutes the load is 33.3 volts. In order to obtain 6.67 amp. at 33.3 volts with this system, the line must supply 13.3 amp. at 100 volts. The efficiency of this system is therefore very low.

Figure 197 (a) shows a transformer Fio. 196.—Currents in a dropwhose primary ac is connected across wlre system 100-volt alternating-current supply. The secondary b'c' has just half the number of turns of the primary ac, and therefore the voltage across the secondary is 50 volts. This secondary b'c' supplies a 2.5-ohm resistance so that the secondary current is 20 amp. Instantaneous directions of currents are indicated. Neglecting the magnetizing current, the primary current Iac is 10 amp. flowing downwards as indicated. It will be noted that the secondary current /CY is 20 amp. flowing upwards. i0 Amp?"'--2CAmp.-t-i0 Amp. C --20 Amp.

(o) Eegular Transformer («) Antotran.former

Fio. 197.—Currents and voltages in an auto-transformer supplying load at 50 per cent, voltage.

If the secondary winding b'c' be combined with the part be of the primary winding, where 6 is the mid-point of the winding ac, no disturbance will occur, as the voltage Vcb is equal to the voltage Vcb and the two are substantially in phase. (The current flows against the emf. in winding be and with the emf. in winding b'c'.) Assume that the windings be and b'c' are in contact at every point giving a single winding, as shown in Fig. 197 (b). The current Icb will now be the algebraic difference of the original primary current he and the secondary current Ic'b, or 10 amp. as shown. Therefore, instead of having two windings, one of which carries 10 amp. and the other 20 amp., a single winding only is necessary and its rating need not be greater than 10 amp. The copper represented by the 20-amp. secondary, Fig. 197 (a), may in this case be eliminated and yet there is sufficient copper to transfer the same power from one circuit to the other. Such a transformer is called an *auto-transformer or compensator.*

The primary voltage is Eac and the winding ac receives the power; the secondary voltage is Ebc', the ratio of transformation is Ebc/Eac the magnetizing current flows through winding ac. Therefore, the winding ac could properly be considered as the primary and the winding be as the secondary. When discussing the *current* and *power* relations within the transformer itself, the treatment is simplified by considering the winding ab as the primary and the winding be as the secondary, the magnetizing current being neglected.

The coil be supplies power to the load and is the secondary of a transformer of which a6 is the primary. Neglecting losses and magnetizing current, both of which are small:

The power delivered to the load is 50 X 20 = 1,000 watts.

The power in the primary ab is 50 X 10 = 500 watts.

The power in the secondary be is 50 X 10 = 500 watts.

Only 500 watts are transformed, but 1,000 watts pass to the load.

The extra 500 watts are *not transformed,* but merely flow *conductively* from the line a'a to the line *bd.* In this case but half the total power is *transformed.*

In the drop wire, the current flowing from a to 6, Fig. 196, undergoes a drop in potential. The power, represented by the product of this drop in potential and the current, goes to heat the wire *ab.* In the auto-transformer, however, the power represented by the current undergoing a drop in potential from a to *b,* Fig. 197 *(b),* is not wasted, but is transferred to the magnetic field. This power, transferred to the magnetic field, appears in the winding *be* where a current of 10 amp. is raised 50 volts in potential. That is, by transformer action, power is transferred from winding *ab* to winding *be.*

Although diagrammatically the auto-transformer looks like a drop wire, its operation is entirely different. The auto-transformer is superior, both as regards efficiency and voltage regulation.

Figure 198 *(a)* shows a regular transformer, which transforms 1,500 watts from 100 volts and 15 amp. to 75 volts and 20 amp. That is, the voltage is stepped down in the ratio of 4 to 3. The primary current *Iac* is 15 amp., and the secondary current *Icv* is 20 amp. as shown. When the windings *be* and *b'c'* are combined to make an auto-transformer, Fig. 198 (6), the net current in *be* is but 5 amp. The winding *b'c'* in (a) may be eliminated entirely and winding *be* in (6) need be only one-third the cross-section of -U Amp. *c c, -«— 20 Amp.— 15 Amp. c-t— 20 Amp.*

(a) Regular Transformer *(l)* Auto-Transformer
Fig. 198.—Currents and voltages in an auto-transformer supplying load at 75 per cent. voltage.

the winding *be* in (a). Hence there is a very considerable saving in copper in the auto-transformer over the regular transformer.

In Fig. 198 (6),

Primary power in *ab* = 25 X 15 = 375 watts.

Secondary power in *be* = 75 X 5 = 375 watts..

Transformed power = 375 watts.

Power *conducted* must then be 1,500

— 375 = 1,125 watts.

Only *one-fourth* the total power involved is now *transformed,* whereas in Fig. 197, one-half the total power was transformed.

The auto-transformer is sometimes called a *compensator.* It is a type of transformer which transforms a portion of the energy and allows the remainder to flow conductively through its windings. Its action is analogous to the balancer set. (See Vol. I, Page 391, Par. 249.) The current /,,& in dropping through the voltage *Eab* raises the current /, to the voltage *E&.* Winding *ab* corresponds to the motor and winding *be* to the generator of Fig. 352, Vol. I, page 391. In fact, a compensator can be used to obtain the neutral of an alternating-current three-wire system in the same manner as a balancer set is used to obtain the neutral in a direct-current three-wire system. The connections of a compensator used in this manner are shown in Fig. 199. The compensator is superior to the balancer set both in efficiency and in maintenance.

For moderate ratios of transformation, the compensator is much more economical in the use of materials and has a much higher efficiency than a transformer which transforms all the power. With the higher ratios of transformation, more and more of the power is transformed and less and less conducted. So the auto-transformer is economical only for small ratios of

Fio. 199.—Compensator used to obtain Fio. 200.—Lighting transformer used a 3-wire lighting system. as a booster.

transformation. Also the low side and the high side are connected together conductively. Therefore, in commercial systems the low side should be grounded at the proper point for reasons of safety, if the high-side voltage is sufficiently high to be dangerous. An ordinary lighting transformer can be used as an auto-transformer to change the voltage by a moderate amount. Figure 200 shows a 20-kv-a., 2,200 to 220-volt transformer. The rated primary current , 20,000 2,200 and the rated secondary current 20,000 220

The high and low sides can carry 9. 1 and 91 amp., respectively, without exceeding their ratings. The low side may be connected to raise the voltage, as shown in Fig. 200. Ninetyone amperes can flow to the load without overloading the lowtension coil. This requires 9. 1 amp. in the high-side coil which is now acting as primary. The line current from the supply must be 100.1 amp. If the transformer losses are neglected, the power supplied

Pi = 2,200 X 100.1 = 220,220 watts power delivered

P2 = 2,420 X 91 = 220,220 watts power transformed = 91 X 220 = 20,000 watts.

Assume 97 per cent. efficiency for the transformer. This means that the loss is 0.03 X 20,000 = 600 watts. The efficiency of the system is 220,220 nn 0 220.220 + 600"

It is to be noted that a device of this type is very similar to the series booster described in Vol. I, page 304, Par. 204, but that it is much simpler and much more efficient. When an ordinary lighting transformer is used in this manner the low-side winding should be grounded at one point as the insulation between the low side and core is not designed to withstand full high-side potential.

91. Transformer Connections. — There are several methods of connecting three-phase transformer banks, as for example, Y-Y, A-A, A-Y, Y-A, V-V, T-T, etc.

Fio. 201.—Y-Y connection of transformers.

Figure 201 shows a Y-Y connected transformer bank, which may be either a step-up or a step-down bank. Unless the primary neutral is connected to the generator neutral, this connection has the objection of having a "floating" neutral. An extreme case is illustrated by attempting to place a load from wire 2 to the neutral, on the secondary side. This power must be supplied by primary coil 2. This primary coil cannot supply the power because it is in series with the primaries 1 and 3 whose secondaries are open-circuited. The two primaries 1 and 3 under these conditions act as very high impedances so that the primary 2 can obtain but very little current through them from the line. Therefore, transformer 2 can supply no appreciable power. In fact, the secondary of 2 may be short-circuited and only a small current will flow. The short-circuit merely pulls the primary and secondary neutrals over to wire 2.

Primaries. Secondaries

Fig. 202.—Delta-delta connection of transformers.

This difficulty of the "floating" neutral may be obviated by connecting the primary neutral back to the generator so that the primary of transformer 2 can take its power from between its line and the neutral. Another objection to Y-Y connection is the fact that the secondary coil voltages contain large third harmonics.

Secoudaries

Fig. 203.—Delta-Y connection of transformers.

The delta-delta bank shown in Fig. 202 is often used, especially for moderate voltages. Its chief advantage is that if one transformer becomes disabled, the system may operate in "V" or open delta. In both the Y-Y and the delta-delta connections, the ratios between the primary and secondary line voltages are the same as the individual transformer ratios.

The delta-Y connection shown in Fig. 203 is a very useful connection for stepping up the voltage. It is not ppen to the objection of a "floating neutral" and of wave distortion, such as the Y-Y connection involves. Another distinct advantage of delta-Y connection over the delta-delta connection is that for high voltages the transformers need not be so well insulated. For a 100,000-volt system, the Y-connected 'transformers need be insulated only for 58,000 (100,000//3) volts, whereas deltaconnected transformers must be insulated for 100,000 volts. The Y-delta system is often used for stepping down the voltage.

The ratio between line voltages in these two systems is not the individual transformer ratio, for the line voltage on the Y-side is /3 times that given by the transformer ratio. A delta-Y bank can-

not be paralleled with a Y-Y or a delta-delta bank, even although the voltage ratios are correctly adjusted, as there will be a 30 phase difference between corresponding voltages on the secondary side.

Transformer primaries may be connected in either Y or delta without any attention being paid to phase relations. The secondaries must be phased like the alternator coils in Par. 59, page 125. The primaries of three-phase transformers, however, must be correctly connected as regards phase relations. The actual phasing is often avoided as the primary and secondary connections are brought out of the case symmetrically.

92. The V-connection.—It was pointed out in Par. 59, page 125, that line voltage must exist between the open ends of the

Primaries Secondaries

Fig. 204.—V or open-delta connection of transformers.

two coils of the delta before the third one is connected. At no load, with only two transformers, three equal three-phase voltages exist around the secondaries and a three-phase transformation is therefore possible with only two transformers. This is called the "V" or open-delta connection, Fig. 204. Under balanced loads, the voltages may become slightly unbalanced. This is not serious in commercial transformers, as their regulation is seldom greater than 2 or 3 per cent.

At first thought it might appear that the V-connection would have two-thirds the capacity of the delta-connection. Both transformers work at a reduced power-factor when connected in V, even though the power-factor of the load remains fixed. Therefore, the kv-a. capacity of the V-connection is less than two-thirds of the kv-a. capacity of the delta-connection having individual transformers of equal rating. The ratio of the Vcapacity to the delta-capacity is l/-/S = 58 percent. rather than 66% per cent. This can be proved as follows:

Let / be the rated current of each transformer and E the line voltage. The power, at unity power-factor, Fig. 204, is

$Pi = V3EI$

As the transformer rating is determined by the *current,* the output of three of these transformers in delta would be

$P2 = 3EI$ Therefore,

Oftentimes in practice a V-bank of transformers is first installed. The third transformer is added when the increase in load on the system warrants it. The rating of the bank is then increased 73 per cent. with an investment increase of but 50 per cent.

93. The Scott or T-connection. — By means of the Scott or T-connection it is possible to transform not only from three-phase to three-phase by means of two transformers, but also from threephase to two-phase or from two-phase to three-phase. The method of connecting for three-phase to three-phase transformation is shown in Fig. 205 (a) and (6). Two transformers having primaries *ad* and 6c and secondaries *a'd'* and *b'c'* are used. The middle point *d* of the winding *be* and *d'* of the winding *b'c'* must be accessible. One end *d* of the primary winding *ad* is connected to the middle point *d* of the primary *be*. The respective ends of the three coils are connected to the three-phase supply *abc*. The transformer *be* is called the *main* transformer and *ad* the *teaser* transformer.

Figure 205 (c) shows the voltage diagram. The three-phase supply is assumed to be 100 volts across lines and the transformers have a one-to-one ratio.

The voltages *Edc* and *Edb* are each equal to 50 volts and are 180 apart, since coil *dc* and coil *db* are both on the same magnetic

Fig. 205.—T-connected transformers, 3-phase to 3-phase.

circuit. Each side of the equilateral triangle, Fig. 205 (c), is equal to 100 volts. The voltage *Eda* is the altitude of the equilateral triangle and is, therefore, equal to 100/3/2 or 86.6 volts. The same relations hold in the secondary coils, so that *a'b'c'* is a symmetrical three-phase system. The full capacity of the transformers is not utilized, however. The teaser transformer operates at only 86.6 per cent. of its rated voltage and in the coils *bd* and *dc* the current lags 30 in one and leads 30 in the other at unity power-factor. This gives a power-factor of 0.866 in the transformer coils and is therefore equivalent to the transformers operating at only 86.6 per cent. of. their rated kv-a. capacity. However, if the teaser is designed for 86.6 per cent. voltage, it operates at full capacity and the capacity of the system is then 100 X 0.866 + 86.6 n noo,,, $4,.$,, 1AA —7-5 = 0.928 of the total transformer capacity. 1UO + oo.D

If the ends b' and d' of the secondaries be connected, as shown in Fig. 206 (a), a two-phase, three-wire system results. The voltage $Ed\,a'$ is equal to only 86.6 volts, whereas the voltage $Eb'c'$ equals' 100 volts. Therefore, the resulting two-phase system has unequal voltages. This may be corrected, however, if the line a be connected to point Oi on the primary of the teaser transformer, the point at being such that dai represents 86.6 per cent, of the total winding of the teaser transformer, as shown in Fig. 206 (6). This will increase the volts per turn in the ratio of 100 to 86.6 and will raise the secondary voltage a corresponding amount. Therefore, a symmetrical two-phase, three-wire system results. By tying the middle points of the secondaries together, a symmetrical quarter-phase, four-or five-wire system may be obtained, as shown in Fig. 207.'

In any of the foregoing connections, d is not the neutral of the primary, as it is not the center of gravity of the voltages. The voltages from the point 0 (Fig. 205 (c) and Fig. 206 (6)) to a, b and c are all equal. Therefore, point 0 is the neutral of the primary system. Point 0 is two-thirds the way down the teaser transformer winding from ai to $d,$ Fig. 206 (6).

In these connections the voltages become slightly unbalanced even under balanced loads. This is due to the unsymmetrical phase relations among the voltages and the currents in the individual coils.

Fig. 207.—T-counected transformers giving quarter-phase, 4-wire system with balanced voltages.

94. Constant-current Transformers.—The transformers heretofore considered are constant potential transformers; that is, the secondary voltage remains substantially constant and a change of load is accompanied by a corresponding change of current. There are instances where a constant *current* is desired, the most common being series street lighting. It will be recalled that constant direct current is obtained from a series generator. Constant alternating current is ordinarily obtained from a constant current or "tub" transformer..

The construction of the transformer is such that the primary and the secondary can move with respect to each other. The primary coil may be fixed and the secondary may move or the secondary coil may be fixed and the primary may move. Both types are found in practice. Figure 208 shows a transformer in which the primary is stationary and the secondary is movable. The load consists of a number of lamps connected in series. The secondary is suspended from a lever which is counter-weighted. A dashpot is provided to prevent rapid fluctuations in the position of the moving coil.

The operation of the transformer is as follows: Assume that the secondary coil is "floating"; that is, it is free to move either up or down and is delivering a certain current to a series load. The currents in the primary and secondary flow in opposite directions (Fig. 209). Therefore, there is *repulsion* between the two coils. Assume that the load changes, for example it decreases. This change of load would be produced by *short-circuit ing* one or more lamps, causing a *decrease* in the load resistance. Because of the decreased load resistance, first the secondary and then the primary current tends to increase. This increases the repelling force between the two coils, resulting in the secondary moving further away from the primary. The leakage flux between the two coils is thus increased and this reduces the secondary induced volts. The secondary coil will move away from the primary until the secondary current is again at its normal value.

The action of such a transformer depends on the change

Fig. 209.—Flux-Path3 in acon-in leakaSe flux of both Primary and stantcurrent transformer, secondary, as is shown in Fig. 209. Because of its large proportionate leakage flux this type of transformer has a very low powerfactor except at or near its maximum load. This is one objection to its use.

Magnetite arcs require a uni-directional current for their proper operation. (See Chap. XIII, page 431, Par. 196.) In order to obtain this current economically by the use of the constant-current transformer, mercury-arc rectifiers are used in connection with the transformer. Ordinarily two rectifier tubes are connected in series. The diagram of connections for a single rectifier tube is shown in Fig. 210. The mercury arc has the property 2200V. A.C.Supply

Fio. 210.—Constant-current transformer and mercury-arc rectifier.

of allowing current to pass in but one direction. When current tends to pass in the opposite direction the mercury vapor acts like a valve and prevents its flow. If there were but one circuit through the rectifier the negative half of the wave would be eliminated and the result would be a number of disconnected waves, as shown in "Fig. 211 (a). The rectifier would not operate under these conditions as the arc would go out between waves and would fail to re-establish itself. By using two anodes, a succession of connected waves, Fig. 211 (6), is obtained. Even under these conditions, the current becomes zero twice each cycle, so that the arc cannot re-establish itself. By using reactance, Fig. 210, the current is held over each half-cycle, resulting in a rippled, uni-directional current wave, Fig. 211 (c). (Also see Chap. XI, page 336, Par. 137(3).)

The starting anode shown in Fig. 210 is an electrode which sends a current at low voltage through the liquid mercury, causing it to vaporize, and in this way establishes the initial arc. The mercury-arc rectifier used under these conditions has a very high efficiency, as the voltage drop across the tube itself is small compared with the circuit voltage. The tubes, however, are fragile and occasional renewals are necessary.

CO Fio. 211.—Rectified current waves.

INSTRUMENT TRANSFORMERS

95. Electrical Measurements at High Voltages.—It is not usually practicable to connect instruments or meters directly to high-voltage circuits. Unless the high-voltage circuit is grounded at the instruments, they may be subjected to highvoltage stresses to ground. This makes it dangerous for anyone to come in contact with the switchboard apparatus. Further, instruments become inaccurate when connected directly to high voltage, because of the electrostatic forces which act on the indicating element. Specially designed instruments may be so constructed that they can be connected directly to high-voltage circuits, but these instruments are usually expensive and are not suitable for commercial work.

By means of instrument transformers, instruments may be entirely insulated from the high-voltage circuit and yet indicate accurately the current, voltage, power, etc., in the high-voltage circuit. Moreover, low-voltage instruments having standard current and voltage ranges may be used for all high-voltage circuits, irrespective of the voltage and current ratings of the circuits.

96. Potential Transformers.—Potential transformers do not differ materially from the constant-potential transformers already discussed, except that their power rating is small. Below 5,000 volts they are usually air-cooled and above this they are usually oil-cooled, the oil being used more for its dielectric qualities than for cooling purposes. As only instruments and sometimes pilot

Fig. 212.—Use of potential transformer on a 13,200-volt single-phase circuit lights are connected to their secondaries, such transformers ordinarily have ratings of from 40 to 200 watts. The low-tension side is almost always wound for 110 volts and the ratio is then determined by the rating of the high-voltage winding. For example, a 13,200-volt potential transformer would have a ratio of $13,200/110 = 120:1$. The ratio of turns may vary a per cent. or so from this value to allow for the trans-

former impedance drop under load. Figure 312 shows a simple connection for measuring voltage in a 13,200-volt circuit by means of a potential transformer. The secondary should always be grounded at one point to eliminate "static" from the instrument and further to insure safety to the operator. Figure 216 shows a potential transformer used in conjunction with a current transformer for measuring power by means of a wattmeter.

97. Current Transformers.—To avoid connecting alternatingcurrent ammeters and the current coils of other instruments directly in high-voltage lines, current transformers are used. In addition to insulating the instruments from high-voltage, they step down the current in a known ratio. This enables a lower-range ammeter to be used than would ordinarily be required if the instrument were connected directly into the primary line.

Fio. 213.—Connections for a series or current transformer on a 13,200-volt circuit.

The current or series transformer has a primary, usually of few turns, wound on a core and connected in series with the line, Fig. 213. When the primary has a large current rating, it may consist of a straight conductor passing through the center of a

Primary

Secondary Flax

Fio. 214.—Construction of one type of current transformer.

hollow core, as shown in Fig. 214. The secondary, consisting of several turns, is wound around the laminated core. The ratio of current transformation is approximately the inverse ratio of turns. For example, if the primary has two turns and the secondary 60 turns, the ratio will be 30:1. The ratio may vary slightly from this value due to the magnetizing current. In Fig. 215 (a), the primary current /i consists of two components, — Iz, the component necessary to balance the secondary ampere-turns, and /o, the magnetizing current. The magnetizing current introduces a slight error in the ratio as well as causing /2 to depart by the angle 0 from the 180

phase relation to I_i. At light loads the magnetizing current may cause considerable error. Figure 215 (b) shows the variation of phase angle and ratio with load for a typical transformer. 10 45

Secondary Current

(6) Phase angle and ratio curves for typical current transformer.

Fio. 215.—Current-transformer characteristics.

The secondaries of practically all current transformers are rated at 5 amp. , regardless of the primary current rating. For example, a 2,000-amp. current transformer has a ratio of 400:1 and a 60-amp. transformer has a ratio of 12:1.

The current transformer differs from the ordinary constantpotential transformer in that its primary current is determined entirely by the load on the system and not by its own secondary load. If its secondary becomes open-circuited, a high voltage will exist across the secondary because the large ratio of secondary to primary turns causes the transformer to act as a step-up transformer. Also, since the counter ampere-turns of the secondary no longer exist, the flux in the core, instead of being due to the *difference* of the primary and secondary ampere-turns, will now be due to the total primary ampereturns acting alone. This means a very large increase in the flux, causing excessive core losses and heating, as well as a high voltage across the secondary terminals.

Fio. 216.—Typical connections of instrument transformers and instruments for single-phase measurements.

Therefore, a current transformer should always have its secondary short-circuited.

Figure 216 shows the method of connecting a typical instrument load, through instrument transformers, to a high-voltage line. The load on the instrument transformers includes an ammeter, a voltmeter, a wattmeter and a watthour meter. Each secondary is grounded at one point. Correction for ratio of transformation must be applied to all the instrument readings, the wattmeter and watthour meter involving the ratio of both the current and the potential transformers. Usually in permanent installations, as on switchboards, the instrument scales themselves are so marked as to take into consideration these ratios. Therefore, the primary power may be read directly.

CHAPTER VIII

THE INDUCTION MOTOR

98. Principle.—The induction motor is the most widely used type of alternating-current motor. This is due to its ruggedness and simplicity, to the absence of a commutator, and to the fact that its operating characteristics are well adapted to constantspeed work.

Motion

Direction of rotation of disc

Fig. 217.—Rotation of metal disc produced by rotating magnet.

The principle of the motor may be illustrated as follows: A metal disc, Fig. 217 (a), is free to turn upon a vertical axis. The disc may be of any conducting material, such as iron, copper, or aluminum. A magnet, free to rotate on the same axis as the disc, is placed above the disc and its ends are bent down so that its magnetic flux cuts through the disc. When this magnet is rotated, the magnetic lines cut the disc and induce currents in it, as shown in the figure. As these currents find themselves in a magnetic field, they tend to move across this field, just as the currents in the conductors of a direct-current motor tend to move across its magnetic field. By Lenz's law, the direction of the force developed between these currents in the disc and the magnetic field producing them will be such that the disc tends to follow the magnet, as shown in the figure.

To illustrate this more in detail, consider Fig. 217 (a), (6), and (c). In *(a)*, the north pole of the rotating magnet is shown as moving in a counter-clockwise direction. The conductor beneath the magnet also moves in a counter-clockwise direction, but more slowly than the magnet. Therefore, the *relative motion* between the magnet and the conductor is the same as if the magnet were *stationary* and the conductor moved in the clockwise direction. This relative motion of the magnet and the conductor is illustrated in Fig. 217 (6), where the

north pole is shown as being stationary and the conductor is moving from right to left. Applying Fleming's right-hand rule (see Vol. I, page 218), the direction of the induced current is toward the observer. The lines of force about the conductor, due to its own current, are therefore counter-clockwise and the resultant field is found by combining the conductor field and the field produced by the magnet. The appearance of this resultant field is shown in Fig. 217 (c) (also see Vol. I, page 309). As the magnetic field is increased in intensity to the left of the conductor and reduced in intensity to the right of the conductor, there is a force developed which urges this conductor from *kft* to *right*. That is, the conductor tends to follow the magnet. Actually, the magnet rotates in a counter-clockwise direction. Therefore, the disc rotates in the same direction but at a speed less than that of the magnet.

The disc can never attain the speed of the magnet, for were it to attain this speed, there would be no relative motion of the disc and the magnet and, therefore, no cutting of the disc by the magnetic flux. The disc current would then become zero and no torque would be developed, which would result in the disc speed becoming less than that of the magnet. Because the disc can not attain the speed of the magnet, there must always exist a *difference* of speed between the two. This difference of speed is called the revolutions *slip*.

It is to be noted that the currents in the disc or armature of this type of motor are *induced* therein, rather than being conducted into the armature as in the ordinary direct-current motor.

A cylinder may be used instead of the disc, as shown in Fig. 218. In the figure are shown four poles, the magnetic lines of which cut the cylinder. If the frame carrying these poles be revolved by mechanical means, the currents induced in the cylinder will cause the cylinder to rotate in the same direction as that of the rotating frame. This cylinder is more representa

Fig. 218.—Rotation of conducting cylinder due to induced currents.

tive of the commercial induction motor

than the disc is, although both operate on the same principle. 99. The Alternating-current Rotating Field.—The rotating fields described in the previous paragraph were produced by rotating the magnetic poles mechanically. This is practically the same as the rotating poles of an alternator field. Rotating magnetic fields may, however, be produced by sending polyphase currents through polyphase windings, such as alternator windings. Such rotating fields are produced entirely by electrical means, there being no mechanical rotation of the pole pieces themselves.

The simplest type of rotating field is that produced by the gramme-ring winding illustrated in Fig. 219. A gramme ring, (5) (6) (7) (8)

Fio. 219. — Rotating field produced by 2-phase currents in gramme-ring winding.

wound for two-phase currents, has two separate windings, one for each phase. Each winding consists of two sections located diametrically opposite each other and each section occupies approximately one-fourth the winding space of the ring. The two windings are called the A-phase and the B-phase, respectively. Care must be taken to connect the two sections of each winding correctly, the correct method being shown in Fig. 219.

Curves *IA* and *IB* show the variation with time of the currents in phases *A* and *B*, respectively. As these are two-phase currents, they differ in time-phase by 90 or one-fourth of a cycle.

At the instant marked (1), the current in phase *A* is zero and that in *B* is negative maximum. With the method of connecting the windings, and the direction of the currents as shown, two S-poles are formed on the upper ends of the B-windings and two N-poles on the lower ends. These four poles combine into two poles, a single S-pole and a single N-pole, each of these last being twice the magnitude of the individual poles which combined to form them. The resultant field is vertical and is directed upwards, as indicated by the arrow *F* beneath diagram (1). In (2) the current in B is still negative, but of lesser magnitude than in (1). The current in *A* has increased positively until its magnitude is equal to that of *B*. Two S-poles and two N-poles again combine to form a single S-pole and a single N-pole, each of double the magnitude of the individual poles forming them. The direction of the resulting field is 45 clockwise from its position in (1). It is to be noted that while the two currents are passing through 45 electrical time-degrees, the resulting field in the gramme ring advances 45 space-degrees. Diagrams (3), (4), (5), (6), (7), and (8) show at different instants the positions of the gramme-ring field resulting from the combined magnetic effects of phases *A* and *B*. The diagram for (9) would be identical with that for (1). The rotating magnetic field has passed through 360 space-degrees while the two-phase currents have gone through 360 electrical time-degrees or one cycle. This constitutes a two-pole rotating field and its speed in revolutions per second is the same as the frequency, or the cycles per second, of the currents. For example, if the currents had a frequency of 60 cycles per second, the field would make 60 revolutions per second, or 3,600 r.p. m.

The gramme-ring winding need not consist necessarily of the two separate windings shown in Fig. 219, but may be a mesh-connected winding, as shown in Fig. 220. This is virtually a continuous winding tapped at four equi-distant points. Two of the diametrically.opposite taps are connected to phase *A* and the other two are connected to phase *B*.

Figure 221 shows a four-pole, mesh-connected winding. This is similar to that of Fig. 220, but is tapped at eight equi-distant points. Two diametrically opposite taps are connected to one line of phase *A* and the two taps at right angles to these are connected to the other line of phase *A*. Another similar set of taps, displaced 45 from the A-taps, connect in like manner to phase *B*. In such a winding, the rotating field completes one revolution during two complete cycles of the current; therefore its angular speed is one-half that of the field in the two-pole machine. Figure 222 shows

the manner of connecting a three-phase circuit to a gramme-ring mesh-or delta-connected winding. Each i of three equi-distant taps is connected to one of the three lines of the threephase supply. This winding produces a two-pole field whose speed in revolutions per second is the same as the frequency of the supply.

Fig. 222.—Delta-connected, *It is to be noted that in each f the 2-poie, gramme-ring winding for foregoing windings, the angle between a p ase circm. e various windings expressed in elec trical space-degrees, is the same as the time-angles between the respective currents in the windings.* (In a two-pole machine one electrical space-degree equals one space-degree; in a four-pole machine two electrical space-degrees equal one space-degree, etc.) 100. Rotating Fields in Drum-wound Machines.—The commercial polyphase induction motor consists of a fixed member called the *stator,* carrying a polyphase drum winding (like that of an alternator), and a rotating member called the armature or *rotor.* As the stator usually receives the power from the line it is called the *primary,* and as induced currents flow in the rotor, it is called the *secondary,* just as in the transformer. The motor will operate, however, if power is supplied to the rotor and the stator acts as secondary. Stator windings are the same as alternator windings for the same number of phases and poles. In fact, the

I linn-*A*

Fig. 223.—Single-layer, 4-pole, 2-phase induction-motor winding, lap-connected.

ordinary alternator winding is entirely satisfactory for an induction-motor winding.

Figure 223 shows a single-layer, drum winding for the twophase, four-pole, induction motor stator shown in Fig. 224. In this machine there are six slots per pole or three slots per pole per phase. Only the one phase, *A,* is shown connected.

Figure 224 shows a section of this two-phase induction motor, taken perpendicular to its shaft. It is wound with

the twophase, four-pole winding, shown in Fig. 223. The time variation of the two-phase currents, I_A and $/B$, is also shown.

At instant (1), the current I_A is zero and $/B$ is negative maximum. By applying the corkscrew rule for determining the relation of magnetic flux to the current producing it, four poles are formed in the stator, two N-poles in the vertical plane and two S-poles in the horizontal plane. At instant (2) the current I_A is positive, and I_B is negative and of the same polarity as in

Fig. 224.—Production of rotating field by 2-phase currents in a 4-pole winding.

(1). The resulting N-and S-poles are again determined by the corkscrew rule and it will be observed that these two poles have advanced 22.5 space-degrees in a clockwise direction, whereas the currents have undergone a change corresponding to 45 electrical 3-Phase Supply time-degrees. Positions (3), (4) and (5) are taken at time-degrees 90, 135, and 180 of the two currents. Between points (1) and (5), the rotating field has advanced only 90 space-degrees, whereas the currents have passed through 180 time-degrees. Therefore, the speed of this rotating field in revolutions per second is onehalf the frequency of the supply in cycles per second.

The N-and S-poles which are produced by the stator rotate in the air-gap and cut the rotor or armature conductors, inducing currents in them. These currents, reacting with these stator poles, produce rotation of the armature, just as in Fig. 217, the induced currents in the disc and the flux producing them react and cause the disc to rotate.

Figure 225 shows a singlelayer, three-phase, four-pole, lap winding adapted to a machine having nine slots per pole or three slots per pole per phase. This winding is very similar to that of Fig. Fig. 225.—Single-layer, 4-pole, 3-phase, 223 and is used in the Ul-induction-motor winding; lap-connected.

duction motor shown in Fig. 226. For simplicity, but one phase, *A,* is shown connected, the other two phases, *B* and

C, being connected in a manner similar to *A.*

Figure 226 shows four successive positions of the rotating field, for corresponding values of the polyphase currents in the stator of this three-phase induction motor. In (1) the current *IA* is zero, so that *Ib* and *Ic* are opposite and equal. The position of the field is shown at this instant. In (2) the currents *IA* and *Ic* are but half their maximum positive values and their positions on the stator are such that their phase belts are on each side of B-belt in which the current is a maximum. Therefore, the field is symmetrical at this position.

It will be noted also that the time-angle between successive values of current in 1-2-3 is 30 electrical-degrees, whereas the field advances but 15 space-degrees between (1) and (2) and also between (2) and (3). Between positions (1) and (4) the currents have advanced 90 electrical time-degrees, but the rotating field has advanced only 45 space-degrees. That is, the advance of the rotating field in space-degrees is equal to one-half the advance of the currents in electrical time-degrees. Therefore, the speed of such a field in revolutions per second is equal to one-half the circuit frequency in cycles per second.

In general it may be stated that in order to produce a two-pole rotating field, the angular space-degrees between the phase belts

Fio. 226.—Rotating field produced by 3-phase currents in a 4-pole, induction-motor winding.

of the winding must be the same as the electrical time-degrees between their respective currents. If the machine has *p* poles, the angular space-degrees between phase belts is *2/p* times the electrical time-degrees between their respective currents. For example, in a six-pole, three-phase machine, the successive phase belts start 40 from each other, that is, % X 120 or 40. In the ordinary drum windings, however, (see Chap. V, pages 108 and 109, Figs. Ill and 112), the coil-sides lap back so that in the above example the *reversed* phase belts would be 20 apart. The currents in such adjacent belts are 60 apart, as

shown in Fig. 226. To reverse the direction of rotation of a two-phase, rotating field, reverse the leads of *either* phase; to reverse the direction of rotation of a three-phase rotating field', interchange any two leads. 101. Synchronous Speed; Slip.—It has just been shown that the angular speed of an alternating-current rotating field depends upon two factors, the frequency of the current and the number of poles for which the machine is wound. The relation between speed, frequency and poles is given by the following equation: (63) where *N* is the speed of the field in revolutions per minute, / the frequency in cycles per second and *P* the number of poles (see equation (2), page 7). This speed, *N,* of the rotating field, is called the *synchronous speed* of the motor. The common synchronous speeds for commercial motors at 25 and at 60 cycles per second are as follows:

Poles 2 4 6 8 12 *Slip.*—If an armature whose conductors form closed circuits be placed in a rotating field, it will develop torque because of the induced currents, acting in conjunction with the rotating magnetic field.

As has already been pointed out, the armature can never attain the speed of the rotating field, for if it did, the cutting of conductors by flux would cease, there would be no rotor current and, therefore, no torque.

The difference between the speed of the rotating field and that of the rotor is called the *revolutions slip* of the motor. For example, if the rotor of a four-pole, 60-cycle motor has a speed of 1,730 r.p. m., its revolutions slip is 1,800-1,730 = 70 r.p.m., where 1,800 r.p.m. is its synchronous speed.

It is more convenient to express the slip as a fraction of the synchronous speed. Denote the speed of the rotor by *Nz* and the synchronous speed by *N.* Then the slip

For example, the slip in the above motor is 1,800-1,730 70,,,..,,..,,.

S--17800-= = 0.039 or 3.9 per cent.

The rotor speed is *f N2 = N(l*-s) (from equation 64) (65)

The full-load slip in commercial motors varies from 1 to 10 per cent., de-

pending upon the size and the type of motor.

102. Rotor Frequency and Induced Emf. — If the rotor of a two-pole, 60-cycle motor is at standstill and voltage is applied to the stator, each rotor conductor will be cut by a north pole 60 times per second and by a south pole 60 times per second, as this is the speed of the rotating field. If the stator be wound for four poles, the speed of the rotating field is halved, but each conductor is then cut by two north and two south poles per revolution of the field and therefore by 60 north and 60 south poles per second, the same as in the two-pole motor. Consequently, the frequency of the rotor currents at standstill (s = 1.0) will be the same as the stator frequency. This holds true for any number of poles. At standstill the motor is a simple static transformer, the stator being the primary and the rotor being the secondary.

If the rotor of the above 60-cycle motor revolves at half speed in the direction of the rotating field (s = 0.5), the rotor conductors are cut by just one-half as many north and south poles per second as when standing still and the frequency of the rotor currents is therefore 30 cycles per second.

By taking other rotor speeds, it can be shown that the rotor frequency i /2 = *sf* (66) v ----.-""' where /2 is the rotor frequency, s the slip, and / the stator frequency. *The rotor frequency is equal to the stator frequency multiplied by the slip. Example.* — What is the frequency of the currents in the rotor of a 60-cycle, six-pole induction motor, if the rotor speed is 1,164 r.p.m. The synchronous speed 60 X 120 *N* =--. --1,200 r.p.m. (Equation 63, page 235)

6

The slip ...«Tii»...« *fz* =0.03X60 = 1. 8 cycles per second. *Ans.*

The rotor frequency has a very important bearing on the operating characteristics of the induction motor.

The induction motor can be used as a frequency changer, provided the rotor is driven mechanically at the proper speed. Current is taken from the rotor, or secondary, through slip-rings. Under these conditions, some of the power is supplied electrically and some mechanically.

103. Alternating-current Torque.—It has already been pointed out, in connection with the direct-current motor, that the torque is proportional to the current and to the density of the magnetic field in which the current finds itself. This same law holds for alternating-current motors, provided the instantaneous values of current and flux are considered.

Figure 227 (a) shows the space distribution of flux from one north pole as it glides along the air-gap of an induction motor. This flux is distributed sinusoidally along the air-gap, as is shown by the flux-distribution curve, t, Fig. 227 (b).

If the slip be small, the reactance of the rotor conductors is low because $/2 = sf$ and $x'z = 2ir/2L2$, where $/$ is the stator frequency, $x'2$ is the rolor reactance at slip s, and L2 is the rotor inductance. Becauseof the rotor reactance the rotor current lags the induced emf. of the rotor by an angle a. At low values of slip, this angle a is very small since tan $a = 27r/sL2/.K2$, where #2 is the rotor resistance.

The induced emf. in any single conductor, I centimeters in length, in a field having a density of B gausses, the conductor moving at a velocity of v centimeters per second with respect to the field, is $e — Blv$ 108 volts, the flux, the conductor and the velocity being mutually perpendicular (see Vol. I, page 217, equation 93). Therefore, when a conductor is cutting flux at a uniform velocity, the flux being sinusoidally or otherwise distributed in space, the emf. in the conductor is zero when it is moving in a region where B, the flux density, is zero; the emf. is a maximum when the conductor is moving in a region where B, the flux density, is a maximum. As the emf. e, is propertional to B at every instant, if v is constant, e will be a maximum when B is a maximum, etc. Therefore, it may be said that e, the emf. per conductor, is in space-phase with the flux. It further follows that the wave shape of the emf. *in a single conductor* is the same as the shape of the space-distribution curve of the flux.

Fig. 227.—Alternating-current torque when current and flux arc in space-phase.

At small values of slip, the angle a, between the induced emf. in each conductor and the current in the conductor, is small and therefore the current in each of the conductors, Fig. 227 (a) is practically in phase with its induced emf. As the induced emf. is a maximum when the conductor is in the field of greatest flux density, the current will be a maximum at practically the same instant. The current is then in time-phase with the emf. and hence in space-phase with the flux. Under these conditions the current in the particular conductor which is under the center of the pole, Fig. 227 (a), is a maximum, and that in the other conductors is less, decreasing sinusoidally as indicated.

Figure 227 (6) shows both the flux distribution in the gap and the current distribution in the conductors of Fig. 227 (a), the current in each conductor being proportional to the flux density of that part of the field in which the conductor finds itself. (For simplicity a smooth current-distribution curve is shown. This would hold true only with a uniform metal sheet about the rotor). The force acting on each conductor is proportional to its current and to the flux density of that part of the field in which the conductor finds itself (see Vol. I, page 310, equation 106). The force due to each conductor, Fig. 227 (a), is indicated in direction by an arrow attached to that conductor. The torque curve is obtained by taking the product of the current and flux at each point, multiplied by a constant. The torque curve for the conductor belt shown in Fig. 227 (a) is given in Fig. 227 (6). This curve is obtained by multiplying the current at each point by the flux density at that point. That is, the ordinate of the torque curve, at any point Fig. 227 (6), is equal to the product of the ordinates of the flux and the current curves at that point, multiplied by a constant. It will be noted that this torque curve is of double frequency, that it is always positive, reaches zero twice every cycle, and

is similar to the power curve of page 22, Fig. 19.

As the value of the slip increases, the reactance of the rotor increases, the reactance being proportional to the rotor frequency and hence to the slip, and the angle a by which the current lags its induced emf. increases, since tan $a = 27r/sL2/2$. The current in any conductor will not reach its maximum value until a time-degrees after the induced emf. has reached its maximum value. In the interval between the time when the induced emf. reaches its maximum and the current reaches its maximum, the maximum point of the flux wave has moved along by the conductor by a electrical space-degrees, as shown in Fig. 228 (a). As a result, some conductors, as a, Fig. 228 (a), find themselves in a reversed field and so exert a torque opposite to that of the other conductors in that belt. Also, conductor a' in an adjacent current belt exerts a torque opposite to that of the other conductors in its belt. The torques exerted by conductors a and a' produce (a) Current belt a space-degrees out of phase with flux wave, when rotor reactance is high.

(6) Space distribution of current and flux, differing in space-phase by angle a. Flo. 228.—Relation among flux, current and torque when current belt is not in space-phase with flux wave. the negative loops of the torque curve, Fig. 228 (6). The torque under these conditions is less than it is in Fig. 227 (6), even with the same values of current and flux. This is due to the negative values of torque, shown in Fig. 228 (6). *Therefore, in order to have maximum torque with fixed values of current and flux, the rotor currents should be in space-phase with the flux.*

The torque $T = Tmaz \cos a$ where $Tmax$ is the torque when the current I and the flux t are in space-phase and a is the space-angle between the current $/$ and the flux t. 104. The Squirrel-cage Motor.—The squirrel-cage motor is the simplest type of induction motor and is the most generally used. The core of the rotor or armature, Fig. 229, like that of the direct

Fig. 229.—Squirrel-cage rotor.

current armature, is usually built up of slotted steel punchings. The winding consists of copper bars placed in slots. These bars have their ends connected together by conducting rings called *end-rings*. The bars are usually bolted to the end-rings and then welded or brazed. Formerly solder was used, but considerable trouble was encountered by its melting and being thrown out of the joint by centrifugal action. Another method is to place the rotor in a mould and cast the ends of the bars in a ring of cast copper. The General Electric Co. manufactures a rotor in which an aluminum grid is cast integral with the end-drings. The methods in which the end-rings are cast integral with the bars of the winding are the best from the operating point of view, as the rotor conductors have no opportunity to work loose. Fig. 230.—Overhung-slot stator showing how ends of coils are taped into slots.

The stator slots in nearly all induction motors of small size, Fig. 230, are of the semi-closed type like those shown in the Fio. 231.—Stator and rotor slots of squirrel-cage induction motor.

rotor of Fig. 231. If open slots are used, magnetic wedges are employed so as to give the effect of semi-closed slots. In the larger sizes of motor, open slots are often used for the stator, as shown in Fig. 231, because of the expense and difficulty of placing the winding in semi-closed slots, and also because the necessity for semi-closed slots is usually less in the larger motors. That is, in the large, higher-speed motors the pole-pitch is large and therefore the ampere-conductors per pole is large. Consequently, the desired flux density in the gap may be readily obtained without an excessive magnetizing current, even if open slots are used.

In practically all motors of the squirrel-cage type, the slots of the *rotor* are semi-closed, as there is little difficulty encountered in placing the rotor bars in this type of slot.

The advantage of the semi-closed slot is that the effective sectional area of the air-gap is increased and the magnetizing current is therefore reduced. Semi-closed slots reduce the pulsations of flux in the individual teeth, and therefore reduce the tooth losses, which otherwise might be serious. On the other hand, the semi-closed slot gives a much higher slot inductance than the open slot and this inductance in the stator and in the rotor lowers the power-factor and decreases the starting and the breakdown torques of the motor.

105. Operating Characteristics of the Squirrel-cage Motor.— The squirrel-cage motor, like the direct-current shunt motor, operates at substantially constant speed. As the rotor cannot reach the speed of the rotating magnetic field, it must at all times operate with a certain amount of slip. At no load the slip is very small. As load is applied to the rotor, more rotor current is required to develop the necessary torque in order to carry the increased load. Consequently, the rotating magnetic field must cut the rotor conductors at an increased rate, in order to produce the necessary increase of current. The slip of the rotor must accordingly increase, so that the rotor speed drops. *The ratio of the slip to the total power delivered to the rotor is proportional to the PR loss in the rotor.* As the resistance of the squirrel cage is very low, the *PR* loss is low and, therefore, the slip for ordinary loads is small. In large motors, 50 hp. or greater, the slip is of the order of 1 to 2 per cent. at full load. In the. smaller sizes of motor, the slip may be as high as 8 to 10 per cent. at full load.

Figure 232 shows the ordinary characteristic curves of a 10hp. squirrel-cage motor. It will be noted that the torque, speed 101112 456789

Horse Power Output Fio. 232.—— Operating characteristics of a squirrel-cage induction motor.

and efficiency curves are very similar to those of a shunt motor.

The power-factor increases with the load for the following reason: At no load the motor takes a current $I0,$ (Fig. 233). *I0* is mostly magnetizing current, although there is a small energy component necessary to supply the no-load losses. The power-factor at no load is cos 0o the value of which may be as low as 0.10 to 0.15. The back electromotive force of the motor remains nearly constant from no load to full load. Therefore, the flux must remain substantially constant, just as it does in the transformer, so that the magnetizing current changes but slightly from no load to full load. As load is applied to the motor, an energy current / is required to carry the load. This current, when combined with /0, gives the total current Ii at this load, and the resulting power-factor is cos *6.* As the load increases, an energy current *I'z* is required. The total current then becomes / 2 and the corresponding power-factor becomes cos 02-It will be observed that the power-factor angle decreases and therefore the power-factor increases as the load on the motor increases. The increased reactance drops in the stator and in the rotor with increase of load tend to oppose this increase of power-factor and when the load exceeds a certain value may even bring about a decrease of power-factor.

Flo. 233. — Increase of powerfactor with increase of load.

As the power-factor increases, a smaller increase of current is required for a given increase of load than would be necessary if the power-factor were constant. Therefore, the current increases more slowly than the load as shown in Fig. 232. At first the efficiency increases rapidly and reaches a maximum value for the same reason that it does in other electrical apparatus. At all loads there are certain fixed losses, such as core loss, friction and windage. In addition there are the load losses *(PR)* which increase nearly as the square of the load. Therefore, at light loads the efficiency is low because the fixed losses are large as compared with the input. As the load increases, the efficiency increases to a maximum, the fixed and variable losses being equal at this point. Beyond this point the *PR* losses become relatively large, causing the efficiency to decrease.

One disadvantage of the squirrel-cage motor lies in the fact that it takes a very large current at low power-factor on starting, and in spite of this large current it develops but little torque. When the motor is at standstill, the squirrel cage acts as the short-circuited secondary of a transformer, causing the motor to take an excessive current on starting, if full voltage is applied.

Figure 234 shows the variation of torque with slip for two different values of line voltage. It will be noted that for small values of slip up to and beyond full load, which is the ordinary range of operation, the torque is substantially proportional to the slip. At higher values of slip, however, the torque curve bends over and finally reaches a maximum. This maximum is called the *break-down torque.* Beyond this maximum point the torque decreases as the slip increases. For most types of load this is a point of *instability,* as an increase in load is accompanied by an increase in slip, and therefore, by a decrease in torque. As the motor now develops a decreased torque with an increased load it must come to a standstill, unless the load is removed. At standstill ($s= 1.0$), the torque is comparatively small.

The underlying cause of this small starting torque is the *reactance of the stator and of the rotor.* The rotor reactance is proportional to the rotor frequency ($x'z = 27r/2L2$). The rotor frequency /2 is proportional to the *slip.* As the rotor slip increases, the rotor reactance increases proportionately, whereas the resistance does not change materially. The effect of this increased reactance is to produce a greater phase difference between the rotor *currents* and their induced voltages (tan a = . As these currents at the same time differ in space 0 Clip 1.0

Fio. 234.—Slip-torque curves for squirrel-cage motor.

phase with the *flux,* less torque per ampere is developed (see Par. 103). In fact the current and the flux may get so far out of space-phase with each other that, even with four or five times the rated current, only a small fraction of the full-load torque is developed. It can be shown that the break-down torque of an induction motor is decreased by an increase in the rotor reactance ($x2 — 2ir-fL$) where x2 is the rotor reactance at

standstill. Therefore, it is desirable that the rotor reactance, *xz,* and hence rotor inductance, be as low as possible (see page 247, equation 67). It can also be shown that the *torque of an induction motor for a given slip is proportional to the square of the line voltage.* If the line voltage is halved the *flux* is halved, neglecting the stator impedance drop-, and the rotor *current* for a given value of slip is halved. Therefore, the torque is quartered, the torque being proportional to the current times the flux, other factors remaining constant. In general, it may be said that the torque for a given slip is proportional to the *square* of the line voltage. For this reason a 10 per cent. drop in voltage may cause a 19 per cent. reduction in the break-down and starting torques. The effect of line voltage upon torque is shown in Fig. 234, the torque at one-half line voltage being one-quarter the torque at full-line voltage for each value of slip.

The stator impedance also reduces the break-down torque. A high stator impedance means a comparatively large impedance drop in the stator for a given current. This decreases the back emf., *E,* hence the air-gap flux becomes less, and therefore the value of the rotor current at any given slip is reduced. This results in a reduction of torque for each value of slip.

The effect of each of these various factors upon the break-down torque is shown in the following equation: The break-down torque __ $KV2\ Tmax = $ (67) where K is constant, V is the terminal voltage, r is the stator resistance, x is the stator reactance, and xz is the rotor reactance at standstill.

The above equation shows that: *The break-down torque is proportional to the square of the line voltage. The break-down torque is reduced by an increase in the stator resistance, and by an increase in the stator and rotor reactances. The break-down torque is independent of the rotor resistance.*

The stator and rotor reactances are proportional to the frequency and to their respective inductances. Therefore, it is desirable that the stator and the rotor inductances be kept low and that the

frequency be not too high.

As the squirrel-cage motor is ordinarily started at low voltage, it develops but little starting torque, because the flux is small and the rotor currents are considerably out of space-phase with the flux.

It is desirable that the stator and rotor inductances be as low as possible. This is accomplished by having the slots partially open and thereby reducing the value per ampere of the leakage flux which links the individual conductors. Ordinarily it is not desirable that the slots be entirely open, as this increases the reluctance of the air-gap and more magnetizing current is required. This in turn reduces the power-factor. Also with open slots the tooth losses may become excessive, particularly in large motors. The rotor-slot design is actually a compromise among these conflicting factors.

Because of the lower reactance accompanying a lower frequency, a 25-cycle motor will in general have greater starting

Fig. 235.—Westinghouse squirrel-cage induction motors as headstock motors for wood-working lathes.

torque and break-down torque than a 60-cycle motor. On the other hand, the magnetizing current in general is higher, because of the higher flux densities employed in the 25-cycle design.

Because of its low rotor resistance, the squirrel-cage motor has excellent operating characteristics for constant-speed work. The slip is small and the speed regulation is good. In addition, the motor is simple, rugged, and requires but little attention. Some of its fields of application are in machine shops, in woodworking shops, in cement mills, in textile mills; in fact it is used in most cases where the load requires constant speed with but little starting torque. Figure 235 shows the application of small squirrel-cage induction motors to the wood-working industry, the motors serving as the headstocks of the lathes.

As this type of motor develops very little starting torque, it cannot be used where it must be started under any considerable load. Another disadvantage is that its speed is not adjustable.

106. Starting Squirrel-cage Motors.— Small induction motors up to 5 hp. can usually be connected directly across the line without undue disturbance of the line voltage. Special starting devices should be used for motors of 7.5 hp. and greater.

Fio. 236.—Switching connections when motor is connected directly across line at starting.

Figure 236 shows the connections often used for the smallersized motors where special starting devices are not required. A double-throw switch, when in the starting position, puts the motor in series with three high-capacity fuses, one in each line. Because of the action of a spring, the switch can make contact on this side only while it is held in position. When the switch is thrown to the running position, the current is supplied through three fuses designed to carry only the safe operating current of the motor. This gives the motor over-load protection that would not otherwise be obtained if fuses sufficiently large to carry the starting current were used during normal operation. Resistances (carbon rods or other types) are sometimes inserted in the starting circuit to limit the starting current. Also, highcapacity fuses at the motor are sometimes omitted, the line fuses giving the required protection at starting.

As the squirrel-cage motor at starting is equivalent to a shortcircuited transformer, it is necessary to reduce the starting current in the larger sizes. One simple method (Fig. 237) is to use a delta-connected motor. By means of a tripleTpole, doublethrow (T.-P. D.-T.) switch the windings are first thrown in Y across, the line, thus applying only $l/3$ or 58 per cent. of the normal voltage to each coil. This makes the *line* current onethird the value it would have, if the motor were directly across the line. When the motor has attained sufficient speed, the switch is thrown over, connecting the motor in delta across the line.

(a) Starting (6) Running

Fio. 237.—The Y-delta method of starting an induction motor.

The most common method of starting the squirrel-cage motor, however, is to use an auto-starter or starting-compensator, similar to those shown in Figs. 238 and 239. In the General Electric compensator shown in Fig. 238, the three coils of a three-phase auto-transformer are connected in Y. When the switch is in the starting position, the compensator is connected across the line with only the line fuses for protection. Under these conditions the three motor lines are connected to three taps, one in each phase of the auto-transformer. Hence the motor voltage is reduced, usually to one-fourth or to one-half its rated value. When the switch is in the running position, the compensator is entirely disconnected from the line and the motor is connected directly across the line through the running fuses. In Fig. 238 the heavy lines show the path of the current when the compensator is in the running position. It should be remembered that a compensator supplying a motor with half voltage reduces the line current to one-fourth its normal value. The motor being at half voltage takes one-half the current that it would take if directly across the line. As this current is supplied by the secondary of a 2:1 transformer, the line current is but half the motor current and is, therefore, one-fourth the cur

Fio.' 238.—General Electric 3-phase starting compensator.

rent that would have been taken had the motor been directly across the line. It is not necessary to use a three-coil autotransformer. In the Westinghouse starting compensator, two coils are mounted on the two outer legs of a transformer core, Fig. 239 (a), very similar to the core used for three-phase core-type transformers (see page 205, Fig. 194 (6)). On starting, these two coils are (c) Running

Fio. 239.—V-oonnected starting compensator.

connected in V across the line and two motor taps are taken off as shown in Fig. 239 (b). The motor is thus supplied at a reduced three-phase voltage. When the starting handle is placed in the run-

ning position, the two motor taps are connected directly to their corresponding lines, Fig. 239 (c), and at the same time the compensator is entirely disconnected from the line. One advantage of this type of starter is that it can be readily used on two-phase as well as on three-phase circuits.

Practically all starting compensators have a no-voltage release as indicated in Fig. 238. When the line voltage decreases to a low value, a solenoid plunger drops, releasing the starting handle which springs back to the "off" position.

107. The Wound-rotor Induction Motor. —If resistance be introduced in the rotor circuit of an induction motor, the slip for any given value of torque will increase.

A given value of torque requires a definite value of flux and a definite value of current. The flux of the induction motor is practically constant, since the back emf. is practically constant. If resistance be introduced in the rotor circuit, the rotor impedance is increased. (At slips which give the ordinary values of torque, the armature reactance is small as compared with its resistance, hence the armature impedance is practically all resistance.) If the slip remains constant, the induced emf. of the rotor does not change. The armature current, which is equal to this emf. divided by the rotor impedance, decreases. The torque therefore decreases.

To bring the torque back to its original value, the armature current must be increased. To increase the armature current, the armature induced emf. must increase. Since the flux is constant, the increase in the induced emf. may be obtained only by this flux cutting the rotor conductors at a greater rate. Therefore for a given value of torque, the slip must increase when resistance is introduced in the rotor circuit.

The slip-torque curve will be changed from curve (1) to curve (2), Fig. 240. It will be noted that full-load torque is obtained at increased values of slip as the rotor resistance is increased. The value of the maximum or breakdown torque will not be affected, but the point of maximum torque moves toward the point of zero speed. (s = 1.0.) That is, the maximum torque occurs at a greater value of slip. The rotor now runs at reduced speed, but the reduced speed is obtained at the expense of efficiency, for the rotor PR losses are increased.

It is evident that speed control may be obtained by the introduction of resistance in the rotor circuit. This method of speed control is very similar to the armature-resistance method of speed control in the direct-current motor (see Vol. I, page 339, Par. 222). The lowering of the speed is accompanied by a material lowering of the efficiency and by poor speed regulation.

0 Slip 1.0

Fio. 240.—Effect on the slip-torque curve of inserting resistance in rotor circuit.

The electrical efficiency of the rotor is equal to the ratio of actual speed to synchronous speed. For example, at 25 per cent. slip, the rotor efficiency is 75 per cent. That is, of the power transmitted across the air-gap, 25 per cent. is lost as heat in the rotor resistance. The other 75 per cent. is converted into mechanical

Fio. 241.—Wound rotor of 100 hp., 440-volt induction motor.

power, although this is not all available at the pulley, because of rotor friction and core losses.

If sufficient resistance be introduced in the rotor circuit, maximum torque may be made to occur at standstill, as shown by curve (3) Fig. 240. That is, break-down torque is obtained at starting. In order to obtain break-down torque at starting, the rotor resistance per phase, Ti, should be approximately equal to the rotor reactance per phase at standstill, xz.

An adjustable resistance cannot be readily placed in the squirrel-cage rotor, so that three-phase rotors requiring external resistance are usually wound either two-phase or three-phase. The two-phase windings may be connected either star or mesh and the three-phase windings may be connected either Y or delta. Such rotor windings are in every way similar to stator windings. The three ends of the three-phase winding

are

St"r Botor Slip-rings
Kotor Bbeostat

Fig. 242.—Connections for a wound-rotor induction motor.

brought out to three slip-rings, as shown in Figs. 241, 242, and 243. Brushes, bearing on each of these three rings, Fig. 242, connect to Y-connected external resistances, usually through a controller. The entire resistance of each phase is in circuit on starting. This causes the rotor current to be more nearly in space-phase with the air-gap flux, so that a large torque is obtained with a moderate value of current. In addition to producing a very good starting torque, the starting current of the motor does not greatly exceed the rated current. As the motor comes up to speed, the external resistance is cut out. The motor then operates on curve (1) Fig. 240.

Even without the controller, the wound-rotor type of motor is more expensive than the squirrel-cage motor, due to the greater cost of winding and connecting the rotor coils. The controller and resistors further add to the cost. In the running position, this type of motor has a greater slip than the ordinary squirrelcage motor, because it is not possible to secure the very low resistance obtainable with the squirrel-cage winding. As has been pointed out, such external resistance may be used to obtain speed control at reduced efficiency and with poor speed regulation. Hence, this type of motor has better starting characteristics, but poorer running characteristics than the squirrel-cage motor.

Fio. 243.—Slip-ring induction motor, assembled.

Wound-rotor induction motors are used where considerable starting torque is required, and frequently where speed adjustment is desired. Common applications of this type of motor are in cranes, elevators, pumps, hoists, railways, calenders, etc. Figure 244 shows a Westinghouse 1,200-hp., 580-r.p.m., woundrotor induction motor driving a Henry A. Worthington centrifugal pump at Minneapolis, Minn.

Another recent use of these wound-

rotor induction motors is in the electric propulsion of battleships. The motors are connected directly to the propeller shafts. Two synchronous speeds are obtained by changing the number of poles. Intermediate speeds are obtained by changing the frequency of the generator.

Where a rheostat is used for starting duty only, the rotor conductors may be connected to resistance grids within the rotor itself. Such grids can be short-circuited by copper brushes

Fig. 244.—Westinghouse 1,200 hp. , 580 r.p.m. induction motor driving a Worthington centrifugal pump.

operated by pushing a rod which protrudes from the center of the rotor shaft. Such a rotor is shown in Fig. 245. This type

Fig. 245.—Rotor of induction motor having starting resistance within rotor.

of rotor cannot be operated with the grids in circuit continuously because of the difficulty of dissipating the heat which is developed within the rotor.

108. The Induction-motor Air-gap.—The air-gaps of directcurrent generators and motors, and-of alternators, are much greater than is necessary for mechanical clearance. This is due to the fact that with too short an air-gap, the effect of armature reaction becomes too great, that is, the field is relatively Iweak as compared with the armature. On the other hand, the air-gap of the induction motor is made just as short as mechanical clearance will permit. The back emf. of the stator varies only a few per cent. from no load to full load. This back emf. is induced by the air-gap flux cutting the stator conductors. As the speed of the rotating field is constant, the flux in the gap must be substantially constant from no load to full load. Therefore, in a given motor, the magnetizing current is practically constant at all loads. If the length of the air-gap be increased, the reluctance of the magnetic circuit is also increased. As the back emf. does not change except slightly, the flux changes also but slightly. Therefore, with a fixed flux the greater air-gap reluctance will necessitate a greater magnetizing current. This

increased magnetizing current lowers the power-factor (see Fig. 233, page 244).

Large slot openings increase the reluctance of the air-gap and so lower the power-factor. Therefore, from the standpoint of the magnetizing current it is desirable to use semi-closed slots or open slots with magnetic wedges. The disadvantage of closing the slot too much is that both the stator and the rotor inductances increase and the breakdown and starting torques are reduced (see page 247, equation 67). The increase of inductance also tends to lower the power-factor.

The small mechanical clearance between the rotor and the stator makes it necessary to have a heavier shaft and heavier and stiffer bearings in the induction motor than are required in other types of rotating machinery of the same speed and size.

109. Speed Control of Induction Motors.—The speed of the rotor of an induction motor is given by f X 120 $N2 =$ —p— (1-s) (pages 235 and 236, equations 63 and 65) where Nz is the rotor speed in revolutions per minute, / is the frequency of supply in cycles per second, P is the number of poles and s is the slip.

Obviously, there are three factors, frequency, slip and number of poles which determine the speed of the induction motor. In order to change the speed, it is necessary to change at least one of these factors.

Changing the Slip.—The slip may be changed by introducing resistance into the rotor circuit. This has already been discussed in connection with the wound-rotor type of motor. At a given slip, any value of torque up to the breakdown torque may be obtained by this method. Its disadvantages are lowered efficiency and poor speed regulation.

These disadvantages may be avoided by introducing counter emfs. instead of resistance into the rotor circuit, either at line frequency, which requires that the rotor have a commutator, or by means of an auxiliary commutating machine which introduces counter emfs. at rotor frequency through slip-rings. This last

method necessitates the use of a commutating type of machine which produces emfs. at rotor, or slip frequency. It must therefore be excited by the rotor currents themselves. The Sherbius1 method of speed control is the most common example of this counter-electromotive force method.

When a current flows against a counter-electromotive force in a rotating machine, mechanical power is developed. This occurs, for example, when the current flows against the counter-electromotive force in a direct-current motor. The counter-electromotive force machine accordingly develops mechanical power, which is available for various purposes. Unlike the voltage drop in a resistance, the counter-electromotive force is practically independent of the current (see "Counter-electromotive Force Cells," Vol. I, page 401). Therefore, a motor employing this method of speed control has good speed regulation and efficiency.

Figure 246 shows the connections employed in obtaining speed control by the foregoing method. *A* is the main induction motor having a slip-ring rotor. *B* is a three-phase commutator motor on whose stator are shunt-field windings *Fi, F2,* and *F$,* spaced 120 electrical space-degrees apart. The rotor or armature of *B* is wound three-phase and the coils are connected to a commutator in the same manner as they are in a direct-current armature. The three-phase slip-ring currents of the main motor are carried into the armature of *B* through the brushes and commutator, the brushes being spaced 120 electrical space-degrees apart as shown. Series compensating windings Ci, Cj and *3* act along their respective brush axes and assist commutation.

i "Theory of Speed and Power-factor Control of Large Induction Motors by Neutralized Polyphase Alternating-current Commutator Machines," by John I. Hull, *Journal of the A. I. E. E.,* May, 1920.

Abo, "Some Methods of Obtaining Adjustable Speed with Electrically Driven Rolling Mills," by K. A. Pauly, *General*

Electric Review, May, 1921, page 422.

Neutralized Shunt 3-Pbase Commutator Motor

Fio. 246.—Connections for adjusting speed of an induction motor by means of a neutralized three-phase commutator motor connected to slip-rings of induction motor.

A three-phase auto-transformer *B* is Y-connected across the slip-rings. A tap on each phase of this transformer feeds one of the shunt fields *Fi, Fz, F3,* the shunt fields being connected in *Y.*

A constant speed generator *G,* usually of the induction type, is mechanically connected to the motor *B,* and delivers back into the line the power received from the motor.

A detailed analysis of the operation of this apparatus involves a somewhat complicated vector diagram and is beyond the scope of this book. However, it can be shown that the commutator motor develops an electromotive force in each phase which is nearly in phase-opposition to its respective phase current. This emf. is practically independent of the slip, if the speed of the commutator motor be held constant by its load, such as the generator *G.* Therefore, neglecting impedance drops, the rotor of the main motor will slip until its emf. is equal to the counter emf. of the commutator motor. As the rotor emf. is proportional to the slip, the rotor slip will be constant at all loads if the impedance drops be neglected.

When a current flows in opposition to an emf. (as in a direct-current motor) it *gives up* energy. Therefore, the current delivered to motor *B* gives up energy, some of which is returned to the line through generator *G.* Ordinarily, in the wound-rotor type of motor this energy is lost in heating a resistance. The fact that this counter emf. is constant gives the motor a practically constant-speed characteristic for any one adjustment. Speed adjustments are made by changing the positions of the taps of the auto-transformer *B.* Because of the cost of two extra machines, this method has been but little employed in this country except in the very large units used in steel mills, where the method is now coming into general use.

Change of Frequency.—Commercial power systems operate at constant frequency and it is impossible to control the speed of induction motors by change of frequency when the motors take their power from such systems. In a few special instances, such as in the electric propulsion of battleships, *(General Electric Review,* April 1919), the motors are the only loads connected to the turbo-alternators. Therefore, it is possible to obtain speed control by changing the speed of the turbines themselves. Even here the range of speed variation is limited, because the efficiency of turbines decreases very rapidly when their speed departs from the speed for which they are designed. *Change of Poles.*—By means of a suitable switch, the stator connections may be changed in such a manner that the number of poles is changed. This changes the synchronous speed of the motor and therefore the speed of the rotor. If the poles be changed in the ratio of three to two, the winding will probably be designed for pitch at the higher speed making it a fullpitch winding for the lower speed. In such a motor the best possible design is not usually obtainable at both speeds. That is, desirable characteristics, such as high power-factor, etc., are sacrificed at one speed in order that a reasonably good motor may be obtained at the other speed. Sometimes the stator connections are changed from delta to *Y* at the same time that the pole connections are changed. This changes the voltage per phase and makes possible a better motor at each speed. Because of the complications involved in changing the connections, it is not desirable to obtain more than two speeds by changing the number of poles. To avoid these complicated switching connections, induction motors sometimes have two distinct windings, the two windings being connected for a different number of poles. The 7,500-hp., wound-rotor induction motors used, to drive the electrically-propelled battleship *Tennessee* have this type of winding. One winding is connected for 36 poles and the other for 24 poles.

In the electrically-propelled battleship *New Mexico,* the motors are direct-connected to the propeller shafts. The stators can be connected for 24 poles or for 36 poles, giving a speed change of three to two. In wound-rotor types of motors it is necessary to change the rotor as well as the stator connections. Otherwise negative torque will be developed by certain of the rotor conductor belts.

Speed Control by Concatenation.—This method requires two motors, at least one of which must have a wound-rotor. The speed is changed by changing the slip of one motor, which changes the frequency supplied to the other motor. The two rotors are connected rigidly together as indicated in Fig. 247. Line frequency is supplied to the stator of one motor, as No. 1, Fig. 247. This first motor should have a one-to-one ratio of transformation between stator and rotor. That is, at standstill, and with the external circuit of the rotor open, the voltage across the rotor slip-rings should be equal to line voltage. Assume that the two motors are similar and that the rotors operate at slightly less than half the synchronous speed of the first motor. The *rotor* frequency of No, 1 motor is slightly greater than half line frequency, as the slip is slightly greater than 50 per cent. (see page 236, equation 66). Therefore, the synchronous speed of No. 2 motor is practically half that of No. 1 motor. The rotors each operate at a speed which is slightly less than half the synchronous speed of the first motor. The rotors so adjust their speeds that their combined torque is just sufficient to carry the load. It is not necessary that the two motors have the same number of poles. The various speeds for combinations in which the two motors have a different number of poles may be determined as follows: If the stator of the second motor is so connected that its rotor tends to turn in a direction opposite to that of the rotor of the first motor, equation (68) becomes

Fig. 247. — Concatenation of induction motors.

Let N be the speed of the combination, /i and /2 the stator frequencies, Pi and P2 the number of poles and *s* and s2

the slips. The speed of the first rotor f X 120

--"-

(1 — S) (page 236, from equation 65)

The speed of the second rotor 2 z

As the two rotors are rigidly coupled, the speed N equals the speed N2. from which

"Pi + P2

PiS2 is very small in comparison with Pi + P2, when the combination is operating near its synchronous speed, and it may be neglected. Then

P2

Again neglecting the term PiS2, the slip becomes

Sl = (70)

The set will not start of itself if connected in concatenation with the rotors tending to turn in opposite directions. It must first be brought up to speed either by an auxiliary motor or by one motor alone, before the second one is connected.

As an example of the speeds obtainable with two motors having a different number of poles, consider two 60-cycle motors, one having 4 poles and the other 20 poles. The following synchronous speeds are obtainable:

Four-pole motor alone: 1,800 r.p.m.

Twenty-pole motor alone: 360 r.p.m.

When the 4-pole and 20-pole motors are in concatenation, aiding, the slip of the first motor, from equation (69), 20 _ 20

Sl " 20 + 4 24

The synchronous speed of the set is N = (1-S,)l,800 = 7 1,800 = 300 r.p.m.

When the 4-pole and the 20-pole motors are in concatenation, opposing, the slip of the first motor, from equation (70), 20 20 Sl 20-4 " 16

The synchronous speed of the set is N = (1-si) 1,800 =-4 (1,800) =-450 r.p.m., or the set now rotates in the opposite direction.

Four different synchronous speeds are obtainable with these two motors, 1,800 r.p.m., 360 r.p.m., 300 r.p.m., — 450 r.p.m.

It is to be noted that the synchronous speed resulting from connecting the motors in concatenation *aiding* is equal to that of a 24-pole motor, or a motor whose poles are equal in number to the *su n* of the poles of the two individual motors. When the two motors are connected in *opposition,* the resulting synchronous speec is equal to that of a 16-pole motor, or a motor whose poles are equal in number to the *difference* of the poles of the two individual motors.

It will; be recognized that the concatenation method of speedcontrol is very similar to the series-parallel method of speed control for direct-current motors (see Vol. I, page 345, Par. 223). In concatenation, at starting and for intermediate speeds, resistance is introduced in the rotor-circuit of the second motor. When the motors are connected in parallel across the line, resistance is introduced in each rotor circuit and is gradually cut out. Motors operating in concatenation are used abroad to some extent, particularly in railway work. Because of its rather complicated connections, this system of speed-control is not used to any extent in this country.

110. The Induction Generator.—If an induction motor be driven-above synchronous speed, the slip becomes negative. The rotor conductors then cut the flux of the rotating field in a direction opposite to that which occurs when the machine operates as a motor. The rotor currents are then reversed with respect to the direction which they had when the machine operated as a motor. By transformer action these rotor currents induce currents in the stator which are substantially 180 out of phase with the *energy* component of the stator current which existed when the machine operated as a motor.

The induction motor, therefore, can be used as a generator, but it has certain limitations which the synchronous alternator does not possess..,

The machine does not have a definite speed for a given frequency as the synchronous alternator has, but the speed with constant frequency varies with the load. The load is practically proportional to the slip. Because its speed is not in synchronism with line frequency, the machine is often called an *asynchronous* generator. The frequency and volt-age of the induction generator are *that of the line to which it is connected,* irrespective of its speed.

An alternator, by itself, cannot deliver power unless its field is excited. The same is true of the induction generator. The alternator usually receives its excitation from a direct-current source or its equivalent and the resulting north and south poles are rotated mechanically. The flux in the induction generator is produced by the polyphase exciting currents in the stator windings and the resulting north and south poles rotate in the air gap at synchronous speed. The currents which excite these north and south poles *come from the line.* Therefore, the induction generator does not receive its exciting current from a separate source, but from the same lines that conduct away the energy that it generates. The induction generator cannot generate its own exciting current but *the exciting current must be supplied by the line.* For this reason it is necessary to have either a static condenser or synchronous apparatus in parallel with the induction generator for supplying its excitation. A static condenser, however, is seldom practicable.

Fio. 248.—Vector diagram of induction motor and induction generator.

Moreover, the induction generator can deliver only leading current. If a load requires a lagging current, a *synchronous* machine in parallel with the induction generator must supply the lagging component of this current.

The reason for this is as follows:

Let *V,* Fig. 248, be the terminal voltage of an induction machine operating as motor. The counter or generated electromotive force *E* is approximately 180 from *V.* Let *I,,* be the quadrature exciting current lagging 90 behind *V* and let *I'm* be the motor energy current. The total current taken by the motor is the resultant current *Im,* lagging behind *V* (see page 244, Fig. 233). When the rotor speed is increased by a sufficient amount, the machine passes from motor to generator action. The magnitude and the phase of the air-gap flux alter by only a slight amount during this transition, just as the flux of a shunt motor does not change in sign and changes in magnitude by only a small amount, if at all, when the machine passes from motor to generator action through the speeding up of its armature. Therefore, the exciting current *I,,,* which produces the flux, remains substantially constant in both magnitude and phase, just as the exciting current of a shunt motor does not change much when the machine is speeded up by mechanical means so that it becomes a generator.

However, as the rotor speeds up, it cuts the flux in a direction which is opposite to that occurring when the machine operates as a motor; that is, the slip becomes negative. Therefore, the induced electromotive force in the rotor conductors reverses in sign, as has already been pointed out. As the rotor reactance is very low at these low values of slip, the currents produced in the rotor are nearly in phase with their induced electromotive forces and they flow in a direction opposite to that which they had as motor currents. These induced currents react on the primary in the same manner that the secondary current of a transformer reacts on the primary. As a result, an energy current *I,,,* 180 from the motor current *I,,,* is induced in the primary. The currents in the rotor are nearly in phase with the induced electromotive force of the rotor. The *induced* electromotive force of the stator is in phase with the rotor electromotive force. The stator ampere-turns, excluding the effect of the magnetizing current, are equal and opposite to the secondary or rotor ampereturns. These stator ampere-turns are represented by the vector *I g* in Fig. 248. Since the rotor ampere-turns are nearly in phase with the secondary-induced electromotive force, the opposite and equal primary component *Ig* of the generator current must be nearly in phase with that component of the primary voltage which balances the secondary emf. *E,* and the generator terminal voltage *V.* In the vector diagram *I'g* actually leads *E* by a small angle which, for simplicity, is neglected in Fig. 248. Therefore, *Ig* is practically all energy current. For any particular kilowatt load, corresponding to the energy current $I'g$, the total generator current is I_{nn} the vector sum of Ig and $I0$. Ig is a *leading* current, as the generator terminal voltage V is nearly in phase with its induced electromotive force E. The generator phase angle 6 is not determined by the load, therefore, but by the generator itself.

If the load requires a lagging current, this machine cannot supply it. This is illustrated by Fig. 249. A certain load requires a current I, lagging a degrees behind the terminal voltage V. It is desired to supply as much of this current as possible by means of an induction generator and to allow an alternator to supply the remainder. Resolve the load current I into two components, an energy component i and a lagging quadrature component iz. The induction generator can by proper speed adjustment supply the energy current i. However, its leading exciting current $I0$ is fixed, as has already been demonstrated. Therefore, I_{nn} the resultant of i and $I0$, is the total induction generator current at this load.

Obviously the alternator must supply that part of the load current which the induction generator cannot supply. That is,

Fio. 249.—Currents supplied by an alternator and by an induction generator in parallel.

the alternator must supply the *difference* between the load current and the induction generator current. To obtain the difference between two vectors, reverse one and add (page 12, Par. 7). As I_{nn} is subtracted, it is reversed and the resulting alternator current is I_{nn} which is equal in magnitude to the arithmetical sum of $i2$ and $I0$. It will be observed that the alternator in this case supplies no power. Its entire current is lagging quadrature current and is equal to the *exciting current* of the induction generator plus the *lagging quadrature* current of the load.

If the load were such as to require a leading current, the quadrature component of which was just equal to $I0$, theoretically the induction generator could of itself supply the entire load. Even then it would be necessary to have synchronous apparatus on the system to se-

cure satisfactory operation.

The inability of the induction generator to deliver lagging current is the principal objection to its use. Considerable kv-a. in synchronous apparatus is required to supply the total quadrature current required. The distinct advantage of the induction generator is the fact that it does not hunt or drop out of synchronism; it is simple and rugged, and when short-circuited it delivers little or no power because its excitation at once becomes zero. Its principal use seems to be in the development of small water powers, where the cost of attendance would prohibit the use of synchronous apparatus. The induction generator connected to the water wheel does not need to be synchronized, requires no direct-current excitation, and does not fall out of syn .14.12.10.08.06.04.02 0.02.01. 06.08.10.12.14 Fig. 250.—Operating characteristics of induction machine as motor and as generator.

chronism. It delivers power if there is sufficient water; if not, it merely runs idle as an induction motor. Such machines would feed into a main generating station located in the vicinity and so could be under the occasional inspection of an operator.

The induction generator is also very useful for braking purposes in railway work. If the induction motors be left connected across the line on a down grade, any tendency of the train to drive them above synchronism will be accompanied by generator action. In addition to braking the train, the generators pump power back into the line and so relieve the main generating station of some of its load. The machine therefore requires no complicated control apparatus when used for regenerative braking, such as is required by direct-current motors operating under similar conditions.

Figure 250 shows the variation of current, efficiency, slip, and torque of an induction machine as it passes from motor to generator. It will be noted that the current does not pass through zero, although the power does. The point of minimum current is the exciting current of the machine, shown by $I0$-At synchronous speed, the line supplies the

core losses, and the friction losses are supplied mechanically. The generator must be driven somewhat above synchronous speed before it supplies its own core losses.

111. The Circle Diagram.—The operating characteristics of an induction motor may be determined experimentally without $o A J$

Fig. 251.—The circle diagram for an induction motor.

actually loading the motor, just as was done in the case of the alternator and of the transformer. It can be shown that with constant impressed voltage and constant frequency, the locus of the primary current I, Fig. 251, as the load on the induction motor is varied, is an arc of a circle. That is, with change of load, the end E, of the current vector I, moves along the arc of a circle $PEHK$. This diagram is approximate in that it neglects the impedance drop and the copper loss in the stator due to the magnetizing and core-loss currents. To obtain data for the construction of this diagram an open-circuit and a short-circuit (or blocked) run are made, as is done with the alternator and the transformer. Using the data obtained from these two tests, the operation of the motor may be determined with a very fair degree of accuracy by the use of such a circle diagram.

The motor is first run at rated voltage without load and the line voltage V, the line current $I0$, and the total watts $P0$ are measured. The no-load power-factor angle 00 can then be deter r mined (cos $00 =$ or for a three-phase motor). The voltage per phase V *is* laid off vertically, Fig. 251, and the no-load current I_{nn} (per phase) is laid off at an angle 60 from V and lagging. The rotor is then blocked. In order that the current may be kept within reasonable limits, the supply voltage per phase, V, is reduced to voltage v', which should be of such value as to give a short-circuit current approximately equal to the rated motor current. The phase current Ia, the total power P' and the phase voltage v' are measured under these conditions. Let V be the rated phase voltage of the machine. $V = V$ for a deltaconnected machine and V

= F/yl for a Y-connected machine. The measured current IB is increased in the ratio of the rated motor voltage V (per phase) to the reduced voltage v'. This gives $IB = 0H$, the current per phase which would exist were the rated line voltage V impressed across the motor when blocked. This current lags V by an angle 6B.

P' cos $Qb = i = r?$. v, where n is the number of phases. OL is drawn making an angle of 90 with $0V$ in a clockwise direction. $IB = 0H$ is laid off' making an angle 6B with OF'. Points P and H on the circle are therefore determined.

Line PH is drawn. PK is drawn parallel to $0L$. It is not necessary to know point K in order to construct the diagram.

With PK as a diameter, a semi-circle is drawn through points P and H. The center, M, of this semi-circle is found by erecting a perpendicular MM, at the center of PH. The intersection of MM' with PK gives the center M of the circle. With MP as a radius and M as a center, the semi-circle $PEHK$ is drawn. PK is the diameter of the semi-circle and its length in amperes is $V\ PK = $ —-j-—, where V is the phase voltage and x and $x2$ are #1--#2 the respective stator and rotor reactances per phase, referred to the stator.

A perpendicular HJ is then dropped from H to OL. The line HF is divided by G into two segments such that $HG: GF = IRz: Ii'Ri$, that is, in proportion to the secondary and primary resistances, respectively, as a one-to-one ratio of rotor to stator turns is assumed. Line PG is then drawn.

At any load current I, $/ 2 (= PE)$ is the secondary current, being equal to $I - I0$ vectorially. EA is the energy component of the current I, and therefore the total power input per phase,

Pi $= EA$ X V The core and friction losses $Pc = BA$ X V per phase

The primary copper loss $Ii2Ri = BC$ X V per phase

The secondary copper loss $IRz = CD$ X V per phase

The output $P = DE$ X V per phase

Draw $P'G'$ parallel to PG and tangent to the circle at E'.

Break-down torque $TB = C'E'$ (to scale).

The above diagram is drawn for but one phase of the motor. The values of power, losses, and torque must be multiplied by n if the motor has n phases.

The torque scale may be found as follows:

The torque is equal to a constant times the power, divided by the speed, the value of the constant depending on the units adopted. The power output per phase is $P = V$ X DE. The rotor speed $N2$ $= N (1 — s)$ where N is the synchronous speed in r.p.m.

$/ \_ CD \_ N(CE-CD) N)$
$2\ CE''\ CE\ CE$

The torque developed per phase K is CE V X $C£$ z's *the total power per phase delivered to the rotor.*

The total power delivered to the rotor by n phases, $P2 = n$ X F' X CE watts. The horsepower output

„ n X $D£$ X F' -33,000 where T7 is the *total* torque.

— ,, $NXDE$,

But $A2 = $ — $££$— from (I)

Substituting in (III) n X-DE X $V =$ $2Tr(NxDE)T$
$746 = CEX\ 33,000$
r-7. M»-X-!?-X pound-feet.

X = 7.04 X n (II)

As the phases n, the voltage F', and the synchronous speed N are usually fixed, the torque $T = K,CE$ where $K' =$ 7.04-— 112. The Measurement of Slip. — There are various methods for measuring slip. The slip may be determined by measuring the rotor speed and subtracting this speed from that of the rotating field as determined from the frequency. As the slip is but a very small percentage of either the synchronous speed or the rotor speed and is the difference of two nearly equal quantities, it is not possible to determine it accurately by the measurement of each of these quantities and so finding their difference.

A simple method of measuring slip is shown in Fig. 252. A "target" or disc is fastened to the end of the shaft or to the pulley of the motor. This disc has the same number of black and the same number of white sectors as the motor has poles. This disc is illuminated by an arc lamp which is fed from the mo-

tor mains. When the current in the arc is passing through its zero values, the arc emits but little light. Therefore, during these periods the sectors on the disc are but dimly illuminated. In one-half cycle the armature of the motor would advance one pole if there were no slip. During this time each black sector would advance to the position just occupied by the adjacent black sector which preceded it. The same is true of the white sectors. During the period of advancement the sectors are but faintly visible because the current in the arc is passing through zero. Each black sector and each white sector is not, therefore, clearly visible until it has reached the position just occupied by the sector of the same color just preceding it. As the disc is

Fig. 252.—Stroboscopic method for measuring slip.

intensely illuminated twice every cycle, while the arc current is passing through its maximum values, all the sectors are clearly visible twice every cycle. Therefore, if the disc rotated at synchronous speed it would *appear* stationary. Due to the fact that each conductor on the rotor does *not* advance one pole each half cycle, the sectors will not reach the position of the next adjacent sector of the same color, but will fall short of this distance, due to the slip. The sectors on the disc will then appear not stationary, but will seem to be rotating slowly backward. The number of revolutions per minute that they *appear* to rotate is the revolutions slip of the rotor. Figure 252 shows a stroboscope for a four-pole machine. Occasionally, black and white stripes are painted on the face of the pulley, Fig. 252, to serve the same purpose.

A mechanical-electrical method of measuring slip is shown in Fig. 253. Two cylinders of insulating material are driven, one by the induction motor shaft and the other by a small synchronous motor having the same number of poles as the induction motor. Each of these cylinders is fitted with a slip-ring, to which a small contact piece is connected. The synchronous motor always runs at the speed of the rotating field. Therefore, every time the induction motor slips one revolution, the contact pieces touch each other, closing the circuit between the two slip-rings. This is indicated by a flash of the light connected in series with the rings through the brushes *b*, Fig. 253.

In the Electrical Engineering Laboratories at Harvard University, the induction motor and the synchronous motor jointly

Fig. 253.—Measurement of slip by means of synchronous motor.

drive a differential through gears, a method developed in these laboratories. The speed of the differential is the revolutions slip of the induction motor. If desired, the speed of the differential, and hence the slip, may be measured with a speed counter with considerable accuracy. By changing gears, the apparatus is adapted to machines having any number of poles. 113. The Induction Regulator.—Without auxiliary apparatus, it is practically impossible to maintain the proper voltage at all the distribution points of a system, because with a fixed voltage at the station busbars, the voltage at the ends of short feeders will ordinarily be greater than the voltage at the ends of long feeders. Owing to the ohmtc and reactive drops in the lines, the voltage at the load end of the feeder may vary considerably with the load on the feeder. In order to maintain a more constant voltage at the distribution point, without using an excessive amount of copper, an induction regulator is often connected to each feeder. This maintains the voltage at the distribution point practically constant.

The induction regulator is a transformer having a movable secondary. In this way it closely resembles the induction motor. The general principle of the single-phase type is shown in Fig.

Fio. 254 (6).—Connections of a single-phase induction regulator.

254. An ordinary winding is placed in the slots on the stator and a drum winding is placed in the rotor slots. When the secondary is in the plane of the primary, the maximum electromotive force is induced in the secondary, because the mutual inductance of the windings is a maximum when the secondary is in this position. When the secondary is at right angles to the primary, the primary flux does not link the secondary so that the induced electromotive force in the secondary is zero. As the mutual inductance of the windings is zero under these conditions, the secondary acts like a choke-coil of very high impedance. To prevent this, a short-circuited tertiary winding is placed on the stator. This acts like a short-circuited transformer secondary, and therefore reduces the inductance of the regulator secondary to a very small amount. The primary winding is shunted across the line as shown in Fig. 254 (6) and the secondary is connected in series with the line (compare with Fig. 200, Chap. VII, page 210). When the secondary is in the plane of the primary in one

Fig. 255.—General Electric 2,300-volt. 60-cycle, feeder voltage regulator, disassembled.

position, its induced electromotive force is a maximum and it is connected to act as a booster. When the secondary is turned 180 degrees from this position, its electromotive force is also a maximum, but it now bucks the line voltage. Any value of voltage between that corresponding to these two positions is obtainable by varying the position of the secondary.

The secondary is turned by a small motor, controlled by relays, Fig. 256. The relays are actuated by a contact-making voltmeter. If the voltage is too high, one set of contacts causes the motor to turn in such a direction as to make the secondary reduce the line voltage. If the voltage is too low, another set of contacts causes the motor to reverse its direction and the secondary boosts the line voltage. Figure 255 shows a General Electric induction regulator of this type, disassembled. In this type, the rotor is the primary and the stator the secondary. The short-circuited winding at right angles to the primary winding is plainly shown on the rotor.

The three-phase induction regulator closely resembles the three-phase, wound-rotor induction motor. The three stator windings or primaries are connected across the line in either Y or delta. The three secondaries, which correspond to the three phases of a rotor winding, are insulated from one another and each is connected in series with one of the three-phase lines. As the stator produces a uniform rotating field, the induced electromotive forces in the secondaries are constant and are independent of the position of the rotor. Their boosting and bucking effect, however, depends upon the phase relations existing between each induced secondary electromotive force and its' respective line voltage. The three-phase regulator requires no short-circuited tertiary winding.

Fio. 256.—General Electric feeder voltage regulator assembled, with panel board.

CHAPTER IX

SINGLE-PHASE MOTORS

114. The Series Motor.—It will be remembered that the direction of rotation of either the direct-current shunt motor or the direct-current series motor is the same irrespective of the polarity of the line voltage. If the line terminals be reversed, both the field current and the armature current are reversed and the direction of rotation remains unchanged. If such motors be supplied with alternating current, the *net* torque developed acts in one direction only.

With alternating current, the shunt motor develops but little torque. The high inductance of the shunt field causes the field current and therefore the main flux to lag nearly 90 in time-phase with respect to the line voltage. The armature current cannot lag the line voltage by a large angle if the motor is to operate at a reasonable power-factor. Therefore, there will be considerable phase difference between the main flux and the armature current. Consequently, such a motor will develop but little torque per ampere (see Par. 103, page 237). This particular type of alternating-current shunt motor is therefore not practicable.

In the series motor, the armature current and the field current are in phase with each other. The main flux is practically in phase with the field current. Therefore, the armature current is substantially in phase with the flux, and the torque curve has no negative loops (see Fig. 227, page 238). Consequently, the series motor develops approximately the same torque per ampere with alternating current as it does with direct current. Fundamentally, the series motor has possibilities as an alternatingcurrent motor.

The *ordinary* direct-current series motor does not operate satisfactorily with alternating current for the following reasons: *(a) The alternating-field flux sets up eddy currents in the solid parts of the field structure, such as the yoke, cores, etc., causing excessive heating and a lowering of efficiency.*

In the alternating-current series motor this difficulty is eliminated by laminating the field structure. Even with laminated field-cores, however, losses in the iron occur with alternating current which do not occur with direct current. (6) *There is a relatively large voltage drop across the series fields, due to their high reactance.* This limits the current and also reduces the output and power-factor to such low values as to make the motor impracticable.

In the alternating-current motor this difficulty is partially overcome as follows:

A low frequency is used, since reactance, X, is $2\pi fL$, where f is the frequency and L the inductance. Even when the field

Fig. 257.—Windings of an alternating-current series motor.

inductance, L, is made as low as is practicable, the field reactance, X, will be considerably too high unless the frequency is made low. The usual lighting frequency of 60 cycles is much too high, except for motors of fractional horsepower rating. Difficulty is experienced in designing a series motor for a frequency of 25 cycles, even. To obtain satisfactory operation, frequencies of 12 and 15 cycles are commonly used abroad for this type of motor.

The inductance varies as the square of the number of turns. The turns per pole must therefore be reduced to a minimum in order to keep the inductance, and therefore the reactance, low. To obtain sufficient flux with few ampere-turns per pole, the reluctance of the magnetic circuit must be reduced to a minimum. This is accomplished by operating the iron at low flux densities and therefore at high permeabilities, and by using a very short air-gap. Because of the small number of field ampereturns and the very low flux density, a very short pole of large cross section is necessary, as indicated in Fig. 257.

(c) *The armature of an alternating-current series motor of a given rating has an unusually large number of conductors.* A motor of fixed horsepower and speed must develop a corresponding torque. The torque developed by a motor is proportional to the product of the field flux and the armature ampere-conductors. Therefore, if the total flux of the alternating-current motor is less than the total flux of a direct-current motor of the same rating, the armature ampere-conductors of the alternatingcurrent motor must be correspondingly increased in order to obtain the required torque. This is one reason why the armature of the alternating-current motor is larger than that of the direct-current motor of 'the same rating. *(d) The alternating-current motor has a lesser number of field ampere-turns and a greater number of armature ampere-turns than the corresponding direct-current motor.* That is, the motor has a strong armature and a weak field. This means that the armature, reaction is unduly large. Therefore, the effect of the armature cross-magnetizing turns, unless compensated, is to produce unusually great field distortion. As this distortion of the field by the cross-magnetizing armature ampere-turns would make commutation practically impossible, this cross-magnetizing action must be neutralized. This is accomplished by means of a compensating winding placed between the main poles, as shown in Fig. 257, this winding being embedded in the pole faces (also see Thompson-Ryan winding, Vol. I, page 275, Fig. 247). In order to reduce the *reactance* of the armature, also, this compensating winding should not only neutralize the cross-magnetizing field of the armature as *a whole,* but it should neutralize it *at every point.* Although it is impossible to secure complete neutralization at every point, a close approximation to this is obtained by distributing the compensating winding over the pole faces, Fig. 257, and by making each group of pole-face conductors carry a current equal and opposite to the current in the Winding 258.—Conductively-compensated series motor.

group of armature conductors directly under it, as is also indicated in Fig. 257.

The compensating winding may be connected in series with the armature, Fig. 258, in which case the motor is said to be *conductively* compensated. When it is necessary to use the motor on a direct-current system as well as on an alternatingcurrent system, conductive compensation is necessary.

If the compensating winding be short-circuited on itself, Fig. 259, the winding is linked with the cross-magnetizing flux of the armature and therefore becomes the shortcircuited secondary of a transformer, the armature ampereturns being the primary. As the secondary ampere-turns of a transformer are practically opposite in phase and equal in magnitude to the primary ampereturns if the magnetizing current be small, the ampere-turns of the compensating winding nearly neutralize the ampereturns of the armature. It is not possible to eliminate entirely the crossmagnetizing flux by this method, any more than it is possible to eliminate the mutual flux in a short-circuited transformer, but it may be reduced to a very small value.

The cross-magnetizing flux links the armature turns, causing the armature itself to have large self-inductance. Therefore, the reactance of the armature alone *(Xa = 2\pi fLa)* is large, which in turn would produce a large reactance *(IXa)* drop and so lower the power and the power-factor of the motor. The compensating winding, acting like the short-circuited secondary of a transformer, reduces this armature reactance to a small value.

This is analogous to the ordinary transformer, which on opencircuit is a very high impedance. When the secondary is shortcircuited, the impedance is reduced to a very low value. The

Fig. 259.—Inductively-compensated series motor.

-Short-Circuited Coll

necessity for reducing this armature reactance drop to a small value is the principal reason for using a distributed compensating winding rather than a more concentrated one. *(e) In the alternating-current series motor a commutating difficulty occurs which is not present in the direct-current motor.*

Figure 260 shows a coil in the neutral plane undergoing commutation. The coil is therefore short-circuited by the brushes. The plane of this coil is perpendicular to the direction of the main field, which is alternating, so that the alternating flux of this field links the coil.

The short-circuited coil Fig. 260.— Transformer electromotive force acts as the secondary of a in coil undergoing commutation, transformer, of which the main field-winding is the primary, and therefore has voltage induced in it. As this coil is shortcircuited by the brushes, and has a low impedance, a large current flows. This current causes severe sparking at the brushes. In addition, it opposes the main flux and so lowers the torque. To reduce this induced current to as low a value as possible, Load Current

Fig. 261.—Resistance leads inserted to improve commutation.

resistance leads are often inserted between the armature coils and the commutator segments, as shown in Fig. 261. Such leads, by increasing the impedance of the short-circuited coil, reduce the short-circuit current. It will be noted, Fig. 261, that so far as the *short-circuit* current is concerned, two such leads are in *series,* while so far as the *external* or *load* current is concerned, they are in *parallel.* This makes the resistance of these leads to the short-circuit current four times as great as it is to the load current. Except when starting, such leads are in the circuit but a small part of the time. If the starting period is too long, the leads in circuit at that time may overheat.

Reactances for reducing this transformer current have been suggested in place of resistances, but the difficulty of finding room for such reactances on a rotating armature has prevented their use. The induced voltage per turn in the armature coil undergoing commutation is proportional to the flux per pole. In order to keep this voltage within allowable limits, the total flux per pole must be made as small as possible. Therefore, the number of poles must be increased in order that there be sufficient total flux to develop the required torque. For this reason an alter IRS

Fig. 262.—Vector diagram for alternating-current series motor.

nating-current series motor ordinarily has more poles than a corresponding direct-current motor.

In order to improve commutation still further, the voltage between commutator bars is kept down to a low value. This requires a large number of commutator segments and a correspondingly large commutator. The voltage between commutator segments is still further reduced by operating the motor at low voltage, usually not over 250 volts.

Figure 262 shows the vector diagram for this type of motor. The resistance drop, $IR,$, of the main field, is in phase with the current I. The reactance drop, $IX,$, of the main field, is in quadrature and leading the current I. IRa and $IRC,$ the resistance drops of the armature and compensating field, are in phase with the current. IXa and IXe the reactance drops of the armature and compensating field, are in quadrature with the current and leading. The reactance drop of the series field is much greater than that of either the armature or the compensating field. The armature cross-magnetizing flux is not essential to the operation of the motor and as far as possible can be neutralized by the compensating field ampere-turns. In fact, the motor operation is improved by the reduction of this crossmagnetizing flux. On the other hand, the main flux is essential to the operation of the motor and cannot be reduced in value without reducing the torque per ampere. Hence, it is not practicable to neutralize the main

flux and consequently the seriesfield reactance drop must be large, even after the turns per pole, etc., have been reduced to a minimum.

When the alternating-field flux is at its maximum value, the armature conductors are cutting the maximum flux, and the back emf. is therefore a maximum. When the field flux is at its zero value, the back emf. is zero. Therefore, the back electromotive force is in time-phase with the flux, and practically in time-phase with the current, as shown in Fig. 262.

The terminal voltage, $V,$ is the vector sum of the back emf., $E,$ and the IR and IX voltage drops in the series field, the compensating field and the armature. The product of the back emf., $E,$ and the current, $I,$ is the power developed in the armature. The power at the pulley is less than this by the amount of the rotational losses. The cosine of the angle θ is the powerfactor of the motor. In order to have high power-factor, the reactance drops must be low and the back emf. high. The reactance drops are lowest and the back emf. is highest at light loads, and therefore the power-factor of the single-phase series motor is highest at light loads, as shown in Fig. 263. This is the reverse of the power-factor relations which exist in the induction motor and in the transformer.

The single-phase series motor has practically the same operating characteristics as the direct-current series motor. This is illustrated in Fig. 263, which gives the operating characteristics of a typical railway motor. The torque or tractive effort varies nearly as the square of the current and the speed varies inversely as the current, or nearly so.

If conductively compensated, the motor operates satisfactorily with direct current and at increased output and efficiency. When the motor is operated with alternating current, the speed may be efficiently controlled by taps on a transformer. This efficient speed control is not possible with direct current.

The single-phase series motor operates satisfactorily in railway work, notably on the New York, New Haven &

Hartford Railroad. From New Haven to Harlem the locomotives take power at 11,000 volts, 25 cycles, from an overhead trolley wire, by means of a pantograph trolley. An auto-transformer on the locomotive reduces this voltage to 250 volts, the rated voltage of the series motors. The electric locomotives run from Harlem into the Grand Central Station, New York City, over the New York

Amperes

Fig. 263.—Characteristic curves of 430-amp., 235-volt. 25-cycle, single-phase Westinghouse railway motor. Continuous rating, 200 amp., 235 volts.

Central 600-volt, direct-current system. The same motors are used for both direct-current and alternating-current service, the control devices switching over automatically when transition is made from one to the other. The motors which operate at 250 volts each on alternating current are connected two in series for direct-current operation.

116. The Repulsion Motor.—If an ordinary direct-current armature be placed in a single-phase magnetic field and the brushes be short-circuited, a simple repulsion motor is obtained. In order to develop torque, however, the brush axis must be displaced from the axis of the main field by about 18 or 20 electrical space-degrees, as will be shown.

The principle of operation of such a motor is as follows: Figure 264 (a) shows a gramme-ring armature and its commutator. This is the same type of armature as would be used for a direct-current machine. This armature operates in a bi-polar magnetic field, the field structure being laminated. The fields are excited by a winding connected directly to a single-phase alternating-current line. At the instant shown, the upper wire is positive and the current is increasing in a positive direction. The flux which is substantially in phase with this current is also increasing and by the corkscrew rule is directed upwards. This flux divides, half going through each side of the ring armature.

It is clear that the winding on each side of the ring armature

Fio. 264.—Currents and cinfs. in the

windings of a repulsion motor, brushes in geometrical neutral.

acts as the secondary of a transformer. Therefore, the alternating flux produced by the field winding, as primary, induces an emf. in each half of the armature. By Lenz,s law, this induced emf. has such a direction as to oppose the inducing flux. The direction of this induced emf. at the instant indicated in Fig. 264 (a) is given by the arrows on the windings. It will be noted, by following through the winding, that the resultant direction of this induced emf. is *upward* in each side of the armature. This is indicated diagrammatically in Fig. 264 *(b),* where the arrows show the general direction of these induced emfs. through the armature. Were there no brushes, it is evident that no current would flow in the armature winding, as the emf. in one-half of the winding is equal and in phase opposition to that in the other half.

I

In Fig. 264, the brushes are shown as being in the geometrical neutral and short-circuited. Each brush is at the mid-point of its transformer winding. As the total emfs. in each winding are the same and the windings are connected in parallel, each midpoint must be at the same potential. Therefore, the brushes short-circuit two points at the same potential and no current flows between brushes.

It is clear that *without* brushes there is no armature current, and even *with* brushes there is no armature current, provided the brush axis is at right angles to the pole axis. Therefore, under both these conditions there is no armature current and, hence, no torque.

00

Fio. 265.—Currents in the windings of a repulsion motor, brushes along pole

Figure 265 (a) shows the same condition existing in the field and armature as was shown in Fig. 264 *(a),* except that the brushes now lie along the pole axis. As the general direction of the induced emfs. has not changed, the brushes are now shortcircuiting the points of the armature winding across which the maximum potential difference exists. There-

fore, current flows between the brushes from both sides of the armature, and in this brush position, the current in the armature is a maximum. But the motor develops no torque with the brushes in this position for the following reason: Two conditions are necessary for the development of torque. *The angle between the space-position of the flux axis and the brush axis must be greater than zero.* For maximum torque this angle should be 90. For example, in a direct-current motor with fixed flux and armature current, the maximum torque occurs when the brushes are in the neutral plane, that is at right angles to the flux. No torque would be developed were the brush axis parallel to that of the flux. *There must be a component of the current in time-phase with the flux* (see Par. 103, page 237). If there is 90 time-lag between the current and the flux, the current is a maximum at the instant the flux is zero, etc., and the average torque is zero. With flux, armature current and brush position all fixed, the maximum torque occurs when the flux and armature current are in timephase with each other.

Under the conditions shown in Fig. 265, the brush axis is parallel to the resultant flux. That is, the angle between the flux and the brush axis is zero. A consideration of Fig. 265 (a) shows that the current flows in opposite directions in the two equal conductor belts on each side of the brush axis. Although it can be shown that the armature current is nearly in time-phase with the flux, no torque is developed because of the space position of the brushes.

Hence in this type of motor, no torque is developed when the brush axis is at right angles to the flux, for then there is no current; no torque is developed when the brush axis is parallel to the flux, because the ampere-conductors under each pole develop opposite and equal torques.

It is obvious, however, that if the brushes be placed in some intermediate position, they will be short-circuiting points of the winding between which a difference of potential exists and therefore currents will flow in the winding, and also the net ampereconductors un- der each pole cannot be zero. It can be shown by a close analysis that the ar- mature current is substantially in time- phase with the flux. Therefore, under these conditions the motor develops torque, and if allowed to do so, the ar- mature will rotate.

Figure 266 (a) shows the brush axis making an angle *a* with the pole axis. The arrows in this figure show the di- rection of the armature *current* at the instant when the upper wire is positive and the current is increasing positively. Figure 266 (6) shows diagrammatically the general direction of these currents through the armature and brushes. It will be observed that the current direc- tion in the conductors under each pole is such as to develop torque. Figure 266 (c) shows the direction of the induced *emfs.* in the armature, neglecting the distorting effect of the armature mmf. on the field flux. The emfs. in each half of the armature act in conjunction, as shown in Fig. 264 (6). Assume for the time being that angle /3 equals angle *a,* Fig. 266 (c). The *current* paths through the winding are *abed* and *afed.* In path *abed* the emfs. *Ecd* and *Ecb* included in angles a and /3 respectively, each equal to the brush displacement angle, are equal and act in opposition. Therefore, they cancel each other, leaving *Eab* as the net emf. through path *abed.* Like- wise in path *afed,* the emfs. *Eta* and *Eff* cancel, leaving *Eed* as the net emf. through this path. The net emfs., *Eab* and *Eed* are effective in sending the cur- rent through the armature.

Fio. 266.—Brush position in a repul- sion motor which gives both current and torque.

The foregoing is not a rigorous analy- sis of repulsion motor operation, but rather a statement of the general prin- ciples on which the operation depends. A rigorous analysis involves vector di- agrams of considerable complexity and is beyond the scope of this book.1

In this type of motor, the direction of rotation depends on the brush posi- tion. For example, in Fig. 266 (c), the direction of rotation may be reversed by moving the brushes so that they cross

the pole axis, the brush axis then mak- ing an angle *ft* with the pole axis. Angle *ft* must be less than 90.

In the foregoing discussion a gramme-ring winding has been consid- ered, as it is a simple matter to follow the winding since JFor more detailed analysis of single-phase motors see "Principles of Alternating Current Machinery," by Prof. R. R. Lawrence; McGraw-Hill Book Co. the conductors do not cross one another, etc. However, it is well known that a drum winding, for the same number of poles, has the same electrical character- istics as the gramme-ring winding. The preceding analysis applies equally well to a drum-wound armature. Also, the foregoing principles apply to motors of more than two poles. Figure 267 shows the brush positions for a four-pole mo- tor.

Instead of displacing the brushes from the geometrical neutral so that a potential difference exists between them, which results in a current, giving rise to torque, the same effect may be obtained by using two field windings displaced at right angles to each other, as shown in Fig. 268. A compensating or transformer field, acting along the brush axis, induces emfs., which in turn cause currents, shown in Fig. 265, and these currents react with the flux of the main field winding to produce torque. This type of motor should not be con- fused with the four-pole type of Fig. 267.

Practically all repulsion motors are made with non-salient poles, rather than with the salient poles shown in the di- agrammatic illustrations just given. The windings are usually of the distributed type, such as are used for induction mo- tors. The fact that the reluctance to the main-field flux and to the transformer- field flux must be kept as low as possi- ble makes it desirable to use non-salient poles and to make the air-gap as short as possible. Otherwise, the magnetizing currents for these fields will be high, lowering the power-factor.

Repulsion motors have characteris- tics similar to those of series motors and have large starting torque. The sparking

is very small at synchronous speed (3,600 r.p.m. for a two-pole, 60-cycle motor) but at speeds differing greatly from this, the sparking may be exces- sive. It will be noted that the motor of Fig. 268 is similar to the inductively compensated series motor of Fig. 259, with the connections of the compensat- ing winding and of the armature inter- changed. There are several types of re- pulsion motor on the market which, while differing in detail from the motor just described, involve identical princi- ples.

116. Single-phase Induction Motor.— Figure 269 shows a twopole motor whose magnetic field is produced by single-phase current flowing in a simple field winding. The current in this

Fig. 269.—Single-phase, alternating field. Fio. 270.—Time-variation of a single-phase alternating field.

field is assumed to vary sinusoidally with time and if the iron be assumed to operate at moderate flux densities, the flux through the armature will vary practically sinusoidally with time. The variation of this field with tune may be represented by the projection of a rotat- ing vector *tmax* upon a vertical axis *XX,* shown in Fig. 270. The vector *tmax* is equal to the maximum value of the flux and its speed of rotation in revolutions per second is equal to the line frequency in cycles per second.

It may also be assumed that this single- phase field is made up of two equal and oppositely rotating fields represented by two equal and oppositely rotating vec- tors, Fig. 271 (a), the maximum value of each of these fields or vectors being equal to one-half *tmax.* The' resultant of two such vectors always lies along the vertical axis and is equal in magni- tude at any instant to the field actually existing at that instant. The same thing is represented in Fig. 271 (6), which shows the flux distribution curves of two fields 0i and *fa,* each of which is equal to one-half the maximum field. These two fields glide around the air- gap in opposite directions and with equal velocities. Their algebraic sum 0 at any instant is the value of the resul- tant field at that instant and this resultant field is *stationary in space.*

The single-phase field may be considered therefore as

Fig. 271.—Representation of a singlephase alternating field by two oppositelyrotating fields.

made up of two equal rotating fields, revolving in opposite directions. (Experiment shows that two such fields actually exist.) Each field acts independently upon the rotor and in the same manner as the rotating field of the polyphase induction motor. One field tends to cause rotation in a clockwise direction and the other field tends to cause rotation in a counter-clockwise direction. Figure 272 shows the sliptorque curve due to each of the two fields. The torques act in opposite directions as shown. At standstill (slip = 1) the two torques are opposite and equal, and the rotor has no tendency to start. If the rotor in some manner be caused to rotate in the direction in which the torque T is $(2-,)$o i o (r2) acting, T will immediately

Fig. 272.—Two opposing torques in a single-exceed the COUnter-torque phase induction motor. and armature wfll begin to accelerate in the direction of T. As the armature speeds up, Ti predominates more and more over Ti and the armature approaches synchronous speed without difficulty. The counter torque due to Ti always exists, however, although it has little effect near the synchronous speed of the field which produces Ti.

When the rotor operates near synchronous speed and rotates in the direction of Ti, its slip is nearly two as regards TV Therefore, the rotating field which produces Ty induces double frequency currents in the rotor at this speed. These doublefrequency currents, however, produce but little torque because of their high frequency. This frequency is double the stator frequency. Therefore, the rotor reactance is many times its value at slip frequency. Consequently, these currents are small in magnitude and make a considerable space-angle with the airgap flux, developing little counter-torque (see Par. 103, page 237).

It is obvious that the single-phase induction motor rotates in the direction in

which it is started.

117. Reactions in a Single-phase Induction Motor.—Although the foregoing treatment of the single-phase induction motor gives some idea of its method of operation, it is not a rigorous analysis nor does it give a physical conception of what actually occurs in the motor.

The reactions occurring in the rotor of a single-phase induction motor are not simple and several factors must be considered if an exact analysis is to be made. In Fig. 273 the main flux £« due to the stator winding passes, at the instant shown, down into the armature from the north pole N. In so doing it links the rotor conductors and due to transformer action, currents are induced in these rotor conductors. These induced currents in the rotor conductors must flow in such a direction as to *oppose* this flux in the same manner as the secondary ampere-turns of any static transformer oppose the primary ampere-turns. The effect of the rotor conductors is the same as if they were connected as shown in Fig. 273, each conductor being connected with one on the opposite side of the armature to form a closed turn. To oppose the flux iM, the current must be flowing inward on the right-hand side of the armature and outward on the left-hand side of the armature, as indicated in the figure.

Assume that the armature rotates in a clockwise direction. There will be an emf. induced in the rotor conductors due to their cutting the flux tM-This induced emf. is called the *speed electromotive force* because it is induced entirely by the cutting of the flux tM due to rotation. Applying Fleming's right-hand rule, this emf. acts inwards on the upper half of the armature and outwards on the lower half, as shown in Fig. 274. This emf. is alternating and is a maximum when tM is a maximum. As the rotor conductors are short-circuited upon themselves, alternating currents flow in them as a result of this induced emf. The rotor reactance being high as compared with its resistance, these currents lag the induced emf. by very nearly 90. Moreover, these currents produce a flux tA, at right angles to $(m,$ as shown in Fig.

274, just as the ampere-conductors of a direct-current motor produce a field at right angles to the pole axis when the brushes are in the geometrical neutral. In practice, the stator completely surrounds the rotor, the air-gap being uniform. At synchronous speed, !a is substantially equal to tM but is 90 from $(pa$ in space.

The speed electromotive force EA is obviously a maximum when tM is a maximum.' The current IA does not reach its maximum until nearly 90 later in time, because the rotor reactance is high as compared with its resistance. In Fig. 273, 0M is shown as having reached its maximum and acting vertically downwards. After a quarter period, $4A$ reaches its maximum and is acting 90 in space from the flux tM, as shown in Fig. 274. It will be recognized that two such fields, acting along axes 90 from each other in space and differing in timephase by an angle of 90, will produce a rotating magnetic field. This field rotates clockwise in Figs. 273 and 274. As the rotor slip increases, tA decreases in magnitude because the speed is reduced. Therefore, the horizontal field becomes less than the vertical field and a so-called elliptical field results.

At standstill, tA is zero and the rotating field becomes a pulsat ing field, which has already been described.

It might be supposed that the above rotating field would react on the rotor in the same manner as the rotating field in the polyphase induction motor. Since tA originates in the armature and also because of its quadrature position, it cannot of itself react on the stator to cause a power current to flow in the stator, and therefore it cannot of itself contribute power to the rotor. However, due to the resultant rotor currents produced by the combined action of the two fields ja and $4M$, it can be shown that the resulting torque acting on the rotor under the above conditions acts in a clockwise direction, and so produces rotation.

118. The Operation of the Polyphase Motor as a Single-phase Motor.—The single-phase induction motor is distinctly inferior to the polyphase motor. For the same weight, its rating is about 50

per cent. of that of the polyphase motor, it has a lower powerfactor and is less efficient.

If one phase of a polyphase motor be opened, the motor will operate as a single-phase motor, although it will not start under these conditions. The rating and the break-down torque of a polyphase motor, operating single-phase, are considerably reduced and if rated polyphase load is applied continuously, the motor may overheat.

Ordinarily in starting a polyphase motor, all three lines are closed when the compensator is in the starting position and the motor starts as usual. When the compensator is thrown to the running position, however, a phase may become open through the compensator. This would occur if one of the fuses were blown, Fig. 238, page 251. The motor then operates singlephase and the only indication that it may give of this condition is overheating if the load is near the rated value. The best test for an open phase is to insert an ammeter in each line.

119. Starting Single-phase Induction Motors.—As the singlephase induction motor is not self-starting, auxiliary means must be used to supply initial torque. One method is to split the phase by the use of inductance, resistance or capacitance.

Figure 275 shows one method of splitting the phase, a two-pole motor being shown. The main winding, which is highly inductive, is connected across the line in the usual manner. Between the main poles are auxiliary poles which have a high-resistance winding and this winding is also connected across the line. As

Anilliary Poles Fig. 275.—Split-phase method of starting a single-phase induction motor.

the auxiliary winding has a high resistance, its current will be more nearly in phase with the voltage than the current in the main winding. For the best conditions, the two currents should differ in phase by 90, but this condition is not readily obtainable, Motor Minor

Fig. 276.—Splitting the phase with resistance and inductance.

and in fact is not necessary. These two sets of poles produce a sort of rotating field which starts the motor. When the motor comes up to speed, a centrifugal device in the rotor opens the switch S and disconnects the auxiliary winding. Another method of splitting the phase is to use a three-phase winding, as shown in Fig. 276, and to connect resistance and inductance as shown. Resistance and capacitance may also be used. Either a delta-connected stator, or a Y-connected stator may be used. The resistance and inductance, when connected as shown, displace the phase relations of the currents in the different phases of the stator with respect to one another and so produce a sort of rotating field. All these phase-splitting devices produce an elliptical rotating field. Because of the characteristics of the field combined with the squirrel-cage characteristics of the rotor, the resulting torque is barely sufficient to start the motor, even without load.

The *shaded-pole method* is shown in Fig. 277. A short-circuited coil of low resistance is connected around one pole tip. When the flux is increasing in the pole a portion of the flux attempts to pass down through this shaded tip. This flux induces a current in the coil which by Lenz's law is in such a direction as to oppose the flux entering the coil. Hence, at first the greater portion of the flux passes down the right-hand side of the pole, as shown in Fig. 277. Ultimately, however, the main flux reaches its maximum value, where its rate of change is zero. The opposing emf. in the shading coil then becomes zero, and later the opposing mmf. of the short-circuited coil ceases, the current in this coil lagging its emf. Considerable flux then penetrates the shortcircuited coil. After the main flux begins to decrease, the induced current in the shading coil tends to prevent the flux then existing in the shaded portion of the pole tip from decreasing. Therefore, the flux first reaches its maximum value at the right-hand or non-shaded side of the pole, and later reaches its maximum at the left-hand or shaded side. The effect of the shading coil is to retard in time-phase a portion

of the flux, so that there is a sweeping of the flux across the pole face from the right-hand to the left-hand side in the direction of the shading coil. This flux cutting the rotor conductors induces currents,

Fio. 277. — Shaded-pole method of starting a singlephase induction motor.

which in turn produce a torque sufficient to start the motor. The shaded pole is not a common method of starting single-phase induction motors and is used only in motors of very small size. It will be remembered that this same shaded-pole principle is

Fio. 278 (a).—Rotor of a Wagner single-phase, type BA motor.

used as the light-load adjustment in the induction watthour meter (see page 64).

The preceding methods of starting the single-phase induction motor produce very weak starting torques which are insufficient to start the motor except under the lightest loads. The Wagner Single-Phase Induction Motor starts as a repulsion motor and has a large starting torque. A cross-section of the motor and a view of the armature are shown in Fig. 278. The armature is similar to the type used in the ordinary direct-current motor, except that the brushes J press on the end of the commutator L rather than radially on its surface. These brushes are shortcircuited on themselves and are set in a position corresponding to those in Fig. 266 or Fig. 267, pages 290 and 291, so that the motor starts as a repulsion motor. It has a large starting torque and comes up to speed rapidly. As it approaches synchronism, a centrifugal device V, Fig. 278(6), is thrown outward and pushes the brushes away from the commutator, while at the same time a metal ring K presses against the commutator bars on the inside and short-circuits them. The motor now operates as a single-phase induction motor.

120. The Induction Motor as a Phase Converter.—If a three-phase induction motor be operated single-phase, as shown in Fig. 279, three-phase voltages exist across its three terminals. The reason for this is as follows:

The back emf. in each phase of a

polyphase induction motor is induced by the rotating field cutting the stator conductors. If the stator is wound for two-phase, the induced emfs. at the stator terminals are twophase; if the stator is wound for three-phase, the induced emfs. at the stator terminals are three-phase. The induced emf. in each phase of a polyphase induction motor is slightly less than the terminal voltage (per phase) by the amount of the stator impedance drop.

It was shown in Par. 117, page 294, that in a single-phase induction motor, a rotating field exists. At small values of slip this field departs but slightly from a true rotating field such as is produced by polyphase currents in polyphase windings. There v.

To 3-Pbase

Dductiou Motors

Stator of thaae Converter

Fig. 279.—Method of obtaining 3-phase power from single-phase supply, by means of squirrelcage induction motor operating as phase-converter.

fore, when a single-phase voltage is applied to one phase of a two-phase or of a three-phase motor, the rotating field is almost identical with that which exists when polyphase voltages are applied to the terminals. Consequently, if a single-phase voltage be applied to one phase of a two-phase stator, a quadrature emf. exists across the terminals of the other phase. If a single-phase voltage be applied across one phase of a three-phase stator, the voltages across the three terminals will very nearly equal one another and will be approximately 120 apart. As the induced emfs. are less than the applied terminal voltage by the amount of the stator impedance drop and as the rotating field is somewhat elliptical, the terminal voltages will not be exactly balanced. For example, in Fig. 279, 220-volts, singlephase, is applied across one phase of a three-phase motor, and voltages of approximately 210 volts and 200 volts are found to exist across the other two phases.

Polyphase induction motors are often used in this manner to produce polyphase voltages from single-phase supply. That is, single-phase voltage is supplied to one phase of the polyphase stator and polyphase voltages are obtained from the stator terminals. When so used the motor is called a *phase converter.*

The phase converter is used to some extent in railway electrification. Although the three-phase induction motor is adapted to railway work, there is considerable disadvantage in using the two trolleys which are required if three-phase power is to be supplied to the locomotive. By using a phase converter, the advantages of the three-phase motor for driving may be secured and at the same time all the advantages of a single trolley are retained. The phase converter receives single-phase power, which is pulsating, and delivers three-phase power, which is substantially steady. This is made possible by the kinetic energy stored in the rotating armature of the phase converter, this energy supplying the power during those times when the single-phase power is negative or is less than the average value of the polyphase power. The armature accelerates and so stores kinetic energy during the periods when the single-phase power exceeds the average power. The armature slows down and so gives up some of its kinetic energy during the periods when the singlephase power is less than the average power. In practice the actual speed variations of the armature are slight.

The electric locomotives of the Norfolk and Western Railway are operated by the use of a phase converter. A two-phase converter is used, as only half the power need be converted under these conditions, the other half flowing conductively from the transformer secondary to the motors. The power is received single-phase from an 11,000-volt trolley and stepped down by a transformer on the locomotive. Special transformer taps are

Fig. 280.—Connections of locomotive phase converter.

used to keep the phases balanced. The general diagram of connections is shown in Fig. 280. It will be recognized that the converter and transformer connection is equivalent to a Tconnection.

This is used in order that three-phase power may be obtained by supplying single-phase power to the two-phase stator of the converter. The phase *ab* to the driving motors is supplied directly from the transformer. The winding *a,b"b',* tapped to winding *ab,* is the main winding of the phase converter (see Fig. 207, page 217). The winding *c'c"c* is the teaser winding tapped to the transformer at c', giving the third wire *c* of the three-phase System. Ordinarily, the teaser winding would be tapped to point *b",* the center of the main converter winding. For convenience, however, the teaser winding is tapped to point *c ',* the center of the transformer winding instead. Under balanced conditions, however, *c '* and *b "* are at practically the same potential, so that as far as voltages are concerned, connecting the teaser winding to *c '* is equivalent to connecting it to *b".* 121. The Repulsion-induction (R I) Motor.—This motor is similar in principle to the simple repulsion motor of Fig. 268, page 291, except that a compensating winding is used, supplied by auxiliary brushes at right angles to the main brushes, as shown in Fig. 281. To reverse the motor, an auxiliary winding *be* is used, Fig. 282, at right angles to the main winding. The points *b* and *c* are interchanged for reversing. This motor has a starting torque of from 200 to 250 per cent. full-load torque and has a drooping speed characteristic similar to that of a compound, direct-current motor. By using a transformer and by shifting the brushes, considerable speed variation may be obtained. 122. The Wagner Type B K, Unity Power-factor Motor.— The Wagner Electric Co. has placed on the market a single-phase motor which has a leading power-factor up to about half load, while for greater loads, the power-factor is almost unity. The motor is a constant-speed motor, the regulation being from 1.5 to 4 per cent. There are two windings on the motor armature, as is indicated by the slot section in Fig. 283 *(b).* The upper winding is an ordinary drum winding connected to a commutator. Beneath this is a squirrel-cage winding separated from the other by a magnetic separator consisting of a steel wedge. 1. Initial field magnet-(6) One slot of rotor, showing conization winding. struction. 2. Auxiliary winding which controls powerfactor. (a)

Fio. 283.—The Wagner Electric Co.'s unity power-factor, type BK motor.

The diagram of connections is shown in Fig. 283 (a); 1 is the main winding and 2 the compensating winding. When the motor starts, the switch 9 is open and as the squirrel cage has little effect at starting, due to the screening effect of the magnetic wedges, the motor starts as a series motor. As it approaches synchronism, the squirrel cage has more and more effect. As synchronism is approached, switch 9 closes by centrifugal action and throws in the compensating winding. The manufacturers claim many advantages for this type of motor, such as high power-factor, constant speed, light service required of the commutator because of the assistance of the squirrel cage, suppression of short-circuit coil currents due to the proximity of the squirrel cage, etc.

CHAPTER X THE SYNCHRONOUS MOTOR

123. The Synchronous Motor.—It will be remembered that the direct-current generator operates satisfactorily as a motor. Moreover, there is practically no difference in the construction of the direct-current generator and the direct-current motor, and there is no substantial difference in the rating of a machine whether it is operated as motor or as generator.

Similarly, an alternator will operate as a motor without any changes being made in its construction. When so operated, the machine is called a *synchronous motor.*

The design of a synchronous motor and of an alternator, each of the same rating and speed, may differ somewhat in details owing to the desirability of securing the best operating characteristics for each. Moreover, synchronous motors are almost always salient-pole machines, whereas alternators may be either of the salient-pole or of the non-salient-pole type.

124. Principles of Operation.—Figure 284 shows a conductor a under a north pole and carrying a current flowing towards the observer. By the well-known law of motor action, a torque develops tending to drive the conductor from left to right. If the current be alternating, it will reverse its direction for the next half-cycle and the torque then acts from right to left. Therefore, the net torque over any given number of complete cycles is zero and no continuous motion can result. This is the condition existing in a synchronous motor when at standstill. The armature conductors carry alternating current and the poles have fixed polarity, being excited with direct current. Therefore, the synchronous motor, as such, develops no starting torque.

Fig. 284.—Torque developed by synchronous motor.

If, however, conductor a can in some manner be brought under the next pole, which is a south pole, for the half-cycle during which the current is in the reverse direction, the resulting torque will.still be from left to right and a tendency toward continuous motion will result. Therefore, in a synchronous motor a given conductor must move from one pole to the next in each half-cycle, if the machine is to operate continuously. This applies to the rotating-armature type of machine. If the machine is of the rotating-field type, any given conductor must be passed by one pole every half-cycle. In any event the synchronous motor must operate at constant speed, if the frequency is constant. There may be momentary fluctuations of speed, but if the *average* speed differs by even a small amount from this constant value, the average torque will ultimately become zero and the motor will come to a standstill. The relation of speed, number of poles and frequency is the same as for the alternator and for the rotating field of the induction motor. That is, the speed $S =$ —p— r.p.m., where / is the frequency and P the number of poles (see Pars. 3 and 101, pages 7 and 235).

Example.—A 500-kv-a., 2,300-volt, 10-pole synchronous motor operates on a 60-cycle three-phase system. What is its speed? 120 X 60

$S=$ =720 r.p.m. *Ans.*

125. Effect of Loading the Synchronous Motor.—If a load be applied to a direct-current shunt motor, the speed is slightly decreased. This reduces the back emf. , —E. The line must supply a voltage $+E,$ equal and opposite to the back emf. $— E$ and in addition, must supply the voltage necessary to overcome the IRa drop in the armature.

That is, $V = E + IRa$ where V is the fixed terminal voltage, I the armature current and Ra the armature resistance. The current $j = V + (-E) = V-E\ Ra\ Ra$

When the back emf. $— E$ decreases, more current I flows into the armature. This increased current supplies the extra torque and power required by the increased load.

When load is applied to a synchronous motor, its *average speed* cannot decrease since the motor *must* operate at constant speed.

It cannot draw the increased current from the line in the same manner that the shunt motor does, that is by operating at decreased speed. Figure 285 (a) shows two poles of a rotatingfield type of synchronous motor. Neglecting any flux distortion, the emf. induced in conductor a is a maximum when conductor a is opposite the center of a pole. It is zero when the pole reaches such a position that conductor a lies midway between the poles. The value of this emf., e, for any position of the pole, axis Y-Y, is shown by curve e.

Assume that a load is now applied to the motor shaft. This must result in momentary slowing down of the rotor, since it requires time for a motor to take increased power from the line. Therefore, the rotor instead of being in the position shown by the solid lines in Fig. 285 (a) will occupy a given position in space at a later time on account of the effect of the load torque. The relations under this condition are shown by the dotted lines. Because of the application of load, the pole center is now at $Y'\ Y'$ instead of being at YY. Therefore, the induced emf. will not reach its maximum value at the same instant that it would have reached it had no load been applied. This maximum value now occurs later in time, due to the slight backward angular displacement of the rotor. This is shown by a new curve of induced emf., *e'*, lagging *e* by an angle *a* where *e* is the emf. which would have been induced had no load been applied to the rotor shaft.

This is further illustrated by the use of vectors. Assume that the motor is running without load and that the current is so small that the back emf., — *E*, Fig. 285 (6), is sensibly equal to the terminal voltage *V* and is 180 out of phase with *V*. *(E* is the component of the terminal voltage necessary to balance the back emf., —*E*). The vector sum of *V* and —*E is* zero, practically.

Now apply load. The terminal voltage *V* is assumed to be constant and so is not affected by the load. The induced or back emf., —*E*, will be shifted backward by an angle *a* because of the backward angular displacement of the rotor caused by the load. Let this new value of back emf. be —*E'* and let the component of terminal voltage necessary to balance it be *E'*. The vector sum of *V* and —*E'* is no longer zero. Therefore, a vector *difference exists* between *V* and *E'*.

In the direct-current motor, the armature current is given by dividing the armature resistance into the difference between the terminal voltage and that component of the terminal voltage *E* which balances the back emf. — *E*. In the synchronous motor, the armature current is given by dividing the armature impedance *Z* into the *vector* difference between the terminal voltage F and the emf. *E'*.

That is / = = ?. (71)

Where *E0* is the vector difference of *V* and *E'*.

Therefore *E,* = *IZ*.

The above equation for the armature current in the synchronous motor is similar to the equation for the armature current in the direct-current motor (see Vol. I, page 317, eq. 109).

As a rule the reactance of the armature of a synchronous machine is high as compared with its resistance, and the current *I* lags the voltage *E0* which produces it by nearly 90. This brings the current *I* very nearly in phase with *E'*

and nearly 180 from the back emf. — *E'*. Therefore, *I* is largely energy current with respect to — *E'*, which means that it supplies considerable internal *power* to the motor.

The rotor, by *shifting its phase backward when load is applied, causes the motor to take an energy current from the line which supplies the power demanded by the increased load.*

The total power supplied to the motor per phase is

P = *VI* cos *8*

The total mechanical power developed is

P' = *E'I* cos *(B + a)*

The *net power at the pulley* is less than *P'* by the amount of the frictional losses and the rotational core losses.

The difference between *P* and *P'* is the armature copper loss.

It should be remembered that the *average* motor speed remains constant.. The rotor merely takes an angular position slightly back of its no-load position, without altering its average speed. This angular displacement of the rotor may be observed by means of a stroboscope (see page 274).

126. Effect of Increasing the Field Excitation.—When the field of a direct-current shunt motor is strengthened, there is a temporary increase in the armature induced emf. This decreases the armature current and the torque is lowered, since the change in armature current is much greater than the corresponding change in the field. As a result, the motor slows down and its back electromotive force accordingly decreases. The armature current then increases until it is again of sufficient magnitude to enable the motor to carry the load.

When the field of a synchronous motor is increased, the motor cannot slow down', except momentarily, for it must run at constant average speed. Since its speed is constant, its back ernf. must increase when the field is strengthened. It might seem then that the motor would stop, for its induced emf. must apparently become greater than its terminal voltage. In the direct-current motor, an induced electromotive force exceeding the terminal voltage would mean gener-

ator action with the result that the machine would cease to operate as a motor. ,7 (Lags Induced cmf. and leads motor terminal voltage)

Vq(Generator terminal voltage) £,'

Fig. 286.—Relation of current to voltage in motor and in generator.

The synchronous motor, however, may operate as a motor and at the same time its back emf. may exceed its terminal voltage in magnitude. Under these conditions, the motor is said to be *overexcited.* Two reactions occur which enable the motor to operate with an overexcited field. First, the motor takes a *leading* current. A leading current in a motor corresponds to a lagging current in a generator. This is illustrated by Fig. 286. A current / is shown lagging the induced emf. — *E* by 90. This current *I* is *lagging* with respect to both the induced emf. and the generator terminal voltage and is therefore a lagging current if the machine is considered as a generator. Such a current weakens the field through the effect of armature reaction (see Par. 63, page 134).

When the machine is considered as a motor, the emf. *E,* which is the component of the terminal voltage that balances — *E,* is opposite and equal to the induced emf., —*E*. The terminal voltage *Vm* differs from *E* only by the armature impedance drop. Therefore, the current *I* is *leading* with respect to the terminal voltage of the motor. It follows then that a current which is lagging when a machine is considered from the point of view of a generator, is leading when the same machine is considered from the point of view of a motor. *In a generator a lagging current weakens the field.* Consequently, *in a motor a leading current must weaken the field.*

This is further illustrated as follows: Figure 287 shows a motor coil moving from left to right. When its axis is in the position *Y,* shown dotted, the coil sides *'.Terminal* Toltage

J a Maximum *Y*

Fig. 287.—Demagnetizing effect of leading current on the field of a synchronous motor.

are under the centers of the poles and the induced emf. is a maximum. As the terminal voltage is substantially 180 from the induced emf., it also will be a maximum at this instant, its direction being indicated in the dotted coil. If the current leads this terminal voltage by 90 it will reach its maximum value one fourth of a cycle ahead of the voltage, or at a time when the axis of the coil is in position X. It will be observed that for this position of the axis, the ampere-turns of the coil act in direct opposition to those of the N-pole. Therefore, the effect of the leading current in the synchronous motor is to weaken the field. In other words, *the armature reaction tends to annul the effect of the increased field current on overexcitation.*

The second effect is illustrated by the vector diagram in Fig. 288. V is the terminal voltage and I is the armature current leading V by an angle *6*. The resistance drop in the armature is laid off in phase with the current I and the IX drop in the armature is laid off at right angles to the current / and leading, in the usual manner. The impedance drop IZ is the vector sum of IR and IX. The voltage E, necessary to balance the back emf., is found by subtracting IZ vectorially from V, just as in the shunt

Fio. 288. — Induced armature voltage greater than terminal voltage when synchronous motor current leads terminal voltage.

motor the component of terminal voltage which is necessary to balance the back emf. is found by subtracting the IR drop from the terminal voltage.

To subtract IZ from V, $—IZ$ is added to V. It will be noted that the emf. E is numerically *greater* than the terminal voltage V. That is, by taking a leading current, the synchronous motor is able to operate with an induced emf. greater numerically than the terminal voltage. This is analogous to the alternator delivering leading current with its induced emf. *less* than its terminal voltage. In each case the flow of power is towards the higher voltage.

127. Effect of Decreasing the Field Excitation.—When the field of a direct-current shunt motor is weakened, the motor speeds up until its back emf. reaches a value which gives the proper

armature current for the particular load condition.

Current a
Mailman
Fig. 289.—Magnetizing effect of lagging current on the poles of a synchronous motor.

When the field of a synchronous motor is weakened, it cannot speed up permanently for it must run at a constant average speed. However, it takes a *lagging* current. This current has two effects.

Figure 289 shows a coil, dotted, whose axis is in position Y. In this position the coil sides are opposite the centers of the pole faces and the back emf. is therefore a maximum. The terminal voltage, which is nearly 180 from the back emf., has its maximum value also for this position of the coil, its direction being indicated in the dotted coil. If the current is lagging the terminal voltage by 90, it will not reach its maximum value until the coil axis reaches position X. The current under these conditions is in such a direction as to strengthen the S-pole. Therefore, in a synchronous motor a lagging current strengthens the field through the effect of armature reaction. When the field of a synchronous motor is weakened, the motor takes a *lagging current which strengthens the field by armature reaction and tends to annul the effect of the weakening of the field.* A lagging current when a machine operates as a motor is a leading current when the machine is considered as operating as a generator. It will be remembered that a leading current in a generator strengthens the field through the effect of armature reaction (see Par. 63, page 135). That is, a lagging current in a motor has the same effect on the magnetic field as a leading current in a generator.

A synchronous motor under any given operating conditions requires a certain excitation. If its field is weakened, its excitation becomes inadequate. This deficit is in part made up by the motor taking a lagging current from the line. A lagging current is ordinarily associated with inductance and therefore with the excitation of a magnetic field. When the

motor takes a lagging current, some of its excitation is therefore obtained from the alternating-current line. In this respect it is similar to an induction motor except that the induction motor takes *all* its excitation from the alternating-current line. This lagging current required by the synchronous motor to help excite its own field weakens the field of the alternators supplying it, and as a result their field excitation must be increased to maintain the line voltage. Therefore, when the field of a synchronous motor is weakened, a part of the excitation which it requires is supplied indirectly by the fields of the alternators supplying the system.

On the other hand, when a synchronous motor is over-excited it x has a surplus of excitation. It _,,,».», ; Fig. 290.—Induced armature takes a leading current. As a lead-voltage less than terminal voltage ing current Will neutralize a por-when synchronous motor current-leads terminal voltage.

tion of the lagging current of inductive apparatus (see Par. 131, page 322) connected to the system, or else will strengthen the fields of the generators supplying the system, the synchronous motor under these conditions indirectly supplies excitation to other parts of the system. Figure 290 shows the vector diagram when the motor takes lagging current. The IR and the IX drops are laid off with reference to the current in the usual manner and the IZ drop obtained. When — IZ is added to V, however, E, which is opposite and equal to the back emf., becomes numerically much less than V. That is, the phase shift of the IZ drop is in such a direction that the machine runs as a motor with a very considerably reduced back emf.

The synchronous motor with salient poles will usually operate even if the field current is reduced to zero. The alternating current in the stator winding will produce a rotating field, just as in the induction motor. Figure 291 shows such a rotating field for a four-pole machine without a rotor. At the particular instant shown there are two N-poles vertically opposite, and two S-poles horizontally opposite. If a four-pole, salient-pole rotor (a) (6)

Fio. 291.—Interlocking action of salient poles with rotating magnetic field.

without excitation be placed in this field, the magnetic lines from the stator will attempt to make the rotor take such a position that the magnetic reluctance is a minimum or the flux is a maximum. In order to accomplish this result, the pole pieces of the rotor when running become locked in with the poles produced by the stator winding, as shown in Fig. 291 (6). These rotating stator poles pull the salient poles of the rotor around with them and in this manner enable the motor to carry a limited load without directcurrent excitation. Although the motor may carry a limited load without any direct-current excitation, its power-factor will be very low and the current will be lagging, which is undesirable. It is to be noted that under these conditions, in the absence of direct-current excitation, the motor takes its entire excitation from the alternating-current lines in the same manner as an induction motor, as has already been pointed out. That is, if sufficient excitation is not supplied by the direct current in the field winding, the motor will take lagging exciting current from the alternating-current line to make up the deficit.

The power-factor of the *induction motor* for a given load cannot be altered without changing the motor design and the ordinary induction motor always takes a lagging current. The power-factor of the *synchronous motor* can be altered at will, and the current can be changed from lagging to leading by simply changing the field excitation.

128. Synchronous Motor V-curves.—If the power, P, delivered to a three-phase synchronous motor be kept constant and the field current // varied, the power-factor of the motor will change. The power for a 3-phase motor is $P = \sqrt{3}VI$ cos 6 where V is the terminal voltage, I the line current and cos 6 the power-factor of the motor. As both P and V are constant, any decrease in the power-factor (cos 6) must be accompanied by a corresponding increase in the current

/. Likewise, any increase in the power-factor must be accompanied by a decrease in the current /.

Therefore, a change in the field current at constant load changes the line or armature current /. In order to determine the relation between the field current and the armature current and also the characteristics of a synchronous motor as regards its ability to correct the power-factor of a system, the so-called V-curves of the motor are obtained. These V-curves show the relation which exists between the armature current and the field current for different constant-power inputs. Several curves are usually obtained, each curve representing a constant value of power input.

The connections for making such a test are shown in Fig. 292.

The field current is varied by means of the field rheostat. For each value of field current, as read on the direct-current ammeter, the corresponding value of the alternating line current is noted. The electrical power delivered to the motor is kept constant by adjusting the load applied to the motor shaft. A polyphase wattmeter is desirable for this experiment, as it eliminates the adding or subtracting of individual instrument readings which is necessary when two single wattmeters are used.

Fio. 292.—Connections for obtaining V-curves of synchronous motor.

Figure 293 shows a set of typical V-curves. The curve *AB* is obtained when the motor is running at very light load. At very low values of field current, the armature current is large and is lagging. As the field current is increased, the power-factor increases and the armature current decreases until it reaches its minimum value /t. If the field current be still further increased, the armature current begins to increase and becomes leading. In other words, the motor passes from *under-excitation* to *overexcitation* when the field current is increased from a low to a high value.

The current I is the value of the current at unity power-factor. This is illustrated in Fig. 294. Let /2 be the value of line current for some power-factor, cos 02. The power (for one phase) is,

Pi = y/2 cos 02 where V is the phase voltage. But (/2 cos 02) = /i 72) for all values of 02.

In other words, for constant power Pi, /i is always the energy component of the current regardless of the powerfactor. Therefore, the current vector,, ... (',,,. vv Fig. 294. —Vector diagram

Will always terminate On trie lineAA showing current variation in perpendicular to V. The current synchronous motor with con .. stant power input.

is a minimum at /i, where the current is in phase with V. The power-factor is then unity. The excitation corresponding to the armature current I is called the *normal excitation* of the motor for the load in question. For an excitation less than the normal value, the motor takes a *lagging* current and is said to be *under-excited;* for values of the excitation greater than the normal value the motor takes a *leading* current and is said to be *over-excited.*

By aid of the V-curves, the power-factor for any other value of line current and given input may be obtained. For example, assume that it is desired to obtain the power-factor for some value of leading current /2, Fig. 293. From Fig. 294, the power-factor cos 02 = /i//2. Therefore, the power-factor for any current I, may be found by dividing the current / into the minimum or normal value of the line or armature current /i for the given input P. The power represented by curve *AB* is obviously for a three-phase motor having a *line* voltage V. *CD*, Fig. 293, is a V-curve taken for a value of power P2, which is obviously greater than Pi. *EF* is a third curve taken for a still greater value of power, P3. A curve drawn through the lowest points of the V-curves is a unity power-factor curve. Curves *XX* and *XY,* drawn through the V-curves at the proper points are 0. 8 power-factor curves, *XX* being for *lagging* current and *XY* for *leading* current. Curves for other power-factors may also be found in a similar manner. These curves are called *compounding curves.*

It should be noted that the normal field current varies with the value of power input to the motor.

129. Amortisseur or Damper Windings.

— Figure 295 shows the rotating field structure of a synchronous motor, around which a squirrel-cage winding is built. The conductors of the squirrel cage are embedded in the pole faces of the rotor. This winding serves two purposes.

It assists the motor in starting and it damps out any tendency of the rotor to oscillate or "hunt." Such windings are called *amortisseur* or *damper* windings or simply *dampers*. If the motor is connected to a system which receives its power from a reciprocating engine unit, there may be pulsations in the supply frequency caused by the variable driving torque of the engine. The synchronous motor is very sensitive to phase changes, as has already been shown, and small changes in the phase of the supply voltage may produce considerable changes in the energy current which the motor takes from the line. This produces pulsations in the motor torque. If these pulsations have a frequency nearly equal to the natural frequency of oscillation of the rotor, they may cause it to oscillate periodically about its normal position. That is, the rotor alternately accelerates and retards, although the average speed does not change. This is called *hunting* (see Par. 75, page 171). These oscillations may become so great as to cause the motor to fall out of synchronism.

Hunting may also be caused by system disturbances, such as switching, short-circuits, etc., and also by sudden changes of load on the motor shaft. Hunting due to such causes usually dies out at a rapid rate, but the first oscillations may be great enough to cause the motor to fall out of synchronism.

The action of the damper winding involves the principle of both the induction motor and the induction generator. So long as the rotor is rotating at synchronous speed, the rotating field of the armature or stator does not cut the dampers and they have

Fig. 295.—Rotor of a 600-volt, 60-cycle, 48-pole synchronous motor showing amortisseur winding.

no effect. That is, the armature mmf. rotates synchronously with the field and there is no relative motion between the field flux and the dampers. Assume that the rotor slows down momentarily. For an instant the rotating field due to the armature mmf. is rotating faster than the field structure. This is equivalent to the rotor slipping temporarily, and currents are induced in the dampers. This is induction motor action and the currents in the dampers are in such a direction that they tend to pull the rotor back again towards synchronism.

Again, if the field poles for some reason swing ahead of their normal position, the dampers cut the rotating field in the opposite direction or the slip becomes negative, temporarily. Induction generator action follows, putting a load on the rotor and tending to slow it down. Therefore, these dampers always tend to pull the motor back into synchronism and thus prevent hunting. Such windings are often used on alternators, particularly of the engine-driven type, to prevent hunting.

130. Starting the Synchronous Motor.— As has been pointed out, the *synchronous* motor is not self-starting. It must first be brought nearly or actually to synchronous speed before it can operate. There are several methods of accomplishing this.

The direct-current exciter for the motor is frequently connected directly to the motor shaft. If a direct-current source of power is available, the exciter may be operated as a motor and thus bring the synchronous motor up to speed. The field of the synchronous motor is then excited and the motor synchronized, just as with an alternator.

If an exciter or sufficient direct-current power is not available, a small induction motor, geared or direct-connected to the synchronous motor shaft, may be used for bringing it up to speed. If the induction motor is direct-connected, its synchronous speed must have a higher value than that of the synchronous motor, in order to compensate for the slip of the starting motor. Such starting motors are often disconnected mechanically after the synchronous motor has been connected to the line. The disadvantage of using an induction motor

is the additional motor, the gears where used, etc. This method of starting is practically not used at the present time.

The synchronous motor is often used to drive a direct-current generator. If sufficient direct-current power is available, the generator may be used as a motor to bring the synchronous motor up to speed. After the motor is synchronized, the field of the direct-current machine is strengthened and it then acts as a generator, taking mechanical power from the synchronous motor.

The synchronous motor may start as an induction motor. First, the field circuit is opened. A polyphase alternating voltage is then impressed on its stator and a rotating field is therefore set up about the rotor.,As a rule, it is desirable to use a compensator so that reduced voltage is applied to the stator windings. The rotating field sets up currents in the pole faces of the rotor and in the amortisseur winding as well, if such exists. This is obviously induction motor action. As the paths of the pole-face currents and of the currents in the dampers have considerable inductance (see Par. 103, page 237), only a comparatively weak starting torque can be obtained. On starting, the rotor currents may be large, and the rotor frequency is that of the stator. The rotor reactance, which is proportional to the rotor inductance and to the frequency, is large. This causes the rotor currents to lag the induced emfs. by a considerable angle and hence; the rotor currents make considerable space-angle with the flux (see Fig. 228, page 240). Therefore the motor develops little torque, even with considerable line current. The motor under these conditions is very similar to the squirrel-cage induction motor, which has a very small starting torque.

However, the starting torque, though small, is usually sufficient to start the machine, which then accelerates until it is at or near synchronism. Before the compensator is thrown into the running position, the field switch is usually closed, so as to minimize disturbances to the system. If the rotor is slipping slightly, it will usually pull into synchronism when the field switch is closed, the field poles locking in with the poles produced by the armature mmf., Fig. 291, page 314.

The motor may pull into synchronism before the field circuit is closed. The flux (see Fig. 291) sweeping by the salient poles shows a less and less tendency to leave them as the rotor approaches synchronism, owing to hysteresis. That is, the flux tends to persist in the poles after the magnetizing force is decreased (see Vol. I, page 181, Fig. 157). This action may be strong enough to pull the rotor into synchronism before the field circuit is closed.

When the field circuit is closed, it may excite the motor poles so that their polarity is opposite to that produced by the revolving field, *i.e.,* by the armature reaction (Fig. 291). The rotor is then thrown back one pole, or in other words, it slips a pole. This may cause considerable disturbance to the system and for this reason the field is usually closed when the compensator is in the starting position. This difficulty may be avoided by applying a weak direct-current field to the motor as it approaches synchronism. This causes the armature reaction to act in conjunction with the direct-current field windings, and the poles then come into synchronism with the same polarity as will be produced by the direct-current excitation. After the motor has pulled into synchronism, it is necessary merely to strengthen the direct-current field to the desired value. The starting compensator may then be thrown quickly into the running position.

When voltage is first applied to the synchronous motor, there may be a very high voltage induced in the field winding. The stator acts as the primary of a transformer, the primary having a comparatively few turns. The flux produced by the stator or primary cuts the field winding at synchronous speed, and as the field has a very large number of turns, a very high electromotive force is induced in the field. This electromotive force may be sufficiently high to puncture the field winding. Therefore, the field winding should be insulated for voltages considerably in excess of that which normal operation requires. The

field is sometimes short-circuited, or is shunted by a resistance when starting, in order to decrease this high voltage. The induced emf. in the field decreases as the rotor comes up to speed, until at synchronism it becomes zero.

131. The Synchronous Condenser as a Corrector of Powerfactor.—The fact that the power-factor of the synchronous motor may be varied at will makes it useful in many installations, particularly in those which operate at low power-factor. It will be recalled that a low power-factor means larger generators, more transmission copper, poorer regulation, and reduced efficiency. Factories and mills using induction-motor drive often have an over-all power-factor as low as 0.5, which is very undesirable. If it is possible to use a synchronous-motor drive in any part of the installation, the motor may be operated over-excited and therefore will take a leading current. This leading current neutralizes some of the lagging current of the system and so improves the system power-factor.

This is illustrated in Fig. 296 for single phase or for one of the phases of a polyphase system. Let V be the voltage of the system and let the total current be I, lagging the voltage V by an angle θt. It is desired to obtain the size of synchronous motor necessary to raise the system power-factor to unity. The synchronous motor is to run without load.

Resolve the current I into two components, an energy component Ii = Icos θi and a quadrature component Ih = I sin θ. The energy current of the synchronous motor is small compared with its quadrature current, when the motor is operating without load, and is added at right angles to the quadrature current. Therefore, in determining the total current taken by the synchronous motor, this energy current may be neglected. For unity power-factor, the motor current will then be substantially equal to the quadrature current I2, but leading. Therefore, the

Fig. 296.—Raising power-factor to unity by means of synchronous condenser.

rating of the synchronous motor is VI, =

VI volt-amperes per phase.

If it be desired to raise the power-factor to some value less than unity, a smaller synchronous motor can be used. In practice, it usually does not pay to raise the power-factor above 0.9 or 0.95, as little is gained by any increase above these values. Moreover, these last few per cent. of improvement in the powerfactor require a much greater proportionate increase in motor capacity.

In Fig. 297 the load power-factor is cos θ. The load on the synchronous motor is assumed to be zero and its losses are neglected. The load current I is resolved into two components I and I2 as before. It is desired to determine the size of synchronous motor necessary to raise the power-factor to cos 60.

The resultant current $I\theta$ is laid off $\theta\theta$ degrees behind V, but terminating on line I-Ii, since the power and hence the energy current Ii is fixed. The synchronous-motor current I, has the value, Fig. 297, I, = I sin $\theta1$ — Io sin $\theta\theta$ = Iz — Io sin θo (73)

It will be noted that the resultant current $I\theta$ is the vector sum of the load current I and the motor current I,.

When a synchronous motor is operated without load for the purpose of merely correcting power-factor, it is called a *synchronous condenser.* Such a synchronous condenser should not be I Fio. 297.—Raising power-factor to cos θo by means of synchronous condenser.

employed unless its investment charges and cost of operation are considerably less than the increased charges occasioned by the low power-factor. Other considerations, such as voltage control, however, are important. When a user of electric power buys on either a kilowatt-hour or a kilowatt basis, a low power-factor is not detrimental to him, except possibly to increase slightly the cost of his mains. This low power-factor is, however, detrimental to the power company, which must install larger generators, conductors, transformers, etc. For this reason many power contracts now penalize low power-factor. 132. The Synchronous Motor as a Corrector of Power-factor. The synchronous motor may correct the power-factor of a system and at the same time deliver mechanical power.

Assume in Fig. 298 that a certain system takes / amperes at a voltage V and that the current / lags V by Oi degrees. It is desired to raise the power-factor of the system to unity by means of a synchronous motor, while at the same time the motor is to supply mechanical power requiring $V I$ watts from the line.

The synchronous motor must first take a quadrature leading current I' in order to counteract the lagging quadrature current / 2 of the load.

$$/2' = /2 = / \sin 6t$$

In addition, the synchronous motor must take an energy current I to supply its losses and also the power required by its load. The total synchronous motor current $!2 + (/2')2$ (74) and the power-factor of the synchronous motor *Example.*—A certain machine shop takes 200 kw., at 0.6 power-factor, from Fio. 298. —Raising power-factor a 600-volt, three-phase, 60-cycle system, to unity by means of a loaded It is desired to raise the power-factor of synchronous motor, the entire system to 0.9 by means of a synchronous motor, which at the same time is to drive a direct-current shunt generator requiring that the synchronous motor take 80 kw. from the line. What should be the rating of the synchronous motor in volts and amperes.

The vector diagram is shown in Fig. 299. Assume that the system is V-connected. The problem will be worked for one phase only.

The energy current of the load, $I_i = I \cos 6 = I \times 0.6 = 192.6$ amp.

The quadrature current of the load, $I2 = I \sin 6 = I \times 0.8 = 256.8$ amp.

At 0.9 power-factor, the resultant power-factor angle $00 = 25.8$.

The energy current of the synchronous motor

i 80000 ii. = — -.---= 77.0 amps.

Total energy current = $I_i + I2$, = 192. 6 + 77.0 = 269.6 amp.

The quadrature current of the system, $I/ = 269.6 \tan 25.8 = 269.6 \times 0.4834 = 130.3$ amp. The quadrature current of the synchronous motor

$I2, = I2-I,' = 256.8-130.3 = 126.5$ amp.

Fig. 299.—Vector diagram for synchronous motor which raises power-factor to cos Bo and at same time supplies power.

The total synchronous-motor current $I« = V (Iu)2 + (W2 = V(77.0)2 + (126.5)2 = V21,930 = 148$ amp. The synchronous motor will then be rated at 600 volts, 148 amperes, or will have a rating of 154 kilovolt-amperes. *Ans.*

The resultant current I0 the vector sum of I and I, is shown in Fig. 299.

133. The Synchronous Motor as a Regulator of Voltage.—

Figure 300 shows one phase of a power system, which may be either a single-phase or a polyphase system. A constant voltage Vg is supplied to the system by a generator

Fio. 300.—Synchronous motor taking power or by a power plant. At the

Synchronous

Motor

through resistance and reactance in series.

receiving end of the line is a synchronous motor whose terminal voltage is Vm. Between Vg and Vm are both resistance and reactance in series. These may be the usual resistance and reactance of a transmission line, or they may exist in an impedance coil, having a resistance R and reactance X, inserted between the supply mains and the motor terminals.

Assume first that the synchronous motor is under-excited and therefore taking a lagging current. Along the vector /, Fig. 301 (a), the IR drop is laid off in phase with I; at right angles to I and leading, the IX drop is laid off. The vector sum of the IR and IX drops, is equal to the IZ drop in the line. Obviously, the motor voltage must be equal to the generator voltage *minus* the IZ drop, vectorially considered. Therefore, IZ is reversed and added to Vg, giving Vm, the motor voltage. It will be observed that numerically Vm is considerably less than Vg.

If the motor now be over-excited, I will lead the voltage Vm. By subtracting IZ from Vg, Fig. 301 (6), the motor voltage Vm becomes numerically *greater*

than Vg. (a) Lagging current; motor voltage (6) Leading current; motor voltage less than generator voltage. greater than generator voltage.

Fio. 301.—Effect of line impedance on synchronous-motor voltage.

This gives a method of controlling the voltage at the end of a transmission line. If the voltage at the receiving end of the line tends to change because of a change in the generator voltage or in the line drop, it may often be held substantially constant by varying the excitation of a synchronous motor placed at the receiving end of the line. In practice, synchronous motors are often installed for purposes of regulation only. At the Los Angeles end of the 240-mile Big Creek Line, two 15,000 kv-a. synchronous condensers are installed, their sole function being to hold the voltage in Los Angeles at the proper value. If the load were removed and no such regulating devices existed, this voltage would rise to values considerably in excess of that at the generating station 240 miles away, due to the line charging current flowing through the line reactance.

Even without the adjustment secured by altering the field current, a synchronous motor tends to maintain constant voltage at the end of a transmission line having reactance. If the voltage at the motor terminals drops, its back emf. tends to exceed the terminal voltage and the motor must then take a leading current in order to operate. This leading current, flowing through the line reactance, tends to maintain the motor voltage, as a leading current flowing through reactance tends to produce a rise of voltage from generator to load. On the other hand, a rise of voltage at the motor terminals tends to cause the motor to operate under-excited. This increases the drop from generator to load and tends to cause the voltage at the load to decrease.

Fio. 302.—Synchronous motor for controlling voltage at end of transmission line.

The effect of the synchronous motor on voltage control may be shown by a laboratory experiment, the connections for which are given in Fig. 302. A synchronous motor, running either light or partly loaded, is supplied from constant potential mains through three series reactances, one in each main. A lamp load or an induction-motor load is connected in parallel with the synchronous motor. Vary the lamp load or the induction-motor load and maintain the synchronous motor terminal voltage Vm constant by varying its field current. It will be found that the field current must be materially increased as the load is increased. Figure 303 shows the general trend of the curve giving the relation between the field current and the load.

It is also instructive to keep the lamp load or the induction motor load constant and at the same time to obtain a V-curve and find the relation of Vm to the synchronous motor field current. The results of such a test are shown in Fig. 304. Vm is considerably lower than Vg for low values of field current, but after unity power-factor is reached, Vm exceeds Vg.

Field Current

Fig. 304.—Effect of field current on motor voltage at constant load.

IL.W.Load

Fio. 303.—Relation of field current to load at motor, motor voltage constant.

134. Industrial Applications of the Synchronous Motor.—

Single-phase synchronous motors are rarely used in practice. Like the single-phase induction motor, the direction in which they rotate is determined by the direction in which they are started. Unlike the polyphase synchronous motor, they will not start by induction.motor action but must be brought up to speed by other means. Polyphase synchronous motors are commonly used.

The inherent disadvantages of the synchronous motor are that it requires a direct-current supply for its excitation, its starting torque is very small, and the motor is very sensitive to system disturbances and may fall out of step when these occur. On the other hand, the ease with which its power-factor can be controlled is a distinct advantage, often outweighing all the disadvantages. The fact that its speed is constant is of little moment, since induction motors, especially in the larger sizes, have only 1 or 2 per cent. speed regulation.

The synchronous motor is used only in the larger sizes where the cost of attendance per kilovolt-ampere is low. Moreover, it should not be used where there are sudden applications of the load, as it may drop out of step under such conditions. An important field of use is in connection with motor-generator sets where a large unit is required and where any sudden changes of load are partly absorbed by the inertia of the direct-current armature. A few such motors, situated at various points in a large system, may make it possible to operate the generating station and many of the transmission lines and sub-stations at high power-factor, in spite of low power-factor in the consumers, loads.

Even with these advantages of the synchronous motor, electrical engineers often prefer to use induction motors for motorgenerator sets, because of their simplicity and greater reliability.

Fig. 305.—Synchronous motors driving direct-current generators.

Figure 305 shows synchronous motors driving direct-current generators in a sub-station.

Electric Propulsion.—Synchronous motors are also coming into use for the electric propulsion of cargo and merchant ships. Such ships, when under way, operate at a constant speed and the constant-speed characteristic of the synchronous motor is not a disadvantage, therefore. As such motors can be operated at unity power-factor, the weight of motor, generator and connecting leads is smaller than when induction motors are used. This matter of weight is important in marine work. The air-gap of synchronous motors is considerably greater than that of induction motors and the mechanical difficulties which a short air-gap involves, such as very accurate alignment, etc., are not present when synchronous motors are used. Owing to the salient-pole feature of the synchronous motor, stator coils may be replaced without removing the rotor. Also, the field windings on the salient poles are less subject to injury than the embedded conductors in the rotor of an induction motor. The dampers of synchronous motors used for electric propulsion are designed to give moderately high torque on starting, reversing, etc. Both the speed and the voltage of the generator may be varied, so that the motors have different starting characteristics from those existing at constant frequency and constant voltage. *Frequency Changers.*—It is sometimes necessary to supply electric power from one electric system to another electric system of different frequency. A common method is to use synchronousmotor-alternator sets. The synchronous motor and the alternator must have a different number of poles, the number of poles in each being proportional to the frequency of the system to which the particular machine is connected. For example, if the frequency is being changed from 60 to 25 cycles, the number of poles of the synchronous motor must be to the number of poles of the alternator in the ratio of 60 to 25 or 12 to 5. The highest speed at which this ratio of frequencies can be obtained will require a set having a 24-pole synchronous motor and a 10-pole alternator. The set will operate at only 300 r.p.m. Except in very large units, electrical machines operating at this very low speed would be costly. A 10-pole, 4-pole combination gives either a frequency ratio of 60 to 24 cycles or a frequency ratio of 62.5 to 25 cycles and operates at 750 r.p.m. Because of its greater speed, this combination is often used, even if it does not give an exact 60 to 25 cycle ratio.

It is often difficult to synchronize such a set, as it must be synchronized with both systems. If the alternator voltages are out of phase with their respective line voltages, the synchronous motor must be made to slip a pole at a time until the alternator voltages are in phase with their respective line voltages. The load is shifted either by advancing the phase of the system supplying the power, as by opening the turbine governors, or by retarding in some manner the phase of the voltage in the system receiving the power.

135. Synchronous Motors of Very

Small Size.—Because of their absolutely constant-speed characteristics, synchronous motors are very useful for driving such devices as must be held in absolute synchronism with tho suppty frequency. Such uses involve the measurement of slip in the induction motor (see page 273), the driving of oscillograph mirrors, stroboscopic devices, mechanical rectifiers, etc.

As the power required of such motors is extremely small and the matter of low power-factor is of no moment, they are often made to operate without direct-current excitation. In Fig. 306, (a) and (6), are shown motors of this type. In (a), the four-pole armature consists of a cruciform-shaped piece of iron with the

(a) 4 Poles *b* 16 Poles

Fio. 306.—Miniature synchronous motors.

spaces filled with wood to make the armature cylindrical. The field is made up of U-shaped laminations and is excited from the alternating-current supply

When the armature is brought up to speed, two diametrically opposite armature poles are attracted to the field poles as the flux is increasing. Because of the inertia of the armature, it continues to rotate when the flux is passing through zero. The next pair of poles are then attracted by the flux as it increases in the opposite direction. Such a motor will therefore run at constant speed, provided the frequency is constant.

A 16-pole motor operating on the same principle is shown in Fig. 306 (6).

These motors really operate on the principle of maximum permeance, although it will be recognized that they are salient-pole synchronous motors of the rotating-field type, having no directcurrent field excitation. Their excitation is produced by armature reaction.

CHAPTER XI

RECTIFIERS: THE SYNCHRONOUS CONVERTER

136. Methods of Obtaining Direct Current from Alternating Current.—At the present time, over 90 per cent. of electrical energy is generated and transmitted as alternating current. A very large percentage of this energy is utilized as alternating current; for example, to operate alternating-current motors, electric furnaces, and many other types of electrical appliances, for illumination purposes, etc. However, there are many cases where the electrical energy must be in the form of direct current, even although the available supply of energy is alternating current. For example, direct current must be used for charging storage batteries, for electrolytic work, for telephone exchanges, etc. The direct-current series motor is practically the only type of motor that can be used for street-railway work and it is also commonly used in railway electrification. In the congested city districts, where the consumers loads are large and close together, direct-current power is preferable to alternating-current power, as capacitive effects in the underground cables are not present when direct current is used and inductive effects in the system are also absent with direct current. Furthermore, in such loads, the importance of continuity of service requires that a large storage-battery reserve be available. This, again, is an additional reason for supplying direct-current service in such districts.

As the power supply in the above cases is almost always alternating current, this alternating current must in some manner be changed to direct current. There are several methods of accomplishing this, the most common being the following: 1. Mechanical rectifier—commutating type.

2. Mechanical rectifier—vibrating type. 3. Mercury-arc rectifier. 4. The Tungar rectifier. 5. Electrolytic rectifier. 6. Induction or synchronous-motor-generator seta. 7. Rotary or synchronous converter. 137. Types of Rectifiers and Converters.—1. *The Rectifying Commutator.*—The rectifying commutator is a commutator driven by a synchronous motor. The segments are so connected that when the alternating current reverses, the connections to the direct-current circuit are simultaneously reversed, as shown in Fig. 307. A uni-directional current is thus obtained. As the brushes cannot have zero width, it is difficult to commutate at the point of zero current and the current and voltage are rarely

zero at the same time. Hence, such devices spark more or less, and so are limited to small currents and voltages.

Fig. 307.—Commutating-type rectifier.

2. *The Vibrating Rectifier.*—The vibrating rectifier, Fig. 308, is based on the same principle as the rectifying commutator, except that the circuit connections are reversed by contacts which are opened and closed, synchronously, by alternating-current magnets and a polarized armature. This type of rectifier ordinarily is designed for use on 110-volt, 60-cycle circuits. The circuit voltage is reduced by means of a step-down transformer, the secondary of which has a middle tap. This secondary excites two series-connected, alternating-current magnets, which are so connected that they both have the same polarity on corresponding ends at every instant. The vibrator is a soft-iron bar magnet, pivoted below these alternating-current magnets, each of its two ends being directly beneath one of the alternatingcurrent magnets. This bar magnet is excited by direct current taken from battery terminals and has therefore a fixed polarity. Assume that at some particular instant the right-hand end of the transformer secondary is positive. By following through the circuits in Fig. 308 it is seen that both the lower ends of the alternating-current magnets are north poles. Also, the left-hand end of the bar magnet is a north pole, and its other end is a south pole. This left-hand end is therefore repelled downwards and the right-hand end is attracted upwards. This closes the left-hand contact, which allows current to flow into the left-hand battery terminal, assumed to be positive. During the next halfcycle the left-hand end of the transformer is positive and the right-hand end of the magnet is repelled. This closes the right-hand contacts and current still flows into the positive or

Fig. sos.—Vibrating rectifier.

left-hand side of the battery. Hence, the battery receives a unidirectional current, and may be charged from alternating-current supply. The contact should open when the current is zero. This adjustment is made by means of the resistance

R, which shifts the phase of the current in the alternating-current magnets. Condensers are connected across the contacts in order to minimize sparking. It makes no difference how the battery is connected, as the direction of excitation of the vibrating magnet causes the current always to flow into the positive battery terminal.

This type of rectifier is designed for 8 amp. at from 8 to 10 volts. Owing to difficulties, due primarily to wave-form, it has not been entirely satisfactory in practice.

3. *The Mercury-arc Rectifier.*—The mercury-arc rectifier has already been mentioned in connection with the constant-current transformer (see page 219). The principle is the valve action of mercury vapor. In order to obtain the best operation, the tube containing this vapor must be exhausted to a very high vacuum. Figure 309 shows a mercury-arc rectifier tube having four terminals. The lower terminal is the cathode, to which the current goesf rom the tube. The two terminals, Ai, A2, are the anodes from which the current enters the tube. A3 is a starting anode,

Fio. 309.—Mercury-arc rectifier for low by means of which the mervoltages.

.

cury arc is established. Current then *enters* the tube from either anode, A i, A 2, depending upon which side of the transformer secondary, *ab,* is positive. When the current attempts to reverse its direction, however, the mercury vapor acts as a valve and prevents any current entering the tube at the cathode. If only one anode were used, the negative half of the alternating-current wave would be eliminated in each cycle and the resultant wave would appear as shown in
Auto-transformer and
Inductive Reactance
solid line, Fig. 310(a). This condition of operation could not.be maintained with the mercury arc, because the arc is extinguished as soon as the current becomes zero.

To obtain a continuous flow of current through the tube, two anodes, Ai, A 2, are necessary, one anode being connected to each end of the transformer

secondary. When one end of the transformer becomes negative, the other becomes positive, so that either one anode or the other is always positive. Therefore, current is always entering the tube from either one anode or the other. Were there no inductance in circuit, the rectified wave under these conditions would appear as shown in Fig. 310(6). The portions of the wave marked A i are due to anode A i, and those marked A 2 to anode A 2. Each of these portions reaches the zero value twice for each cycle of current supply. This would cause the arc to be extinguished. By introducing inductance in the circuit, however, the current is held over the zero point and the resulting wave is similar to that shown in Fig. 310(c), being more or less pulsating in character.

The direct current leaves the cathode, enters the positive terminal of the battery to be charged (or other translating device) and flows to the neutral of the auto-transformer.

The operation of the auto-transformer is as follows: Assume that at some particular instant, terminal *b* of the transformer secondary, Fig. 309, is positive and terminal *a* negative. Current obviously attempts to pass from 6 to a through some external circuit. One path is by way of the anode *Az,* the tube, the cathode and through the battery to the neutral *N* of the autotransformer. As some of this current must return to terminal *a* of the transformer secondary, it attempts to pass through the winding *Nd* of the auto-transformer A part of the current does pass through this winding and in so doing creates a flux in the core of the auto-transformer which induces an emf. in the winding *Nc.* The direction of this emf. is such as to cause the remainder of the current to flow from *N* to *c.* This current flows through the local circuit *NcA2.* This, it will be remembered, is the principle of the auto-transformer (see page 206, Par. 90).

The anode *A 3* is for starting purposes only. When the tube is tilted, a conducting stream of mercury is established between *A* and the cathode. The resulting current flow vaporizes some of this mercury and so establishes the arc. A

ballast resistance *R* is necessary in order to limit the current at starting, since there is then a metallic path of low resistance between *A3* and the cathode.

For low-voltage circuits, this type of rectifier has not as yet been developed in large capacities. Figure 311 shows the front and rear views of a complete rectifier panel such as would be used for charging vehicle batteries.

4. *The Tungar.1*—The tungar is based on the following principle: An incandescent filament emits minute negative charges called *electrons.* When the discharge of these electrons occurs in an electrostatic field, the electrons attain considerable velocity. If a gas is present, these electrons collide with the gas atoms and ionize them. That is, when an electron collides with an atom of

'For complete description see "The Tungar Rectifier," by R. E. Russell, *General Electric Review,* 1917, page 209.

gas, that atom is broken up into an electron and a positive ion. The region in which this action occurs then becomes *ionized.* Ionized gas is a conductor of electricity.

Fig. 311.—Single-phase mercury-arc rectifier with panel.

Figure 312 (a) shows a glass bulb containing an inert gas, usually argon, at reduced pressure, and also an ordinary coiled tungsten filament. Near the filament is a graphite anode. A transformer *ab* steps down the supply voltage and the filament is connected across its secondary. The filament then becomes incandescent and emits negative charges or electrons.

One terminal of the transformer secondary *c* and one end of the filament are connected to the transformer primary at *b.* The filament is then at practically the same potential as that of the power-supply line *b,b.* The voltage of the battery being charged is somewhat less than the voltage between line *a'a* and line *b,b.* Therefore, the potential of the graphite anode is different from the potential of point *c,* usually by approximately 5 or 6 volts. Consequently, during one half-cycle the potential of the filament is negative with respect to that of the anode and during the next half-cycle its potential is positive with respect to that of the anode.

When the filament is negative, the negative charges or electrons are repelled by it, because like charges repel each other. These

Rectified

Current

Bulb

Rectified Current

(a) (6)

Fio. 312.—Tungar rectifier.

electrons attain a considerable velocity and break up the gas particles into ions. The region between the filament and the anode becomes conducting and as a result current flows from *a* into the positive terminal of the battery, through the battery to the anode, to c and then to *b*. When the filament is positive, the electrons or negative charges which it tends to emit due to its incandescence are attracted toward the filament, since positive and negative charges attract each other. Consequently, the electrons which produce the ionizing action are withdrawn from the region between the filament and the anode. As a result the gas is no longer ionized and it ceases to be a conductor. No current can flow, therefore, during this half-cycle. The current can flow only in one direction, therefore, from the graphite to the filament, and the device acts as a rectifier. A very small and almost negligible part of the current is due to the electrons themselves, which act as carriers of negative electricity from filament to anode.

Figure 312 (6) shows the connections for one commercial type of low-voltage tungar, the switches and cut-outs being omitted. Both the current to be rectified and the current for heating the filament are supplied by the transformer secondary, the filament being connected between a 2.5-volt tap and one end of the secondary. Current regulation may be obtained by adjusting the resistance and the reactance. Where electrical connection between load and primary mains is permissible, an auto-transformer with taps may be used.

The devices shown in Fig. 312 elimi-nate the negative half of each wave, but this is not a serious disadvantage when ordinary batteries are being charged. However, a two-bulb rectifier supplying a continuous, pulsating current is also manufactured. The efficiency of the tungar rectifier is from 35 per cent. in the smaller sizes to 75 per cent. in the larger sizes. The capacities at present are not much in excess of 750 watts.

5. *Electrolytic Rectifiers.*—Electrolytic rectifiers are based on the following principle: If a lead plate and an aluminum plate be immersed in a sodiumor ammonium-phosphate solution, current can pass from the solution to the aluminum. As. soon as the current attempts to reverse and pass from the aluminum to the solution, a thin insulating film of aluminum oxide is instantly formed over the aluminum plate, and acts as an insulator up to about 150 volts. This prevents the current flowing from aluminum to solution, and such a device may be used, therefore, as a rectifier. Figure 313 shows such a simple rectifier, giving a continuous pulsating current like that shown in Fig. 310(6).

Such rectifiers are of low efficiency, 60 per cent. and lower, and are of small capacity. They are used primarily for charging low-voltage batteries from alternating-current supply. Their advantage lies in their cheapness and simplicity.

Lead Plate Fio. 313.—Electrolytic rectifier.

6. *Induction- or Synchronous-motor-generator Sets.*—None of the foregoing devices is capable of converting alternating to direct current on the large scale required in modern power systems. To convert large amounts of power, induction-motoror synchronous-motor-generator sets may be employed. The capacity of such units is limited only by the size in which it is possible to construct the direct-current generator. The disadvantage of a motor-generator set is that it requires two machines, with corresponding cost and floor space, and the over-all efficiency is not extremely high, being the product of the efficiencies of the individual units of the set.

7. *The Rotary or Synchronous Convert-er* is a single machine which converts alternating to direct current or *vice versa,* and may be built to convert large amounts of power efficiently and economically. Because it has only one armature and one field, the synchronous converter usually costs less than an equivalent motor-generator set. Because the armature current is small, being the *difference* between the alternating and the direct currents, this type of machine has a high efficiency when operating under favorable conditions. 138. Principle of the Synchronous Converter. —It has already been demonstrated that alternating current is generated in the armature coils of the ordinary direct-current generator. If taps be brought out properly from the armature winding to sliprings, alternating current may be taken from this same winding and the machine becomes an alternator. Such an alternator can obviously operate as a synchronous motor.

The synchronous converter is constructed like the ordinary direct-current generator, although the relative dimensions may be different. It has fixed poles, a rotating armature, a commutator, a shunt field, and usually a series field. In addition to the commutator, however, leads are taken from the armature to slip-rings, in the manner shown in Fig. 314 (also see Figs. 315 and 316). Figure 314 represents a two-pole, single-phase converter.

In the synchronous converter, as commonly used, alternating current is supplied to the slip-rings and direct current is taken from the commutator and brushes. If, however, the directcurrent brushes be open-circuited or removed, the machine becomes, under these conditions, a synchronous motor of the rotating-armature type. On the other hand, if direct current be supplied to the brushes and commutator, and the slipring brushes be disconnected, the machine becomes a shunt or compound motor.

If the machine be driven mechanically, and current be taken from the slip-rings only, it becomes an alternator. On the other hand, if current be taken from the commutator only, it becomes a direct-current generator. Both alternating and direct current may be taken from it simultaneously, and it then becomes a *double-current* generator.

Fig. 314.—Two-pole, single-phase, synchronous converter.

In the synchronous converter as ordinarily used, alternating current is supplied to the slip-rings so that the machine operates as a synchronous motor, so far as the alternating-current side is concerned. At the same time direct current is taken from the commutator and brushes, and therefore this side of the machine has characteristics very similar to those of a shunt or compound generator. When operated in this manner, the machine is said to be a *direct* synchronous converter.

The machine, however, may take power from the direct-current supply, operating as a direct-current motor, and deliver alternating current from the slip-rings. When operated in this manner, the machine is said to be an *inverted* synchronous converter. This is not the usual method of operation.

139. Polyphase Converters.—The output of a converter increases materially with the number of phases. For example, the rating of a six-phase converter is more than twice its rating when operated single-phase (see page 355).

The connections of polyphase converters are comparatively simple. For example, the four-phase converter shown in Fig. 315 requires four slip-rings. The points at which the slip-rings are connected to the winding are 90 space-degrees apart in the

Fio. 315.—Two-pole, 4-ring, 4-phase synchronous converter.

two-pole type. If the machine has four poles, two taps from each ring to the winding are necessary. This is illustrated in Fig. 316, in which a three-phase, four-pole converter is shown. Two taps run from each ring to the winding; in this case the taps are diametrically opposite. For example, if the tap from one ring connects to a portion of the winding which at some particular instant is under the center of a north pole, then there must be similar taps running from this same ring to every point of the

winding which lies at that instant under the center of a north pole. (See points o,a, Fig. 316).

A six-phase, six-pole synchronous converter will have six sliprings and three taps from each slip-ring, making a total of 18 taps to the winding.

A simple rule for obtaining the number of taps to the winding is to remember that if the machine has *n* phases, there must be *n* slip-ring taps for every 360 electrical space-degrees, or for every *pair* of poles. (This does not hold for single-phase.) For example, in Fig. 316 there must be three taps for each pair of poles, or six taps in all. Figure 327, page 357, shows how the taps are

Fio. 316.—Three-phase, 4-pole, synchronous converter.

brought out from the armature to the slip-rings in a 14-pole, six-phase converter.

It is to be noted that the slip-ring taps must be brought out at equidistant points along the winding, in order that the alternating voltages may be balanced. Hence, the direct-current windings that can be used for a synchronous converter are more or less restricted, for the number of coils must be divisible by the number of slip-ring taps.

140. Single-phase Voltage Ratios in a Synchronous Converter. In a synchronous converter, both the alternating and the directcurrent electromotive forces are induced by the same system of conductors, cutting the same field. Therefore, there must be a fixed ratio between the direct-current and the alternatingcurrent *induced* electromotive forces.

In a single-phase converter, there are the same number of active conductors between the direct-current brushes as between the alternating-current slip-rings, as will be seen in Fig. 314. The same number of conductors, cutting the same field, gives both the direct-current emf. and the single-phase emf.

It will be remembered that the electromotive force between the brushes of a direct-current generator is the sum of the emf. waves generated in each of the individual conductors connected in series between the brushes. The resulting

electromotive force is the peak value of the resulting wave, as is shown in Fig.

Resultant a.c. e *(a)* (6)

Fio. 317.—Relation between direct and alternating induced emfs. in a synchronous-converter armature.

317(a). (Also see Vol. I, page 219, Pars. 164 and 165). For simplicity, Fig. 317(a) shows only the wave resulting from two coils between slip-rings spaced 90 apart.

In a single-phase machine, there are just as many conductors between the slip-ring taps as between the brushes. Therefore, the resultant alternating emf. wave between slip-ring taps is found by adding together the alternating waves, 90 apart, point by point, as shown in Fig. 317(6). Comparing Figs. 317(a) and 317(6), it will be noted that the direct-current emf. is equal to the *peak* value of the alternating emf.

Therefore, in a single-phase converter, the direct-current induced electromotive force is equal to the /2 times the effective value of the single-phase alternating-current induced electromotive force. This ratio may be modified by wave form as in the split-pole converter (page 361). 141. Polyphase Voltage Ratios in a Synchronous Converter.— It will be remembered that the total single-phase electromotive force generated in an alternator armature is the vector sum of the individual inductor electromotive forces, as shown in Fig. 318. In (a), the several conductors upon the surface of the armature are shown. In (6) are the vector electromotive forces generated in the various conductors, together with their vector sum (also see page 121, Par. 58). The total single u, ,,

Single-Ph. Slip—T
Ring Taps,...

(a) (6)

Fig. 318.-—Relation of induced emfs. to belt span, in a closed armature winding phase voltage is the diameter of a circle drawn to the proper scale, as shown in Fig. 318(6). The three-phase electromotive force is the vector sum of the individual electromotive forces included within a 120 arc, Fig. 318(6). The four-phase electromotive force is the vector sum of the electromotive forces included within a 90 arc, and the six-phase electromotive force is the vector sum included within a 60 arc.

This gives a simple method for obtaining the various electromotive force relations in a converter armature. Draw a circle, Fig. 319, whose diameter is 100 units. Let this represent a singlephase electromotive force of 100 volts, effective. The direct-current electromotive force will then be V2 X 100 = 141. 4 volts, which is shown by extending the diameter. The three-phase electromotive force is the length of a chord subtending an arc of 120, or 86.6 volts. The four-phase electromotive force is the length of a chord subtending 90, or 70.7 volts. The sixphase electromotive force is the length of a chord subtending 60, or 50 volts.

D.C.-U1 Volta

Single-Phase-100 Volta 3-Pha«e-86. 6 Volts 4-PhBse-70.7 Volta 6-Phaae-50 Volto

Fio. 319.—Relations existing among voltages in a synchronousconverter armature.

141.4 100 86.6 70.7 50 Ratio Op A. C. Emfs. To D. C. Emf.

0.707 0.612 0.50 0.354

142. Current Ratios in a Synchronous Converter. — The relations between the direct and alternating currents in a synchronous converter may be determined as follows: *Single-phase:*

If the efficiency is assumed to be 100 per cent. and the powerfactor unity VI = Vi /t

/! V 141.4 r

where V and I are the direct-current voltage and current respec tively and Vi and /i are the single-phase voltage and current respectively.

If the efficiency be T; and the power-factor *P.F.,* VI = $v.Ji$ X *P.F.* X r,

The single-phase current

J.414/ *lrtXP.F. ("*

In practice, the efficiency is from 92 to 96 per cent. and the power-factor is rarely allowed to drop below 0.9. *Three-phase:* At 100 per cent. efficiency and unity power-factor VI = where $V3$ is the three-phase line voltage and Ia the three-phase line current.

If the efficiency be 77 and the power-

factor *P.P.,* the three-phase line current ij X *P.F. Four-phase:*

At 100 per cent. efficiency and unity power-factor VI = $2/2vJ2$ where F2 is the voltage between adjacent lines and /2 is the fourphase line current, Fig. 320.

VI V Vz'

Fig. 320.—Currents and voltages in a 4-phase synchronous-converter armature.

If the efficiency be 77 and the power-factor *P.F.,* the four-phase line current 7 h *Six-phase:*

The six-phase system may be considered as composed of two Y-systems, or two delta systems, each having one-half the capacity of the six-phase system. (See Par. 149, page 364.) Figure 321 shows a six-phase double-Y connection in which the sixphase voltages between adjacent lines and to neutral are Vt-A current It flows in each line. As the six phases are all connected together at the neutral, this system may be split into two equal Y-systems, Fig. 321(6), each having Vt volts to neutral. The output of each Y-system at unity power-factor is *3VtII* watts.

At 100 per cent. efficiency and unity power-factor

VI = 2(3F6/6) = 6F6/6

(a) *(b)*

Fio. 321.—Currents and voltages in a double-Y, 6-phase system.

If the efficiency be i; and the power-factor *P.F.,* 0.471*I*

6 77 X *P.F.*

Summarizing the above, (for unity power-factor and 100 per cent. efficiency): *Example.*—A 500 kw. converter, Fig. 322, has an efficiency of 92 per cent. at full load and operates at a power-factor of 0.94. The direct-current voltage is 550 volts. The alternating-current side is operated six-phase. Find the direct current and all the A.C. line currents and voltages.

Vt = 550 X 0.354 = 195 volts between adjacent lines and to neutral. From equation (83)

Fio. 322.-—Currents and voltages in a 6-phase, 500 kw., synchronous converter and transformers.

143. Conductor Currents in the Armature of a Converter.—It has already

been pointed out that the synchronous converter has a high efficiency because the *net* current in each armature conductor is the *difference* between the alternating current and the direct current which would of themselves exist in that conductor. The reason for this is obvious. The alternating current entering through the slip-rings is a *motor* current, driving the machine as a synchronous motor and is therefore in opposition to the induced electromotive force. The armature current which is delivered by the commutator to the brushes is a *generator* current and is therefore in conjunction with the induced electromotive force. Both the alternating and the direct current utilize the same conductors, rotating in the same field. Under these conditions the two currents must flow in opposite directions. Therefore, the *net* current in each conductor must be the *difference* between the motor current and the generator current.

The wave-form for the resultant current in the various conductors is very irregular and differs for the different armature conductors. The value of the resultant current also differs in the different conductors.

Consider conductor a, Fig. 323, which lies midway between two slip-ring taps. First consider the direct current in this conductor as the conductor moves through successive positions 1, 2, 3, 4. If the load be assumed constant and the width of the brush be neglected, the direct current will be positive, and will not vary as the conductor moves from (1) to (2) to (3). At (3), the brush position, the current reverses abruptly and then remains constant until the conductor reaches position (1). This is shown in Fig.324(a). The conductor a is midway between slip-ring taps, so that it is at the center of the alternating-current phase-belt which is included between these slip-ring taps. The phase of the electro

Fio. 323.—Relative positions of conductors and slip-ring taps.

(6) Resultant Current

Fig. 324.—Current at unity power-factor in a conductor midway between slipring taps.

motive force in a is the same as that of the resultant electromotive force of the entire belt. This is evident from a study of Fig. 318, page 347, although a conductor at the exact center of the winding is not shown in that figure.

Assume that the current is in phase with the induced electromotive force. When a is in position (1), Fig. 324, the alternating current in the entire phase belt is zero; when a reaches position (2), the current is a maximum; etc. This current is plotted in Fig. 324(a), a sine wave being assumed The alternating current is opposed to the direct current, since one is a motor current and the other a generator current for the same induced electromotive force. The resultant current is found by adding the two currents, point by point, the result being shown in Fig. 324(6). This resultant current is irregular in form, and its effective value is small compared to that of either of the component currents.

This resultant current, though periodic, is not a sine wave, and therefore must be made up of a current wave of fundamental frequency and higher harmonics. As the current is assumed to be in phase with the induced electromotive force, the product of this current of fundamental frequency and the induced electromotive force gives the power necessary to supply the rotational losses, which include friction, windage, and core losses.

Next consider conductor b, Fig. 323, at one of the slip-ring taps, but in the same phase belt as a. As this conductor passes through the successive positions (1), (2), (3), and (4), the direct current is the same for each position of 6 as it was for the corresponding position of a. This direct current is plotted in Fig. 325 (a). The alternating current in b must be the same as in a, for the two are in the same phase-belt and so are in series. When conductor 6 is in position (1), a is in position (4), and therefore the current in both a and b is a positive maximum, from Fig. 324. When b reaches (2) the current is zero, etc. This current is plotted in Fig. 325 (a). The resultant current is shown in Fig. 325 (6).

It will be noted that the resultant cur-

rent in b is *distinctly greater* in magnitude than the current in a, Fig. 324 (6). Therefore, the heating in the conductors nearer the slip-ring taps will be greater than it is in the conductors midway between taps. On the other hand, it can be similarly shown that the heating in conductor c, in the same phase-belt as a and b but at the other tap, is different from the heating in either a or b, if the power-factor is other than unity.

The converter rating is determined by the allowable temperature of the hottest part of its armature. Although the conductors midway between slip-ring taps are operating at temperatures lower than the allowable safe values, the converter rating must be adjusted to conform to the safe temperature limits of the conductors whose temperature is highest.

The greater the number of phases, the greater will be the number of slip-ring taps. This will produce a lesser temperature (6) Resultant Current Fio. 325.—Current at unity power-factor in a conductor at slip-ring tap.
range due to difference in position of the various armature conductors, because the resultant of the direct and the alternating current for conductors located near the slip-ring taps, which conductors operate at the highest temperature, is decreased in magnitude. The average heating for all the conductors will be reduced, which will permit an increase in rating for the converter. The rating of a given converter increases rapidly with increase in the number of phases, as shown in Table 144.

Table 144 gives the rating of a converter for different numbers of phases, the output as a direct-current generator being taken as unity.

Table 144.—Effect Of Number Of Phases And Of Poweb-
Factok On The Output Of A Synchronous Converter

The considerable gain in rating obtained by operating a converter six-phase is the reason that six-phase converters aresocom (6) Resultant Current Fig. 326.—Effect of low power-factor on current in conductor at slip-ring tap.
monly used. The advantage of the gain obtained by operating twelve-phase is

usually offset by the added wiring complications. 146. Effect of Power-factor on Converter Rating.—The rating and efficiency of a converter decrease much more rapidly with decrease in power-factor than is the case with other types of alternating-current machinery. This results from the rapid increase in the resultant current in the converter armature with phase displacement between the alternating and the direct-current waves. Assume that in Fig. 325 the alternating current lags the induced electromotive force by 45. This corresponds to a power-factor of 0.71. For the same power and electromotive force, the alternating-current wave must be increased to 1/0.71, or 1.41, times the value shown in Fig. 325. This current wave is shown in Fig. 326(a). It is to be noted that the resultant wave shown in Fig. 326(6) has been increased considerably in magnitude over the value shown in Fig. 325(6). Hence, for the same heating in the two cases, it would be necessary to lower by a considerable amount the output of the converter operating at a power-factor of 0.71. Table 144 shows the large reduction in rating caused by lowering the power-factor from unity to 0.9.

At values of power-factor other than those near unity, the synchronous converter loses most of its advantages over the motor-generator set. Therefore, a converter should be operated at a power-factor which is very nearly unity.
146. Armature Reaction in a Converter. —At unity powerfactor, the resultant current in a converter armature is comparatively small, as shown in Fig. 324(6). Therefore, the armature reaction is correspondingly small and there is practically no distortion of the field. As a result, the machine commutates very much better than when operating as a direct-current generator carrying the same load. When the power-factor decreases, the resultant armature current increases, as shown in Fig. 326(6). As the rotational losses do not change to any great extent with change of power-factor, the power necessary to overcome these losses changes only a small amount with change of power-factor.

Hence the *energy component* of the fundamental of the resultant current changes only a small amount with change of powerfactor, since the power necessary to rotate the armature is equal to this energy component multiplied by the back electromotive force. Therefore, at power-factors less than unity, practically the only current which is added to the energy current existing at unity power-factor, is a quadrature current, lagging or leading the induced electromotive force by 90 time-degrees. Only the energy component or the component of current in phase with the induced electromotive force produces cross *magnetization* (see page 132). When the converter is operating *direct,* any current in quadrature with the induced electromotive force merely strengthens or weakens the field, depending upon whether the current lags or leads. Consequently, there is magnetizing action upon the fields when the current lags and demagnetizing action when the current leads. (See Chap. X, Pars. 126 and 127, pages 309 and 312.) As a result, the added quadrature current merely strengthens or weakens the field but does not distort it. Hence there is little or no sparking in a converter armature due to field distortion.

Fig. 327.—Armature of a 4,000-kw., 625-volt. General Electric synchronous converter.

It will be remembered (see Vol. I, page 280), that in a directcurrent machine, an electromotive force of self-induction exists in the armature coils which are undergoing commutation. It is desirable, therefore, that a counter electromotive force, opposite and equal to this electromotive force of self-induction, be induced in these coils. Otherwise, sparking will exist even if there be no field distortion. In a direct-current generator, this counter electromotive force is obtained either by moving the brushes ahead of the neutral plane or by the use of commutating poles. This counter electromotive force assists the current in the coils undergoing commutation to reverse, and better commutation results. This same electromotive force of self-induction exists in the converter coils which are undergoing commutation. Therefore, commutating poles are used in converters, particularly in those of large capacity, in order to improve commutation. The commutating poles need not be as strong as those which are required for a direct-current machine of the same rating, as there is little or no cross-magnetization to be neutralized.

The resultant current in the armature conductors of a converter, under ordinary conditions of operation, is considerably less than either the alternating or the direct current. Therefore, a much larger commutator, in proportion to the armature, is required, than would be necessary for a direct-current generator having an armature of the same size. Converter armatures have abnormally large commutators, as shown in Fig. 327.

147. Voltage Control.—The ratio of the direct-current electromotive force to the alternating-current electromotive force in a converter armature is fixed, regardless of field excitation. However, the ratio of *brush* voltage to *slip-ring* voltage may be changed a limited amount by varying the field excitation. The brush voltage and the diametrical slip-ring voltage, increased by/2, differ from each other by the *impedance drop* through the converter armature. If this impedance drop changes either in phase or in magnitude, the ratio of brush voltage to slip-ring voltage changes. The impedance drop may be varied in phase and in magnitude by changing the excitation. Weakening the field below the value which gives unity power-factor makes the current lag, increases its value and lowers the induced electromotive force. (See page 313, Fig. 290.) Strengthening the field above the value which gives unity power-factor makes the current lead, increases its value and raises the induced electromotive force. (See page 311, Fig. 288.) The effect of changing the field excitation is therefore to change the power-factor, which in turn changes the magnitude and phase of the impedance drop in the armature, as has already been explained in connection with the synchronous motor (see pages 311 and 313). The ratio of brush voltage to slip-ring voltage can therefore be changed in this manner. This ratio can be varied by only 2 or 3 per cent. above and below normal and the voltage ratio and the power-factor cannot be adjusted independently.

Series Reactance.—It was shown in Par. 133, page 326, that the voltage at the terminals of a synchronous motor can be raised by over-excitation and lowered by under-excitation, provided there is sufficient reactance in the circuit between the motor and the source of constant voltage. As the converter is operating on its alternating-current side as a synchronous motor, it has excitation characteristics similar to those of the synchronous motor. That is, *over-excitation* causes it to take a *leading* current, and *under-excitation* causes it to take a *lagging* current. Therefore, with series reactance in the alternating-current line, the alternating voltage may be raised and lowered by changing the excitation (see Par. 133, page 326). This may be accomplished by hand regulation of the shunt-field rheostat, or automatically by means of a regulator, or by compounding the machine.

Instead of using special series reactances, the transformers, which are usually necessary with a converter, may be designed to have sufficient leakage reactance for this purpose.

The disadvantage of this method of voltage control is that a change of voltage is accompanied by a change of power-factor. Lowering the power-factor by any considerable amount is not desirable, because of the decreased efficiency and output which result. The voltage and power-factor cannot be changed independently. Therefore, this method is usually limited to less than 10 per cent. variation above and below the normal voltage.

Induction Regulator.—The induction regulator has already been described in connection with the induction motor. (See page 275.) This type of regulator may be connected between the transformers and the converter, and the alternating voltage impressed on the converter terminals may be raised and lowered thereby. This changes the direct-current voltage by a corresponding amount. Under these conditions the voltage may be raised independently of power-factor, but the extra equipment is an objection to the use of the induction regulator. *Series Booster.*—A low-voltage alternator is often connected to the shaft of the converter. This alternator has the same number of poles as the converter. The armature of the alternator is connected in series with the alternating current lines supplying the converter, as shown in Fig. 328. By raising the field of the alternator or booster, the alternating voltage of the converter is raised. The converter voltage may be lowered, not only by decreasing the booster field, but by reversing it as well. When the booster voltage is assisting the converter voltage, the booster acts as an alternator and takes mechanical power from the converter armature. This increases the *energy* component of the resultant armature current in the converter and hence changes the cross-magnetizing effect of the armature. When the booster voltage bucks the converter voltage, the booster receives electrical energy and delivers mechanical energy to the converter

Converter Field Fio. 328.—Synchronous converter with series booster.

Booster

Field Rheostat

Converter

Field Rheostat

shaft. That is, it operates as a synchronous motor and tends to drive the converter mechanically. Therefore, the energy current in the converter armature is decreased and may even be reversed. This causes a variation of the cross magnetization which in turn requires that the strength of the commutating poles be changed accordingly. This is accomplished by separate windings on the commutating poles, the current in these windings being controlled by the booster field-rheostat. (For further information, see Standard Handbook, Sec. 9.) The distinct advantage of this method of control is that the voltage may be varied independently of power-factor. The objection to this type of

voltage control is the additional machine. Figures 329 and 330 show converters having booster-generators. *Transformer Taps.*—The converter voltage may be adjusted approximately to the desired value by taps on the transformers. Owing to the arcing and burning of sliding contacts, the use of transformer taps is not common for adjustment during operation. The use of taps for fixed adjustment of voltage is, however, common. *Split-pole Converter.*—The split-pole converter is based on the following principle:

The total direct-current electromotive force generated depends

Fio. 329.—General Electric 3,500-kw., interpole-type, synchronous converter with synchronous booster.

on the total flux between brushes, irrespective of the manner in which this flux is distributed. The alternating electromotive force depends on the form of the flux wave, as well as on the total flux. Therefore, if the distribution of the flux be altered without changing its total value, the alternating electromotive force may be altered in value, but the direct-current electromotive force will not be affected.

In the split-pole converter, the form of the alternating electromotive force wave is varied by means of auxiliary poles adjacent to the main poles. The main poles are excited by the main-field winding, and the auxiliary poles by a separate winding. By changing the auxiliary excitation in conjunction with the excitation of the main winding, the wave form of the alternating electromotive force may be changed, thus varying the ratio of the alternating-current to the direct-current electromotive force.

Fig. 330.—Westinghouse synchronous converter with booster generator.

The brushes in a generator must be moved *forward,* in order that the machine may commutate in the fringe of a leading poletip (see Vol. I, page 285, Par. 197). To balance the electromotive force of self-induction, the brushes of the split-pole converter must be moved forward, in order to commutate in the fringe of a leading pole-tip, Fig. 331. This fringe must come from the main

poles, for their flux is nearly constant in strength, whereas the flux of the auxiliary pole is varied over a wide range and may be reversed even. Therefore, in a *direct* synchronous converter, the armature must rotate from *main* to *auxiliary* pole, Fig. 331, whereas in an *inverted* synchronous converter, the armature must rotate from *auxiliary* to *main* pole.

Fio. 331.—Relation between direction of rotation and position of auxiliary poles in splitpole synchronous converter.

Fig. 332.—General Electric, 8-pole, 750-kw., 375 r.p.m., shunt-wound, split-pole synchronous converter.

Figure 332 shows a General Electric, split-pole converter without commutating poles.

148. Experimental Determination of Voltage and Current Relations in a Converter.—An instructive laboratory experiment is carried out with a converter connected in the manner shown in Fig. 333. The series reactances may be omitted if the transformers themselves have sufficient leakage reactance. Connect instruments to measure the three-phase input, a voltmeter to measure the transformer primary voltage, a voltmeter to measure the slip-ring voltage, ammeters to measure the currents between the transformer secondaries and the converter, directcurrent instruments to measure the converter output, and a direct-current ammeter to measure the field current.

Keep the load on the converter constant at its rated value. Vary its field over the maximum range of operation, reading all instruments. With field current as abscissas, plot as ordinates:

Fio. 333.—Connections for testing a synchronous converter.

1. Voltages V_1, V_2, Fa, and F4. 2. Efficiency of the entire unit. 3. Power-factor.

Also, check the currents by the equations of Par. 142, page 350. Note the effect of power-factor on efficiency.

Other experiments may be performed using these same connections, such as keeping the field current constant at its normal no-load value $(P.F. = 1.0)$ and noting the changes in efficiency and power-factor as the load is increased. Plot efficiency and power-factor as ordinates with output as abscissas.

149. Synchronous-converter Connections.—Transformers are usually necessary with synchronous converters. The directcurrent voltage is always low and the alternating voltage at the slip-rings must be still less. Moreover, transformers are necessary for obtaining a sixphase from a three-phase system.

Usually, the transformer primaries may be connected either in Y or in delta. The most common six-phase connections for the transformer secondaries are the "diametrical," the star, and the double-Y. (See Fig. 321, page 350.) The difference between the diametrical and the star is that the secondaries are connected together at the neutral point in the star, whereas three separate secondaries are connected across diametrically opposite points in the diametrical connection. There is no difference between the double-Y and the star if the neutrals of the two Ysystems are connected together. Other than a slight effect on harmonics, and the fact that a neutral is available in the doubleY, there is little difference in the use of the three connections, except with the split-pole converter.

If the induced electromotive force of the converter armature contains harmonics, there will be no circulatory current within the armature itself, for in the direct-current type of armature such as is used for the converter, any electromotive force induced under a given pole in one part of the armature is opposed by an opposite and equal electromotive force induced under an opposite pole. However, if the *line* voltage is practically sinusoidal and the *induced* electromotive force of the converter contains harmonics, there will be unbalanced harmonic voltages. The current due to these unbalanced voltages will consist entirely of harmonics which contribute no energy but do heat the armature and transformers.

This effect is negligible in the ordinary converter, but in the split-pole type, the voltage control depends on the

introduction of large harmonic voltages into the electromotive force wave. Therefore, when this type of converter is used, the transformer connections must be so chosen that as many as possible of the harmonic currents are eliminated. Most three-phase transformer connections eliminate the third harmonic current and its multiples, with the following two exceptions. The primaries cannot be connected in delta if the secondaries are connected either diametrical, six-phase star or double-Y, with the neutrals of the two Y-systems connected together, for the third harmonic currents in the secondaries, due to unbalanced harmonic voltages, will cause third harmonic currents to circulate in the primary delta, producing extra heating in the converter armature and in the transformers.

If the transformer primaries be connected in Y, with no neutral connection, no third harmonic currents or multiples thereof can flow into the Y, as these currents are all in phase with one another. In order that currents may flow to a common point, there must be phase difference, as the currents flowing toward the point must be equal to the currents leaving the point at any instant, or electricity will accumulate at the point. If no third harmonic currents can flow in the transformer primaries, none can flow in their secondaries, hence there will be no circulatory harmonic currents between the transformer secondaries and the converter armature if the primaries are connected in Y without a neutral connection to the main generator. However, if the neutral of

Fig. 334.—Double-delta connection of transformers to 6-phase synchronous converter.

the transformer primaries be carried back to the main generator, the third harmonic currents and multiples thereof can return to the generator through the neutral. Therefore, the secondaries cannot be connected either diam.etrical, six-phase star, or double-Y with interconnected neutrals, if the primaries are connected in Y with a neutral return to the generator.

The harmonic currents other than the

third and multiples thereof are not eliminated by three-phase connections, but they are reduced to small values by the use of series reactances or by using transformers having high leakage reactance.

Figure 322, page 351, shows the connections for a 500 kw. converter and transformers taking power from 6,600-volt, threephase, 60-cycle mains and delivering 550 volts direct current. The transformer primaries are connected in delta, and the secondaries can be connected either diametrical, star, or double-Y. (If this were a split-pole type of converter, the primaries could not be connected in delta, but they must be connected in Y without neutral return to main generator.) The advantage of the star and the inter-connected double-Y connection is the fact that a neutral is accessible. The voltages and currents at each part of the system are shown. Unity power-factor, 98 per cent. efficiency for the transformers and 95 per cent. efficiency for the converter are assumed.

The double-delta connection of secondaries may also be used. Such a connection for a converter is shown in Fig. 334. The arrows point in the relative directions in which the voltages act. No neutral is available if this method of connecting the transformers is used.

160. The Inverted Synchronous Converter.—When a converter operates from a direct-current source and delivers alternating current, it is known as an *inverted* synchronous converter. The direct-current side has characteristics very similar to those of a shunt or compound motor. The alternating-current side has characteristics very similar to those of an alternator. A converter when operating inverted has the same rating as when operating direct. When operating from the alternating-current supply, the speed of the converter must be in synchronism with the supply, and hence constant. When operating from the directcurrent supply, the speed is determined by the back electromotive force and the flux, just as in any direct-current motor, and the speed may vary. In fact, at times there is a tendency for the inverted converter to race, so that inverted

converters should have speed-limiting devices. An inductive load on the alternating-current side weakens the field through armature reaction, in the same manner that the field of an alternator is weakened under similar conditions. The weakening of the field increases the speed of the converter. This increased speed causes the current to lag still more (tan 0 =-5) because of the increased frequency. As the effect is cumulative, and may cause the armature to reach dangerous speeds, the necessity for using a speed-limiting device is obvious.

A centrifugal device is often used to trip the circuit-breaker when the speed exceeds the safe value. Another method, not often used, is to have an exciter on the converter shaft. As the speed increases, the exciter voltage increases and the converter field is strengthened. This tends to check the increase of speed of the converter.

Inverted converters will operate satisfactorily in parallel on the alternating-current side, any converter being made to take more load by weakening its field.

151. Starting the Synchronous Converter from the Alternatingcurrent Side.— There are several methods of starting direct synchronous converters, some of which are similar to the methods used with the synchronous motor.

If polyphase currents are supplied to the armature, a rotating field is produced about the armature, Fig. 335. This is similar to the rotating field of the induction motor, except that it is produced by a rotating armature about itself. If the armature speed is below syn-' chronism, this field cuts the pole faces and the damper windings (Fig. 339), and induces currents. A reaction results between the rotating field and these induced currents, producing rotation.

When starting the converter in this manner, several precautions are necessary. The armature is the primary, and the shunt field coils are the secondary of a transformer, the secondary having a very large number of turns. The rotating armature field, therefore, induces very high voltages in the field coils on starting and tends to puncture them. To

reduce this voltage, the field is usually split into sections by a field-splitting or sectionalizing switch. Figure 336 shows the connections of a three-pole switch used to sectionalize the field circuit of a four-pole converter into

Fig. 335.—Relative directions of rotation of the armature and of the rotating field produced by the armature.

four parts. *This sectionalizing switch should be open when starting from the alternating-current side.*

If there be a switch short-circuiting the series field, this should be opened, as otherwise the currents generated in the series field by the transformer action of the armature will cause undue heating. If there be a series-field shunt or diverter, this should be opened for the same reason.

The rotating field produced by the armature *cuts* the armature conductors, Fig. 335, just as if the armature were rotating and cutting the flux of a stationary field, as in the direct-current generator. This field induces voltages in the armature coils. Some of these coils are shortcircuited by the brushes, so that sparking results under the brushes, even though there is no direct-current load. This sparking may not be severe, as the rotating field is comparatively Fl-336. —Connections of shunt field and.. ' shunt field-splitting switch.

weak in the interpolar spaces where the brushes are, because of the high reluctance of the air path at these points. However, if interpoles are used, the reluctance of the interpolar space is reduced very materially so that sparking becomes severe. Consequently, brush-raising devices are usually installed on interpole machines, to lift the brushes on starting and so eliminate this sparking. One brush in a positive brush-holder and one brush in a negative brush-holder are usually left on the commutator to supply the field excitation. In order to reduce the sparking caused by these two brushes short-circuitng armature coils in which electromotive forces are induced by the rotating field, these brushes are often beveled so that the time of short-circuiting is reduced to a minimum. Converters are started at re-

duced voltage, obtained from taps on the transformer secondaries, although starting compensators are used at times in the units of smaller size.

Field-Reversing Switch

As a rule, converters excite their own shunt fields. The armature rotates in a direction *opposite* to that of the rotating field which is set up about it, Fig. 335. Therefore, as the armature approaches synchronism, the rotation of this field becomes slower and slower with respect to the field structure, as the rotating field rotates in one direction and the armature in the opposite direction. The field poles themselves, which are magnetized alternately north and south by this field, become more and more slowly magnetized as the armature approaches synchronism. Finally, due to hysteresis action, (see Chap. X, page 321, Par. 130), the poles themselves become permanently magnetized through armature reaction, and the armature pulls into synchronism in a manner similar to that of the salient-pole synchronous motor when started in this manner.

When the shunt-field switch is closed, the field produced by the shunt winding may oppose the field built up in the field poles by armature reaction. Consequently there is a tendency for the armature to slip a pole. Should the armature slip a pole, the direct-current voltage at the brushes reverses. This reverses the shunt-field current, which again causes the converter to slip a pole. This action, unless checked, may continue indefinitely. It may be stopped by reversing the shunt field current by means of the field-reversing switch, Fig. 336.

It often happens that the direct-current field is not strong enough to cause the armature to slip a pole, because the field voltage may be low, due to the alternating voltage being reduced through the starting taps. However, the tendency exists, and due to the resulting distortion of the pole flux, the brushes are no longer in the commutating zone. The brush voltage is thereby reduced, which again lowers the tendency to slip a pole. The converter will continue to run under these conditions, but it will take a large current at low power-factor, will spark at the brushes, and its operation will be unsatisfactory. By reversing the field current, however, normal operating conditions can be obtained.

152. Methods of Obtaining Correct Polarity.—It is important that the converter always come up with the same direct-current polarity, as it may be operating in parallel with other apparatus. As has just been pointed out, the converter may build up with either polarity. If this polarity happens to be wrong, there are several methods of correcting it.

Below are given some of these methods. The starting compensator or transformer taps are assumed to be in the starting positions.

(a) Open the shunt-field circuit and then open the line switch long enough for the converter to slip one pole. This can be determined very readily with a stroboscope. Close the field switch and then throw the alternating-current switch quickly into the running position. With a little practice this operation can be readily performed. (6) Reverse the shunt field by means of the field-reversing switch. This causes the machine to slip a pole and so reverses the direct-current voltage, making it correct. If left this way the machine will continue slipping, one pole at a time, as has just been pointed out. Therefore, the shunt-field switch must be thrown back immediately to its original position. (c) When the converter is first connected across the alternating-current line, the rotating field produced by the armature cuts the armature conductors and generates alternating currents in these conductors, as has already been pointed out. The brushes are stationary and the field rotating, so there is no commutating action. Therefore, there is an alternating electromotive force of line frequency across the brushes at the instant of starting. The armature rotates in a direction *opposite* to that of its rotating field, because of the reaction with the pole-face currents. This is illustrated by Fig. 335. The rotating field about the armature is shown as rotating clockwise. A conductor, such as the pole faces, when placed

in this field, would tend to rotate clockwise. That is, if the armature were held stationary, the field structure would tend to rotate in the direction of the rotating flux produced by the armature, or in a clockwise direction. Therefore, the torque produced by this rotating flux is in such a direction that it tends to cause the field structure to rotate in a clockwise direction. However, the field structure is fixed in position and the armature is free to rotate. The *reaction* between the two remains unchanged. Consequently the armature will rotate in a *counter-clockwise* direction. The relative motion between armature and field structure is the same as if the armature were stationary and the field were free to rotate. -(-Bus

Fig. 337.—Method of obtaining correct polarity by closing equalizer and series-field switches.

As the speed of the armature increases, the field produced by it must rotate slower and slower in *space*, although it does not change its speed relative to the armature. The brushes tend to become more and more nearly stationary with respect to this rotating field, so that their commutating action becomes greater and greater. The frequency of the electromotive force across the brushes becomes less and less, and when the armature finally pulls into synchronism, becomes zero, and a direct-current voltage exists across the brushes.

If a direct-current voltmeter be connected across the brushes, its pointer will tend to oscillate at line frequency when the alternating current is first switched on. As the armature speeds up, this frequency becomes less and less, and the pointer is soon able to follow the slow oscillations. When the frequency of oscillation becomes very low and the pointer is just going through zero in the positive direction, the field switch should be closed. This insures the converter's coming in with the correct polarity. A zero-center type of voltmeter is desirable when this method is employed.

(d) If the converter operates in parallel with others, and equalizers are used, a weak field of the correct polarity may

be produced in the field of the incoming converter by closing a line and an equalizer switch, as indicated in Fig. 337. This tends to make the armature reaction build up fields of the correct polarity and so insures the converter coming in properly. 153. Starting Synchronous Converter by Means of an Auxiliary Motor.—As was pointed out in Chap. X, one method of starting a synchronous motor is to bring it up to speed with an auxiliary motor and then synchronize. (See page 320, Par. 130.) This same method may be used with the converter. The methods of synchronizing are identical with those used with the alternator. (See page 168, Par. 74.) This method of starting is practically obsolete. 164. Starting Synchronous Converter from the Directcurrent Side.—If sufficient direct-current power is available, the converter may be started from the direct-current side, starting as a shunt motor. When started in this manner, the series field should be short-circuited, as it will oppose the shunt field when the machine operates as a motor and will therefore reduce the starting torque. The transformer secondaries are short-circuits on the direct-current armature at starting, as the frequency is zero and their resistance is very low. This is particularly true if the brushes happen to be resting on commutator segments which are connected directly to the slip-ring taps. Therefore, the transformers should be disconnected. The proper speed is obtained by adjusting the shunt field. As there is practically no voltage control in the simple converter when operating in this manner, it is not always possible to adjust the alternating voltage to a value equal to that of the line. To prevent any disturbance which may result from synchronizing at a voltage other than bus-bar voltage, some of the starting resistance is often left in the armature circuit until after the machine has been synchronized. 166. Parallel Operation of Synchronous Converters.—Synchronous converters may be operated in parallel on the directcurrent side, just as shunt and compound generators are similarly Operated. If one series field winding be used on each machine, only one equalizer is necessary. If the machine is a three-wire converter and is compounded, there will be two series fields, as shown in Fig. 338. In this case two equalizer switches are necessary. (See Vol. I, page 376, Fig. 338.) The loads are shifted by changing the voltages of the converters, either by field control or by any of the other methods already described.

Better operation is obtained if each converter has its own transformer bank, rather than by having a single bank supplying all the converters. This introduces more or less reactance between converters and stabilizes their operation. It may even be necessary to install series reactances in the transformer leads.

The alternating side of a converter may be accidentally opened by a circuit breaker or otherwise, while the direct-current side may still be connected to a source of power, such as other converters or a storage battery across the bus-bars. The converter will then tend to operate as a shunt motor, usually with a weakened field, due to the differential action of the compound winding. Under these conditions the converter may tend to race. Therefore, converters are usually equipped with reverse-energy relays on the direct-current side, or else the direct-current breakers are

Neutral

Fio. 338.—220-volt. 3-wire D. C. system obtained from a 220-volt synchronous converter employing double-Y connection.

interlocked with the alternating-current ones, so that the directcurrent side will be opened simultaneously with the alternatingcurrent side.

The *resultant* current in the converter armature conductors produces the torque which overcomes the stray-power losses of the converter. This resultant current is the *difference* of two nearly equal currents, as has already been demonstrated. A small percentage change in either the motor current alone or in the generator current alone produces a large percentage change in this torque current. Therefore, the converter

is very sensitive to line disturbances, such as fluctuations of voltage or of frequency. Accordingly, it has a much greater tendency to "hunt" than has the synchronous motor. For this reason, converters always have amortisseur or damper windings or grids built around and into the poles, as shown in Fig. 339. The action of these windings is the same as in the synchronous motor described on page 318, Par. 129, except that the windings are now stationary in space. The armature which produces the rotating field rotates at synchronous speed in one direction and the rotating field itself rotates at synchronous speed in the opposite direction with respect to the armature. Under normal operation therefore, the field is stationary in space with respect to the amortisseur windings.

Fio. 339.—Main pole with damper winding.

166. The Three-wire Converter.—It is pointed out in Vol. I, Chap. XII, that the neutral of a 3-wire system may be obtained by the use of two or more slip-rings connected to the directcurrent armature. A reactance coil is connected across these slip-rings and alternating current flows through this reactance. The direct-current neutral is connected to the middle point of this reactance and the direct current of the neutral divides and

Fio. 340.—D. C. neutral obtained from neutral of 4-phase, diametricallyconnected transformers.

passes back into the armature through the reactance. The reactance has a low resistance and has practically no effect on the direct current.

It is to be noted that a synchronous converter with the proper transformer connections provides a neutral point for just such a direct-current neutral. For example, if a six-phase, double-Y, or a six-phase star, Fig. 338, a four-phase star, Fig. 340, or a three-phase Y-connection of transformer secondaries be used, an excellent neutral is provided.

In the first two of these connections, the direct current flows in opposite directions through the two-halves of each transformer secondary, so there is no direct-current magnetizing action on the

core. In the Y-connection, however, this is not the case and the magnetizing action of the direct current, acting in conjunction with that of the alternating current, produces an unsymmetrical cyclic magnetization of the iron. This is undesirable, as it results in an increased magnetizing current whose positive and negative values will be unequal and dissimilar. By splitting each transformer secondary into two sections, a,a', b,b', and c,c', Fig. 341, and connecting as indicated, it will be observed that the direct current flows in opposite directions in

Secondaries

Fig. 341.—Zig-zag connection of three-phase secondaries for eliminating D. C. magnetizing of transformer cores. the two-halves of each secondary winding, and consequently, has no appreciable magnetizing effect.

Figure 338 shows the complete connection for a six-phase, 220-volt, three-wire converter, having two series fields and with. the direct-current neutral connected to the neutral of the doubleY connected transformer secondaries. Figure 340 shows a direct-current system supplied by a four-phase converter, the neutral being obtained through two starconnected transformers.

CHAPTER XII TRANSMISSION OF POWER BY ALTERNATING CURRENT 157. Transmission Systems.—To transmit power economically over considerable distances, it is necessary that the voltage be high. High voltages are readily obtainable with alternating current. As high as 15,000 volts may be generated directly. For voltages in excess of this it is desirable to use transformers, as it is difficult to insulate the generators for these higher voltages. The transmission voltage is usually too high for commercial uses, but for purposes of distribution it may be stepped down to the desired value by the use of transformers.

Direct-current voltages for commercial power can be raised and lowered only by machines having rotating commutators. The efficiency of such apparatus is not high, and operating difficulties are encountered in connection with the commutators, even at comparatively low voltages. Hence, alternating current is nearly always used for transmission purposes. (The one exception is the Thury1 System in Europe.) Where considerable power is involved, polyphase systems are used because of the many advantages of polyphase over single-phase systems. For example, polyphase motors are considerably cheaper and lighter than single-phase motors of equal rating and as a rule have better operating characteristics. The output of generators when operating polyphase is much greater than when operating single-phase. (See page 76.)

Of the polyphase systems, the three-phase system is generally used for transmission, although the employment of two-phase for. distribution purposes is not uncommon. The three-phase system has the advantage that it requires the least number of conductors of all the polyphase systems; the voltage unbalancing even with unbalanced loads is not usually serious; and for a given voltage between *conductors,* with a given power transmitted a given distance with a given line loss, the three-phase system i See Vol. I, Page 303, and also "Standard Handbook," Fifth Edition, Section XI.

requires only 75 per cent. as much copper as either the singlephase or the two-phase system.

The single-phase system is used in railroad electrification, where single-phase power is supplied at the trolley. The most notable examples of this are the New York, New Haven and Hartford Railroad and the Norfolk and Western Railway.

When the voltage is so high as to make transformers necessary, the power is usually generated at 6,600 volts. This voltage is not so high as to make difficult the proper insulation of the generators, and at the same time the armature conductors and the leads running to the switchboard do not become too large.

6600-V. Generators Q O 6600-V. Bus-bars 'i I, ,,

Delta-Y A Generating

Transformer Bank L 1 &tati011 110,000-V. Bus-bars "T", TJ 110,000-V.

Transmission Line
110,000-V. Bus-burn
Y-Y Trans. Bank
13,200-V. Bus-bars
13.200-V.
Underground Cable
Delta-Y
Transformer Bank 4000,2300-V.
-Ph. 4-Wire Bus-bars,,.,,. Distributing Lines
(Overhead or underground)
Fio. 342.—Typical connections of a power system.

The transmission voltage is largely determined by economic considerations. Although a high voltage reduces the conductor cross-section, the saving in copper may be offset by the increased cost of insulating the line, by the increased size of transmissionline structures and by the increased size of generating and substations, due to the large clearances required by the high-voltage leads and bus-bars. A rough basis for determining the transmission voltage is to use 1,000 volts per mile of line.

Because of the danger involved, it is not usually permissible to carry high-voltage transmission lines through thickly populated districts in order to reach the distributing sub-stations. The voltage is usually stepped down to about 13,200 volts at sub-stations located at the outskirts of the city and thence carried into the city underground, or occasionally overhead, at 13,200 volts.

Figure 342 shows a typical system. No attempt is made to show switches, circuit-breakers, etc. Power is generated at 6,600 volts and is delivered directly to the 6,600-volt bus-bars. It is then stepped up to 110,000 volts, the transmission voltage, by delta-Y transformer banks whose secondaries are connected to the 110,000-volt bus-bars. The power then passes out over the duplicate transmission lines to a sub-station located in the outskirts of the district where the power is to be utilized. It is then stepped down to 13,200 volts by Y-Y-transformer banks and delivered to the 13,200-volt bus-bars at this sub-station. The power then leaves these 13,200-volt bus-bars for the various distributing substations in the district. One distribut-

ing sub-station is shown. Here the voltage is stepped down to a three-phase, four-wire system. This system has 4,000 volts between conductors, or 2,310 volts to neutral, for distribution to the consumers.

Usually the lighting and the power loads are connected to separate feeders, in order to avoid the annoying flickering of the lamps when motors are thrown on or off the line. The lighting loads are usually supplied by 10:1 transformers located on the poles, from whose secondaries 230-115-volt, three-wire systems are obtained, Fig. 343. The two wires coming from the top crossarm to the crossarm next beneath and going through the fuse cut-outs to the transformer are the 2,300-volt lines. The 230-115-volt secondary wires leave the front side of the transformer and feed three vertically-arranged conductors of the threewire secondary mains, which supply the local lighting loads. The *power* consumers are usually connected to the secondaries of V-connected or delta-connected transformers located at the consumer's premises. In order that the secondary mains may not be too large, 440 and 550 volts are generally used for the power loads. (Also see Vol. I, page 380, Fig. 341.)

Fio. 343.—Typical 2300-230/115-volt lighting transformer and secondary 3wire mains.

In the sub-station, other power-transforming apparatus may be installed, such as constant-current transformers; motor-generator sets or synchronous converters, for obtaining direct current, etc. 168. Transmission Line Reactance; Single-phase.—In making line calculations for the transmission of direct-current power, the resistance alone needs to be considered. In making similar calculations for alternating-current lines, it is necessary to take into consideration not only the line resistance, but the line reactance as well. In cables and in overhead lines operating at high voltage, it is also necessary to consider the capacitance between conductors.

Figure 344 shows the cross-section of a two-conductor, singlephase line. As the current at any instant flows in opposite directions in the two conductors, the circular paths of the magnetic lines set up about one conductor must always go in a direction opposite to that for the other conductor. That is, when one magnetic field is acting in a clockwise direction, the other must be acting in a counterclockwise direction. This causes the two fields to act in conjunction in the area between the two conductors,

Fig. 344.—Magnetic field between the as shown in Fig. 344. Thus, two conductors of a single-phase line. two parallel wires form a FCC tangular loop of one turn, through which flux is set up by the current in the wires. This flux links the loop and the circuit has inductance, therefore. It might appear that this inductance would be negligible, because the loop has but one turn and the flux path is entirely in air. It must be remembered, however, that the cross-sectional area of the flux path is *large,* usually being from 1 to 20 ft. wide and several miles long. Although the flux density is small, the total flux linking the loop is usually considerable.

It can be shown that the inductance of such a loop is $L = 2Z(0.080 + 0.741 \, Iogi0-)$ mil-henrys, (84) where D is the distance between conductor centers, and r is the radius of each conductor, both expressed in the same units. I is the length of the line in miles. The reactance of the loop is $X = 2fL$ (85) where f is the frequency in cycles per second.

It is usually more convenient to consider the inductance of a single conductor only. The inductance per single conductor is obviously one-half the value given in equation (84), which applies to the two conductors of the circuit.

The reactance per mile then becomes $X = 2ir/(80 + 741 \, logTM-) \, 10-6$ ohms per mile. (86)

Table I in the Appendix gives values of the reactance at 60 cycles per second for solid and stranded conductors, at various spacings. The reactance for stranded conductors is slightly less than the corresponding values given for solid conductors. The reactance at other frequencies may be found by direct proportion. (For more complete tables see Sec. XI, Standard Handbook, fifth edi-

tion.) *Example.*—A single-phase transmission line is 40 miles long and consists of two 0000 solid conductors spaced 4 ft. on centers.

(a) Find the inductance of the entire line and the reactance per conductor at 25 cycles per second; (6) at 60 cycles per second; *(c)* if a 200-amp., 60-cycle current flows over this line find the total reactance drop.

The diameter of 0000 conductor is 460 mils; the radius, $r = 0.230$ in. D/r " oSo " 209 log.o 209 = 2.32 (p. 462) The inductance per mile $L, = 2(0.080 + 0.741 \times 2.32) = 3.60$ mil-henrys (from equation 84). (a) The total inductance $L = 3.60 \times 40 = 144$ mil-henrys or 72 mil-henrys per conductor. Ana. The reactance per conductor at 25 cycles $X, = 27T \, 25 \times 72 \, XlO-3 = 11.3$ ohms *Ans.* (6) The reactance per conductor at 60 cycles $Xz = 2ir60 \times 72 \times 10-s = 27.1$ ohms. *Ans.'* (c) The total reactance drop $V = 27.1 \times 200 \times 2 = 10,840$ volts. *Am.* 169. Transmission Line Reactance; Three-phase.—In transmission line problems it is convenient to consider the reactance of the individual conductor, rather than the reactance of the looped line or of the entire circuit. The convenience becomes more apparent when three-phase lines are considered. In Fig. 345 are shown the three conductors of a three-phase line, symmetrically spaced. That is, each conductor is at an apex of the same equilateral triangle. The current at the instant shown is flowing outward in conductor A and inward in conductors B and C. The field produced by each conductor is indicated. These fields are continually changing, due to the cyclic variation of the current in the three phases, and this causes a rotating field in the region between the conductors. This rotating field is similar to the rotating field of the polyphase induction motor, and as it cuts all three conductors, it induces electromotive forces in them.

In treating this problem, however, it is simpler to consider the reactance of each conductor separately. If the spacing is symmetrical, the flux produced by each conductor does not induce any electromotive force in the circuit com-

posed of the other two conductors. For example, Fig. 346 shows the circular field produced by conductor C acting alone. As none of its lines links the circuit AB, conductor C does not induce any electromotive force in loop AB. Likewise, conductor A induces no electromotive force in loop BC, and conductor B induces no electromo tive force in loop CA, provided the conductors are symmetrically spaced.

In the three-phase case, therefore, the reactance per conductor is found by equation (85), page 381, or by consulting the tables, page 465. The distance between the centers of conductors is used for D. *Example.* — A three-phase line consists of three 0000 solid conductors placed at the corners of an equilateral triangle, 4 ft. on a side. Find the reactance drop per conductor per mile when a 25-cycle alternating current of 120 amp. flows in the conductors.

Instead of calculating the reactance X, it may first be found in Appendix I, page 465, for 60 cycles per second, its value being 0.677 ohm. The 25-cycle reactance is 25/60 of this value, and is equal to 0.282 ohm.

160. Transmission Line Capacitance; Single-phase.—If a *direct-current* voltage be applied to a transmission line under noload conditions, no current flows after the first few moments, except the almost negligible leakage current. If an *alternating* voltage be applied to a transmission line, considerable current may flow, even if there be no appreciable leakage and no connected load. This current is the *charging current* of the line, and leads the voltage by almost 90. The line acts as a condenser, the conductors being the plates and the air the dielectric. Each conductor becomes charged, first positively and then negatively, which results in an alternating current.

This is illustrated by Fig. 347, which shows conductors A and B of a single-phase line. At the instant shown, conductor A is positive and conductor B is negative. The electrostatic flux

Fio. 347.—Electrostatic flux between line conductors.

existing in the field between A and B

is shown. The capacitance *between conductors* of such a line can be shown to be approximately where D is the distance between conductor centers and r is the radius of each conductor, both expressed in the same units.

The simplest method of treating transmission-line problems is to work with voltages to *neutral* and with capacitances to *neutral.* (6) Line capacitance replaced by (c) Line capacitance replaced by a single condenser. two series-connected condensers.

Fio. 348.—Substitution of equivalent condensers for transmission line capacitance.

In Fig. 348 (a), an imaginary plane surface *xy* is shown midway between conductors A and B and perpendicular to the plane of the conductors. The electrostatic field between this surface and each conductor is the same. As the plane bisects every electrostatic flux line, the potential difference between conductor A and any point in the plane is equal to the potential difference between conductor B and this same point. That is, the potential of every point on the plane *xy* is midway between the potential of conductor A and that of conductor B. Hence, every point in this surface is at the same potential and *xy* is an equipotential surface. The plane *xy* may be replaced by a thin conducting plate of infinite breadth without disturbing the electrostatic field. Each conductor has the same capacitance to this plate. This capacitance must be *twice* the capacitance between the conductors themselves. That is, the capacitance C between conductors, Fig. 348 (6), may be replaced by two equal capacitances, $C_i, C_i,$ connected in series, Fig. 348 (c), where $C_i = 2C$. The joint capacitance of the two capacitances C_i, C_i in series is obviously just equal to that of the single capacitance, C. The point 0 is the neutral of the system, its potential being the same as that of the plate *xy.*

If the capacitance to neutral is used when calculating the charging current, the voltage to *neutral* must also be used. With half the voltage and twice the capacitance, the charging current per conductor is the same as if the total voltage

and the capacitance between conductors had been used.

The capacitance to neutral may be found by multiplying equation (87) by 2.

$C_i = $ —' —-mf. per mile to neutral. (88) logio

Obviously, the line charging current is $/_{,,} = 2irfC_iEWe$ amperes per mile of line.

where / is the frequency in cycles per second, E is the voltage to *neutral,* and C_i is the capacitance to neutral in microfarads per mile of line.

Appendix J, page 466, gives amperes per mile of line, per 100,000 volts to neutral, at 60 cycles per second, for various sizes of conductor and various spacings.

Example.—A 40-mile, 60-cycle, single-phase line consists of two 000 conductors spaced 5 ft. apart. What is the charging current if the voltage between wires is 33,000 volts?

The diameter of 000 wire is 410 mils. The radius
r = 0.205 in.
$D/r = 60/0.205 = 293$
logic 293 = 2.47
C, = 40 X - = 0.628 mf.

The charging current
Oq finn I, = 260 X 0.628 X-"'""- 10-" = 3.91 amp. *Ans.* £ 161. Transmission Line Capacitance; Three-phase. —Figure 349 shows the three conductors $A, B, C,$ of a three-phase line, these conductors being symmetrically spaced. There is capacitance between each pair of conductors, which can be represented by three equal capacitances $c', c', c',$ Fig. 349(a), connected in delta. In determining the capacitive relations in this type of system, it simplifies the problem to substitute an equivalent Y-system for the delta-system. It is obvious that any delta-load may be replaced by an equivalent Y-load. This is the same as considering that each conductor has capacitance c to a fictitious neutral 0, Fig. 349(6). In the actual line the (6)

Fig. 349.—Delta capacitance of a 3-phase system replaced by an equivalent Y-capacitance.

neutral may be the ground. The voltage across each of these condensers c is $E/$-

/3 where E is the line voltage.

Equation (87), page 384, may then be applied to finding the capacitance to neutral, c, D being taken as the distance between conductor centers. The voltage to neutral £/V3 is used for determining the charging current per conductor.

Example.—Assume that a third wire be added to the system of paragraph 160 to form a symmetrical spacing and that the system is operated threephase, 33,000 volts between conductors. Find the charging current per conductor. $r = 0.205$ in.

$D/r = 60/ 0.205 = 293$
Iogi0 293 = 2.47
, 0.0388 40 v 40 X 2.47 0.628 mf.
Volts to neutral = 33,000/V3 = 19,070 volts.

The charging current per conductor
$h = 2ir60$ X 0.628 X 19,070 = 4.52 amp. *Ans.* This may be checked by Appendix J, page 466.

162. Three-phase System; Conductors Spaced Unsymmetri cally.—If the conductors in a three-phase system are *not* symmetrically spaced, being located at the corners of a triangle whose sides may be of any length, as $A, B,$ and $C,$ Fig. 350(a), the side D of the equivalent equilateral triangle, Fig. 350(6), may be found as follows: $D = Vabc$ (89) B D (a) (b)

Fig. 350.—Unsymmetrical spacing and equivalent symmetrical spacing.

This value of D should be used as the distance between the conductor centers of the equivalent system in transmission line calculations.

163. Single-phase Line Calculations.—In determining the voltage drop in an alternating-current line, both the resistance and the reactance must be taken into consideration. The voltage to supply the resistance drop is in phase with the current, and the voltage to supply the reactance drop is in quadrature with the current and leading.

Fio. 351.—Single-phase line having resistance and reactance.

In making transmission-line calculations, it is convenient in all cases to work to neutral. Figure 351 shows a single-phase line which has a resistance per wire of R ohms and a reactance per wire of X ohms. The load takes a current I amperes at a powerfactor cos θ, and the *total* voltage at the load or receiver is *2Ea*. The voltage to *neutral* at the receiver is therefore *Ea*. The *total* voltage at the sending or generating end is *2Ea*.

If this system be split along the line *CD*, two systems result, one of which is shown in Fig. 352. Each of these two systems transmits one-half the total power and the sending-end and receiving-end voltage of each system is half the voltage between conductors. The voltage at each end is now the voltage to neutral. The ground is assumed to be the return conductor. The return conductor need be merely hypothetical, however, for under balanced conditions, Fig. 351, no current flows back through the -= Ground-=r Ground

Fig. 352.—Single-phase line and voltages to neutral.

ground, as each half of the system acts as a return for the other half. Therefore, the voltage drop through the ground is zero. That is, Fig. 352, for purposes of calculation, the ground may be considered as having zero resistance and zero reactance.

Let it be required, in Fig. 352, to determine the generator voltage *Eg* when the load voltage *ER,* the current *I,* and powerfactor cos θ are given. The vector diagram is shown in Fig. 353(a). The component of voltage to supply the *IR* drop is laid off in phase with the current *I;* the component to supply

Fig. 353.—Vector diagrams for single-phase transmission line.

the *IX* drop is laid off 90 ahead of current *I.* The resultant of these two components is the component to supply the *IZ* drop, or to supply the actual voltage drop per conductor. The voltage at the generator *E0* is the vector sum of *ER* and *IZ.* In Fig. 353(6) the *IR* and *IX* components are added to *ER* vectorially. It will be seen that this figure is similar to Fig. 145, Chap. VI, page 139, and its geometrical solution is identical. $Ea = \sqrt{(ER \cos \theta + IR)^2 + (EK \sin \theta + IX)^2}$ (90) *Example.* — It is desired to deliver 4,000 kw., single-phase, at a distance of 25 miles, the load voltage being 33,000 volts, 60 cycles, and the power-factor of

the load being 0.85. The conductors are spaced 4 ft. apart. The line loss shall not exceed 10 per cent. of the power delivered. Determine: (a) the size of conductor; (6) the resistance drop per conductor; (c) the reactance drop per conductor; *(d)* the voltage at the sending end; *(e)* the line regulation. Neglect capacitive effects. (o) The line loss = 4,000 X 0.10 = 400 kew. = 400,000 watts. The loss per conductor = 400,000/2 = 200,000 watts.

The current $I = $ = 142.5 amp. = $(142.5)^2/?$' = 200,000 watts. $R' = 200,000/(142.5)^2 = 9.85$ ohms. Res. (per mile) = 9.85/25 = 0.394 ohm. From Appendix H, page 464, the wire having the next lowest resistance per mile is 000 A.W.G., the resistance of which is 0.333 ohm per mile. *Am, (b)* Total resistance per conductor $R = 25$ X 0.333 = 8.34 ohms. $IR = 142.5$ X 8.34 = 1188 volts. (c) From Appendix I, page 465, for 000 conductor and 48-in. spacing, the reactance per conductor is 0.692 ohm per mile.

Total reactance per conductor, $X = 25$ X 0.692 = 17.3 ohms. *Ans.* The reactance drop $IX = 142.5$ X 17.3 = 2,470 volts. *Ans. (d)* Applying equation (90), using volts to neutral *(ER* = 16,500 volts), cos $\theta = 0.85$ $\theta = 31.8$ sin $\theta = 0.527$ $Eg = \sqrt{(16,500 \times 0.85 + 1188)^2 + (16,500 \times 0.527 + 2470)^2} = \sqrt{(15,220)^2 + (11,170)^2} = \sqrt{357} \times 10^6 = 18,900$ volts,

The voltage at the generator = 2 X 18,900 = 37,800 volts. *Ans. (e)* The line *regulation* is defined as the *rise in voltage when full load is thrown of the line, divided by the had voltage.* ,,,. 37,800-33,000

Regulation = — -— 5-0-,.-— or 14. 4 per cent. *Ans.* **164. Three-phase Line Calculations.** — The advantage of working transmission line problems to neutral is much more obvious in three-phase lines than in single-phase lines. Figure 354(a) shows a three-phase system, each conductor of which has a resistance of R ohms and a reactance of X ohms. The voltage to neutral at the load is *ER* and the voltage to neutral at the sending end is *Eg.* In order to determine the line characteristics, one phase

is removed, Fig. 354(6), and its characteristics determined. Under the condition of balanced load, which is assumed, the relations in all three phases are similar, so that the results obtained with one phase may be applied to the other two. As each pair of wires is the common return of the third wire, no current returns through the ground under the balanced conditions assumed. As the voltage drop between the load neutral and the generator neutral is zero, the ground may be considered as a return conductor of zero resistance and of zero reactance, as was R x -A/vvww—'TNnreswi R X (a) Three-phase transmission line having resistance and reactance.

(i) One phase of 3-phase line.

Fio. 354.—Three-phase line having resistance and reactance.

done in the single-phase case. The load need not necessarily be Y-connected, as indicated in Fig. 354(a). The same method is used even if the load be delta-connected and there be no neutral. The delta-load is replaced by an equivalent Y-load and the computations are made for one phase only.

Example.—Solve the problem of Par. 163, assuming three-phase transmission, other conditions remaining the same. Power to be delivered, 4,000 kw. ; load voltage, 33,000 between conductors; distance, 25 miles; frequency, 60 cycles; load power-factor, 0.85; spacing of conductors, 48 in.; allowable line loss 10 per cent of power delivered. Find (a), (6), (c), *(d),* and *(e),* Par. 163. (a) The power per phase = 4,000/3 = 1,330 kw. The voltage to neutral $Ea = 33,000/\sqrt{3} = 19,070$ volts.

1 330 000 Current per conductor $I = $ J9 070,x Q85 = 82'3 amp' Allowable loss per conductor = 1,330 X 0.10 = 133 kw. = 133,000 watts.

1 Do A/y) Resistance per conductor $R,- -/go'gz = 19-64$ ohms. Resistance per mile = 19.64/25 = 0.786 ohm.

From Appendix H, page 464, the wire having the next lowest resistance per mile is No. 1 A. W. G., the resistance of which is 0.665 ohm per mile. *Ans.* (6) Total resistance per conductor $R = 25$ X 0.665 = 16.6 ohms

IR = 82.3 X 16.6 = 1,365 volts. *Ana.*
(c) From Appendix I, page 465, for No. 1 wire and 48-in. spacing, the reactance is 0.734 ohm per mile.
Total reactance per conductor X = 25 X 0.734 = 18.35 ohms. The reactance drop IX = 82.3 X 18.35 = 1,510 volts. *Ans.* (d) From equation (90), using volts to neutral, *(En* = 19,070 volts). cos 6 = 0.85 0 = 31.8 sin 6 = 0.527 Eg = V(19,070 X 0.85 + 1,365) + (19,070 X 0.527 +1.510)2 = V(17,580)2 + (11,560)2 = V443 X 10s = 21,000 volts. The voltage between conductors at the sending end E,o = V5 X 21,000 = 36,400 volts. *Ans.* ,. _.,. 21,000-19,070 1,930,,,,, *(e)* Regulation =---or 1(U per cent' *Ans* 165. Lines Having Considerable Capacitance. — Heretofore the line capacitance has been considered negligible in its effect on the regulation. In long lines of high voltage the charging current, due to the line capacitance, may have a very considerable

— Ground-dbr Ground

Fig. 355.—Transmission line having resistance, reactance and capacitance.

effect on the regulation. Its tendency is to cause the voltage to rise from the sending end to the receiving end. The capacitance of the usual line is distributed uniformly along the line. The calculations are very considerably simplified, however, if the total capacitance C to neutral be divided, one-half being concentrated at the sending end and one-half at the receiving end, in parallel with the load, Fig. 355. This assumption introduces little or no error in the results, even for the longest existing 60-cycle lines. The condenser at the sending end has no effect on the regulation, but its charging current $Ic/2$ must be added vectorially to the line current / in order to obtain the total current supplied by the generator. The current /c/2 taken by the condenser at the load must be added vectorially to the load current IR in order to obtain the total line current I. The problem is then treated by the methods already outlined. *Example.*—It is required to deliver 30,000 kw. at 0.80 power-factor at a distance of 100 miles, with a line loss not exceeding 10 per cent. of the power delivered. The volt-

age at the load is 120,000 volts, 60 cycles, and the lines are arranged at the apexes of an equilateral triangle, 12 ft. on a side. Determine: *(a)* the line regulation; (6) the total power supplied by the generating station.
The power per phase, *P.* ??fO = 10,000 kw.
o
The volts to neutral at the load, ER = 120.000/ V3 = 69,300 volts. The current per conductor at the load, / =69(1)786 = 18'5 amp'
The power loss per conductor = 10,000 X 0.10 = 1,000 kw. = 1,000,000 watts.
1 000 000 The conductor resistance K =-/,-0/rvv: = 30.7 ohms. (180.5)2
30 7
Res. per mile =,7, = 0.307 ohm.
1UU
From Appendix H, page 464, the wire having the next lowest resistance per mile is 0000, the resistance per mile of which is 0.264 ohm.
The conductor resistance R = 100 X 0.264 = 26.40 ohms.
From Appendix I, page 465, the reactance per conductor per mile for 0000 wire and 144-in. spacing is 0.810 ohm.
Total reactance = 100X0.810 = 81.0 ohms.
The charging current for 0000 wire with 144-in. spacing and 100,000 volts to neutral is from Appendix J, page 466, 0.523 amp. per mile.
The total charging current for the above line is fiQ OAT) /« = 0.523.X X 100 = 36.2 amp.
As only one-half the line capacitance is assumed at the receiving end, the charging current flowing over the line, $Ie/2$ = 36.2/2 = 18.1 amp.
In order to find the total line current, however, this 18.1 amp. must be added vectorially to the 180.5 amp. of load-current. Therefore, Fig. 356, the load current is resolved into an energy component / cos 9 = i = 180.5 X 0.8 = 144.4 amp., and a quadrature component *I* sin6 = i2 = 180.5 X 0.6 = 108.3 amp.
As the quadrature component, 108.3 amp., lags the load voltage by 90 and the charging current, 18.1 amp., leads the load voltage by 90, the resulting

quadrature component is *i' =* 108.3-18.1 = 90.2 amp. The total line current /' = V(144.4)2 +190.2)' The voltage at the sending end /c/2-18.1
Fm. 356.—Effect of line charging current on total line current.
170 amp. Eg = VT69,300 X 0.8 + 170 X 26.4)2 + (69300 X 0.6 + 170 X 81.0)« = V(3,590 + 3,060) X 10" = 81,500 volts. ,, T-,, 81,500-69,300 12,200 (a) Line regulation --(59300 = 69 300r per ' 8' (6) The total line loss *P'* = 3 X (170)2 X 26.4 = 2,290 kw. *Ans.* The total generator power Pa = 30,000 + 2,290 = 32,290 kw. *Ans.* 166. Corona.—Figure 357 shows a tapered conductor whose diameter at the large end is about a half-inch. This conductor tapers gradually to a point. It is suspended vertically in air with its tip about 18 in. from a conducting sheet or plate, which is grounded. The secondary terminals of a high-voltage transformer are connected, the one to the tapered conductor and the other to the plate.
A low voltage is first applied to the transformer and the voltage is then gradually increased. When the secondary voltage is in the neighborhood of from 3,000 to 4,000 volts, a bluish discharge occurs from the pointed tip of the conductor. This may be plainly seen if the room be darkened. As the voltage is increased, the bluish discharge forms further up on the conductor and surrounds it in a ring. When the voltage reaches the neighborhood of 100,000 volts, this bluish discharge may have formed on the rod up to a point where the diameter of the rod is about in. Meanwhile the discharge from near the pointed end, and the accompanying hissing sound, will have become quite vigorous.
Step-up
Transformer O
O o - Ground Fig. 357.—Formation of corona on a tapered conductor.
This bluish discharge is called *corona.* It occurs when the electrostatic stress in the air exceeds about 75,000 volts maximum per inch, or 53,000 volts effective per inch. At this voltage gradient the number of electrostatic lines per unit area becomes too great for

the air to withstand. (See Vol. I, Chap. IX, p. 200.) This is the reason why corona first appears at the sharp point. The electrostatic flux lines are more concentrated at points. This is illustrated in Fig. 358, which shows a conducting body suspended in air, the potential of the body being considerably above ground potential. The electrostatic lines leaving this body are indicated. They are much more dense at those parts of the surface having a smaller radius of curvature.

When air is so highly ionized that corona forms, its dielectric strength is practically nil, and the air may be considered as broken down or disrupted electrically. Under these conditions, the air becomes a partial conductor and is practically valueless as an insulator.

Fio. 358.—Effect of radius of curvature on distribution of electrostatic lines.

Corona is always accompanied by the production of ozone, the odor of which is readily'detected. In the presence of moisture, nitrous acid forms when corona occurs. The acid and ozone may attack metals and other substances, such as insulating materials. When corona occurs, the resulting ozone is very active chemically.

Corona is accompanied by a dissipation of energy. If a transmission line be operated at a sufficiently high voltage, corona loss occurs. Where a line is long, the loss becomes serious and must be considered when the line is designed. The loss may be reduced by increasing the diameter of the conductors and thus increasing their radius of curvature. This fact favors aluminum for transmission line conductors, other factors being equal. Figure 359 shows the conductors of a highvoltage line illuminated by the corona discharge. (For a more complete discussion see "The Law of Corona and the Dielectric Strength of Air," by F. W. Peek, *Trans. A. I. E. E.*, Vol. XXX (1911), p. 1889.) LIGHTNING ARRESTERS 167. Multigap Arresters.— Abnormal voltage rises occur in power systems due to lightning discharge, switching, shortcircuits, and other disturbances. These voltage rises may damage the system and any connected

apparatus, by puncturing insulation, by producing insulator flash-overs, arcing grounds, etc. It is therefore highly desirable to relieve the line of such disturbances whenever possible. This is done by means of lightning arresters, whose function is to relieve any abnormal voltage rise by passing current to ground and therefore preventing damage to the system and connected apparatus.

Lightning arresters are connected between the line to be protected and the ground. They must have four properties. They 359.—Illumination of transmission line by corona.

should be practically an open circuit when the line is at normal voltage. They should provide an easy path to ground for the discharge. They should be able to absorb the energy of the discharge. They should be able to suppress the dynamic arc which follows the transient discharge and which the power of the system tends to maintain.

One type of arrester for low voltage is shown hi Fig. 360. A number of cylinders made of non-arcing metal are connected between the line and ground. There is a small air-gap between adjacent cylinders. A carbon rod of high resistance is shunted from the conductor across approximately three-fourths of these cylinders, a medium resistance across approximately one-half the

Fig. 360.—General Electric multignp lightning arrester.

cylinders, and a comparatively low resistance across a little over one-quarter of the cylinders. The cylinder spacing is such that the full line pressure which exists across the last five cylinders cannot jump the series gaps. However, any considerable increase of line voltage causes a discharge through the resistance, across these five gaps and thence to ground. If the discharge becomes sufficiently heavy, the voltage drop through the high resistance becomes excessive and increases the voltage across the next four gaps which then break down and assist in the discharge. In a very heavy discharge, the voltage drops across all the resistances become large and the discharge passes to ground through the entire series of gaps. When

the line returns to normal voltage, the cooling effect of the large number of cylinders, combined with the rectifying property of their metallic vapors, tends to prevent the dynamic arc being sustained. Such arresters are suited only for low voltages (up to 5,000 volts), and can absorb only small amounts of energy. 168. Horn Gaps.—For high voltages, the horn gap, Fig. 361, is often used. The gap consists of two horns, each mounted on an insulator, and the gap itself is located between the lower parts of the horns. One horn is connected directly to the line to be protected and the other is connected through a resistance, usually water, and a choke coil to ground. The gap is so set that ordinary operating voltages cannot jump it. When the voltage rises so that it is from 150 to 200 per cent. of its normal value, it jumps the gap and the disturbance passes to ground. The resistance and choke coil limit the current and so prevent the line being grounded by the arc. The function of the horns is to break the arc. An arc tends to rise because of its heat, and also because of the well known law that a current tends to form as large a loop as possible, in order to make the permeance of the magnetic circuit a maximum. (See Vol. I, p. 12, Par. 17.)

Horn gaps are not altogether satisfactory, because they often arc over unnecessarily; the protection which they afford is insufficient because of the resistance and choke coil, and they do not always suppress the dynamic arc which follows the transient discharge. This results either in a permanent arcingground or the destruction of the gap. 169. The Aluminum Cell Arrester.— The aluminum cell arrester has proved to be the most reliable type of arrester and is now universally used on large power systems. It is based on the following principle: If aluminum be immersed in certain electrolytes and a direct-current voltage be impressed on the aluminum and the electrolyte, no appreciable current flows, except for an instant. This is due to the fact that the current builds up a very thin film of aluminum oxide on the plate, which acts as an insulator. This film builds up with alternating current as well as with direct current. The oxide, however, constitutes an insulator and also a dielectric of almost infinitesimal thickness. Therefore, considerable capacitance exists between the plate and the electrolyte. This would result in a considerable charging current through the arrester, with alternating current, if it were connected directly across the line.

This film is an excellent insulator up to approximately 340 volts effective. When the voltage exceeds this critical value, the film breaks down and allows a large current to pass. When the voltage again drops below this critical value, the film re-forms and stops the current flow. Hence, such a device is an electrical safety valve. Its characteristics are therefore ideal for a lightning arrester. Several cells are always connected in series, the number of cells depending on the voltage. The oxide films prevent any discharge so long as the line voltage is normal. If the line voltage becomes abnormally high, the aluminum films are broken down and the discharge passes readily to ground. When conditions again become normal, the films re-form and no power arc can follow the discharge. In practice, the aluminum is in the form of cones, the proper

Fig. 362. — Cross-section of aluminum cell arrester.

number being clamped together in a stack, Fig. 362. Each cone is about half filled with electrolyte. The entire stack is immersed in oil, as the oil acts as an excellent insulator and also absorbs the energy of the discharge.

Were the stack connected directly across the line, the charging current to the stack would cause considerable heating in the cell and would therefore reduce its capacity for absorbing the energy

Fig. 363.—37,000-volt aluminum cell lightning arrester for outdoor service on 3-phase, ungrounded circuit.

of the discharge. Consequently, there is a small horn gap in series with each arrester, as shown in Fig. 363. The gap is very short in comparison with the arcing distance of the circuit, so that it does not interfere to any extent with discharges occurring during abnormal voltage rises. The use of spheres for the gaps, which are shown in Fig. 363, increases the speed of the gaps in discharging high-frequency impulses.

In time, the film dissolves and it is necessary to re-form it about every 24 hours. This is done by closing the auxiliary gap (lower gap, Fig. 363), which allows the arrester to charge through a carbon-rod resistance and so re-forms the film. The purpose of the carbon rod is to limit the charging current and to damp out high-frequency disturbances that might otherwise occur when the arrester is being charged.

In arresters whose rating exceeds 12 kilo volts, there are three stacks for a grounded 3-phase system, each stack being connected

Ground Stack/

Fig. 364.—Connections of aluminum-cell arresters on 3-phase, ungrounded system.

between its respective conductor and ground. In a nongrounded system, the three line stacks are connected *in* a common Y, and a fourth stack, called the ground stack, is connected between the neutral of the Y and ground, Fig. 364. In Fig. 364, No. 3 is the ground stack. By revolving the transfer switch through 180 in a horizontal plane, No. 2 becomes the ground stack and No. 3 is connected between the middle conductor and the system central.

Lightning arresters should be connected to the incoming line where it enters the station, or even outside the station. Choke coils, consisting of a few turns of bare wire, are connected between the arresters and the station bus-bars, Fig. 365. When a surge reaches the station, it has a choice of two paths, the inductive path through the choke coil into the station and the condensive path through the arrester to ground. Obviously, a surge, being of high frequency, will take the path to ground through the arrester, whose condensive reactance is low at high frequencies.

To High Tension
Bus-bara

Fio. 365.—High-voltage entrance and connections to lightning arresters and bus-bars.

TRANSMISSION LINE CONSTRUCTION 170. Pin-type Insulators.—The success of any transmission line depends to a large extent on the insulators. Little or no difficulty is encountered in insulating low-voltage lines. Pintype insulators are always used for such lines, because they are cheap, are easy to install and act as rigid supports for the conductors. Pin-type insulators are made of glass, of porcelain, and of patented compounds. Glass is suitable for lines of light construction, such as telephone lines, and for power lines of moderate voltage. Its advantages up to 10,000 or 15,000 volts are its cheapness and the fact that cracks and flaws are readily detected. On the other hand, it is hygroscopic and breaks readily.

Porcelain has excellent mechanical and electrical characteristics, but is more expensive than glass. Internal flaws are not readily detected and cracks in the porcelain cause rapid deterioration of the insulator. Porcelain is practically the only material used for insulators on high-voltage power lines.

Patented compounds have good mechanical characteristics and are readily moulded to any desired form. They cannot withstand the severe mechanical stresses combined with the electrical stresses and weathering encountered in power lines.

In the larger sizes of pin-type insulator, the insulator is made up in sections cemented together, Fig. 366. Pin-type insulators

Fio. 366.—Typical 77,000-volt. pin-type insulator.

can be safely used for voltages up to about 66,000 volts, but for these high voltages they are large, expensive, and produce excessive torsion in the cross-arms. 171. Suspension-type Insulator.—It seemed at one time as if the insulator would limit transmission voltages, as the pintype had practically reached its limit in size, weight, and cost. The introduction of the suspension-type insulator, however, has raised the limit of transmission voltages to more than double the value possible with the pin-type insulator. With the suspension type of insulator, the conductor is suspended instead of being rigidly supported. A string of suspension insulators is made up of several units in series, the number of units depending on the voltage. A single unit can safely operate at from 16,000 to 25,000 volts, depending on local conditions. Under normal conditions, the insulator string acts as a flexible support for the conductor and offers little or no resistance to horizontal forces. Hence, the stresses in adjacent spans should be nearly balanced or the string will be pulled out of the vertical line. When a span breaks, the string is thrown temporarily into the adjacent unbroken span as a strain or dead-end insulator. Suspension insulators are also used as strain insulators at dead ends, railroad crossings, etc. Figure 367 shows a section of a link-type suspension insulator in which the suspension loops link each other. Figure 368 shows a string of such insulators arcing ovcrundei high voltage.

Fig. 367.—Section of link-type suspension insulator.

Fio. 368.—Arc-over, at 60 cycles, of a 6-unit string of link-type insulators.

172. Transmission Structures.—There are three general types of transmission structures employed in this country, wooden poles, steel poles, and steel towers. Concrete poles are used occasionally.

Wooden poles are used on the lighter lines, especially where the voltage is low. Wooden poles have the advantage of being cheap, particularly when used near wooded sections. They are also light, easily fitted and erected. On the other hand, their life is comparatively short no that they require frequent renewals. They are not sufficiently strong for heavy lines operating at high voltage. Owing to the limited height of wooden poles, the spans must be short.

Steel poles are ordinarily made of four main members supported and braced by lattice work, Fig. 369, and are usually set in concrete. This type of pole is strong and, if painted occasionally, has a long life. It does not re-quire a wide right of way. It is particularly useful in mill yards and along railroad tracks, where the space is limited, Except for moderate heights, however, towers are cheaper than steel poles, especially in this country, where labor costs are high.

Steel towers are a development of the windmill tower so common in this country. They are ordinarily composed of four main members braced by light cross-members. They are stronger and more rigid than either the wooden or the steel pole. As they are made of a comparatively few standard members, riveted or bolted together, the labor costs are comparatively low. Owing to the spread of the four main members, they are able to resist the high torsional stresses such as would result from the breaking of the conductors on one side. Towers may be set in concrete bases. This is necessary if the ground is marshy. A less expensive method is to rivet plates or feet on the bottom and bury the lower supports directly in the ground. The towers are usually shipped "knocked down" and are assembled on the spot by the erecting crew. Figure 370 shows a transmission tower of unusual height which supports the power lines of the Eastern Connecticut Power Company at a river crossing.

A cheaper form of transmission line structure is the flexible tower. This form of tower is based on the principle that if the stresses in two adjacent spans are equal, the structure acts merely as a prop which supports the line but which need not resist longitudinal forces. Flexible towers, Fig. 371, are merely A-frames designed to withstand the maximum transverse stress which may occur, but are not intended to withstand stress in the direction of the line. When these towers are used, an anchor tower about every mile is necessary, in order to take care of any unbalanced longitudinal forces which occur when conductors break. When suspension insulators are used, a steel ground wire is necessary at the top of the structure to give longitudinal support to the tower. The advantage of flexible tower construction lies in the fact that the towers are usually assembled complete in the shop and are easily erected.

Fig. 371.—132,000-volt. singlecircuit. A-frame, flexible tower of SUB-STATIONS 173. Transformer Sub-stations.—The function of the substation is to receive the electrical energy, usually at a voltage too high for commercial purposes, and to deliver this energy at other voltages and sometimes at other frequencies such as may be required for the district served.

The sub-station may be a transformer station only, receiving energy at a voltage of 26,400 volts, for example, and transforming it to 2,300 volts for general distribution. Figure 372 shows the wiring diagram of such a-station. Two distribution lines leave the station at 2,300 volts, one for lighting and one for power. Power loads and lighting loads should be kept separate, if possible, in order to avoid the nickering of lamps when the motor loads are thrown on and off the line. Usually 2,300 to 230-115 volt transformers are used to step down the voltage for lighting purposes, a three-wire system being employed for the secondary. (See page 379, Fig. 343.) Owing to the possibility of the low-voltage wires coming in contact with high-voltage wires, and so exposing the consumer to danger, one wire of the secondary of lighting circuits, usually the neutral, should be grounded at each consumer's premises. As motor loads are usually three-phase, two V-connected transformers, three single-phase transformers, or a single three-phase transformer may be used for stepping down the voltage. In order to save secondary copper, motors are often operated at 440 or 550 volts. Some large consumers, employing a few large motors, may operate them at 2,300 volts and thus eliminate the step-down transformers.

174. Motor-generator and Synchronous-converter Substations.—It is often necessary to obtain direct current, either for power supply to a thickly populated district or for electric railways. As has been pointed out (page 342, Par. 137), either the synchronous-motor-generator set, the induction-motorgenerator set, or

the synchronous converter may be employed

S Ph. Incoming Lines 26,400 V. L.A.

8-Ph. Outgoing Lines 2300 V.

Fig. 372.—Typical connections for a transformer sub-station.

for changing the alternating-current supply into direct current. The advantage of the synchronous-motor-generator set is that its power-factor may be controlled; its disadvantage is its tendency to fall out of step when line disturbances occur. The advantage of the induction-motor-generator set is that the induction motor tends to continue operating even when severe line disturbances occur; the induction motor does not require directcurrent excitation; it is very rugged. Its principal disadvantages are that it takes lagging current and at light loads its powerfactor is low.

Fig. 373(a).—Switch assembly with cells, mechanism, etc.

The advantages and disadvantages of the synchronous converter as compared with motor-generator sets have already been discussed in Par. 137, page 342.

175. Oil Switches.—With an ordinary air-break switch, it is practically impossible to break a high-voltage circuit supplying any considerable amount of power. Special air-break switches are in use for interrupting high-voltage circuits, but the knife blades of these switches are from 4 to 6 ft. long and the switch is provided with horn gaps. Such switches are suited only to outside mounting, where there is ample space for the resulting arc. The power rating of such switches is very limited. To interrupt high-voltage circuits, especially where the power is *Fio.* 373(6).—Details of individual pole.

Fio. 373(o) and (6).—General Electric triple-pole, 15,000-volt. 1200-amp., motor-operated, oil circuit breaker with interlocks between the mechanism, the cell doora, and the disconnecting switch.

large, the switch contacts must be immersed in oil in order to quench the resulting arc. When the voltage is even moderately high, a separate compartment for each phase is necessary. The

switches usually have double breaks, Fig. 373(a) and (6), and the energy concentrated at each break is half the total energy. The effect of the oil is to cool and quench the arc while the circuit is being opened. The heat of the arc tends to carbonize the oil so that it is occasionally necessary to renew the oil. During short-circuits, the switch may be called upon to absorb a large amount of energy in a very short time. The resulting pressure within the switch compartments may be very high, so that it is necessary to construct the tanks of heavily riveted or welded steel. Even so, explosions of switch cells are not uncommon.

Due to the fact that carbonized oil may form a conducting path between switch contacts, there is always a possibility of injury to persons working on the supposedly dead side of the switch. Therefore, it is always desirable to have an air-break disconnecting switch in each phase. The disconnecting switch may form a part of the switch, as in Fig. 373, or it may be installed on a separate outside mounting. (See wiring diagram, Fig. 372.) The disconnecting switch is not called upon to interrupt the circuit under operating conditions, but is opened only after the oil switch has interrupted the circuit.

Practically all oil switches operating at high voltages, or connected in circuits of considerable power, are operated by remote control. Both solenoids and motors, energized from a low-voltage circuit and controlled from the switchboard by low voltage, are used to operate the oil switch.

The switch of Fig. 373 is motor-operated. The motor winds a spring immediately after the switch has operated, leaving the spring ready to open or close the switch, depending on what the next operation is to be. Two separate compartments per pole are used, Fig. 373, one for each contact. This makes the energy per cell half that which would exist if a single tank were used. The oil baffles shown in Fig. 373(6) are particularly important.

176. Arrangement of Apparatus in Sub-stations.—The purpose of the sub-sta-

tion building is to protect the equipment and the operator from the weather. The incoming high-voltage lines are brought in either through the roof, by means of roof-bushings, or through the sidewalls by means of wall-bushings, Fig. 365, page 401. The incoming wires are bent to form drip loops so that water will not run down the wires into the station.

The high-voltage bus-bars are usually located near the roof of the station so as to be out of the way. It is also desirable to place other high-voltage equipment, such as lightning arresters, oil switches, etc., on some form of balcony or else inside an enclosure so that the possibility of personal contact is minimized. 177. Automatic Sub-stations.—In order to eliminate the cost of having an attendant in the smaller sub-stations, automatic sub-stations have been developed. These are particularly adapted to electric railway work. After the trolley voltage in the vicinity of the station has fallen below a predetermined value and remained there for a minute or so, a combination of relays and switches starts up one of the synchronous converters or

Fig. 374.—Automatic railway sub-station, Pacific Railway, showing 1000-kw. 50-cyele, synchronous converter and automatic control equipment.

motor-generator sets and connects it to the trolley line. If the load on the station exceeds the safe load of the machine in service at that time, another machine starts up automatically and after it is connected across the line, the field rheostat operates to make it take its share of the load. Likewise, the machines drop out of service automatically after the load has fallen below a predetermined value. Fig. 374 shows the interior of one of these stations. 178. Outdoor Sub-stations.—When the voltage is high, the clearances required by the high-tension leads and bus-bars within a sub-station may require a large building and hence a considerable investment. The investment in equipment and in buildings situated along transmission lines and supplying small loads may be large compared with the kilowatt-hours consumed. Sub-stations for small loads would not be economically possible were it necessary to place all the apparatus within a building. Transformers, switches, lightning arresters, have been designed

Fio. 375.—Outdoor sub-station, showing transformers, oil switches, bus-bars, etc. *(Dallas Power and Light Co.)* so that it is possible to operate them out of doors. The building needs only to house the switchboard and the operator, if one is necessary. The oil switches, the lightning arresters, the transformers, and the bus-bars can all be placed out of doors. The apparatus must be practically air tight to keep out moisture. Outdoor sub-stations on a large scale are highly developed at the present time. Figure 375 shows an outdoor sub-station of moderate size.

CHAPTER XIII

ILLUMINATION AND PHOTOMETRY

Light is a form of radiant energy and is probably due to vibrations set up in the ether by luminous bodies. It has the property of producing the sensation of vision on the retina of the eye and so enables objects to be seen and distinguished.

Illumination means specifically the light incident on a surface or object, but in a broader sense it has come to signify that branch of engineering having to do with the distribution and utilization of light. The measurement of light and light distribution is called *photometry*. 179. Candlepower.—The brightness of a light source is called its *luminous* intensity. The luminous intensity of a body is measured in terms of the light intensity in a horizontal direction given by a standard candle, and is called candlepower. Candlepower is denoted by /. 1 That is, if a light source, such as an incandescent lamp, were replaced by 14 standard candles without altering either the total light emitted or its distribution, the incandescent lamp would have a luminous intensity in a horizontal direction of 14 candlepower.

Candles of standard dimensions, burning under standard conditions, have in the past been used as standards of luminous intensity. Owing to the difficulty of reproducing such a standard with a sufficiently high degree of precision

and owing to the variation of its luminous intensity with atmospheric conditions, etc., the candle has not proved an acceptable standard, particularly at the present time when a high degree of precision in light measurements is necessary.

No perfectly satisfactory standard of luminous intensity has as yet been devised. At present the Bureau of Standards maintains incandescent lamps which constitute a standard of luminous intensity at some known voltage. These lamps are constant for a considerable time if used only occasionally. By means of these lamps, secondary incandescent lamp standards may be calibrated and used.

i Photometry symbols will be found often to duplicate electrical symbols. For example, / = candlepower and in electrical units / = current. Photometric and electric units are not of the same character. 180. Unit Solid Angle or Steradian. — In order to understand the fundamentals of light emission and distribution, it is necessary to know what is meant by *solid angle. A unit solid angle is the angle at the center subtended by a unit area on the surface of a sphere which has a unit radius.*

Figure 376 shows a sphere whose radius is 1 ft. An area of 1 sq. ft. on its surface subtends a conical solid angle at the center. This angle is a unit solid angle, finq.fi. sometimes called the *steradian.*

As the area of the surface of a sphere is equal to 4Tr2, there must be 4ir units of solid angle about the center of a sphere. This may be seen by letting $r = 1$. If any area on the surface of a sphere be divided by the square of the radius, the result is the solid angle that this area subtends at the center.

Example. — A certain sphere is 3 ft. in diameter. How many unit solid angles does an area of 3 sq. ft. on its surface subtend at its center. The number of unit solid angles 1.33 steradians. *Ans.* 376. —Unit solid angle.

1 Lumen =,.,..2 181. Luminous Flux: Lumen. — Light may be considered as a flux which emanates from a luminous source in the same way that magnetic flux emanates from a magnetic pole.

The amount of illumination emitted by a luminous source may be considered as being the total light *flux* emanating from that source.

Figure 377 shows a candle placed at the center of a sphere whose radius is 1 ft. Assume that this candle emits light uniformly in all directions, the intensity being equal to that in the horizontal plane or one candlepower. (A candle of this type is never met with in practice but is given here merely for purposes of illustration. The ordinary standard candle emits an intensity of one candlepower in the horizontal plane only, the intensity in other directions being much less than one candlepower.) Let *B* be a unit solid angle at the center subtended by an area of 1 sq. ft. on the surface of the sphere. A certain amount of light flux will be confined by this unit solid angle and as light flux is. emitted radially in straight lines, no flux enters or leaves the solid angle through its sides.

Fio. 377.—One lumen, the unit of light flux.

The light confined by this unit solid angle and coming from such a standard candle is the unit of light flux and is called the *lumen.* The number of lumens is denoted by *F.*

As there are *4-ir* units of solid angle at the center of a sphere, it is evident that each standard candle would emit *4ir* lumens if its light intensity were the same in every direction and equal to the horizontal intensity. The difference between candlepower and luminous flux should be clearly understood. The candlepower is *intensity* of light emission and may vary in different directions. On the other hand, luminous flux represents the *total* light emitted in any given region.

In the past, incandescent lamps have been rated on their mean horizontal candlepower, because in the carbon lamps, which were then the only type in general use, the shape of the filament and its distribution in the bulb were practically the same in all lamps. Therefore, all lamps had light distribution curves of the same general form. That is, the ratio of mean spherical candlepower to mean horizontal candlepower was practically constant in the lamps then in use.

With the advent of new types of lamps, the disposition of the filaments became quite different in the various lamps and the mean horizontal candle-power was no longer a measure of the total light output of a lamp. The feeling at the present time is that a lamp should be rated according to the total light flux which it emits, or in other words, a lamp should be rated in lumens and not in mean horizontal candlepower.

The mean spherical candlepower, which is the average of the candlepower emitted in all directions, is also a measure of the total light flux emitted by a luminous source.

A luminous source which has a luminous intensity of one candlepower in every direction has a mean spherical candlepower equal to 1.0 and emits 4rr lumens. *Therefore, the number of lumens emitted by a light source is equal to 4w times the mean spherical candlepower. Example.*—An incandescent lamp has a mean spherical candlepower of 20. How many lumens does it emit? The total light flux $F = 4ir\ 20 = 251$. 4 lumens. *Ans.* 182. Illumination.—Illumination is the amount of light flux or the number of lumens falling on a unit area. This corresponds to flux density in magnetism. It will be remembered that flux density is defined as the number of magnetic lines passing normally through a unit area. (See Vol. I, Page 7, Par. 13.) The unit of illumination is the *foot-candle* and corresponds to one lumen per square foot, the square foot being taken normal to the direction of the light flux. It is denoted by F the symbol E, where $E = -r-A$ is the area of the surface taken normal to the direction of the light flux. For example, in Fig. 377, one lumen is included by the solid angle B. If the sphere be thought of as hollow and having a radius of 1 ft., a square foot on its surface intercepts one lumen and the light flux is perpendicular to the surface at every point. Therefore, as the illumination is assumed to be equal in all directions, the illumination at every point on the inside wall of this sphere is one foot-candle. Such uniform distribution of light seldom occurs in practice.

A sphere having a radius of 2 ft. has four times as great a surface area as a sphere having a radius of 1 ft. With a fixed luminous source at the center, both spheres intercept the same total light flux. The light *intensity* at the surface of the 2-ft. sphere is one-fourth the light intensity at the surface of the 1-ft. sphere. *Therefore, to obtain the illumination* in foot-candles, *on a surface* which is normal to the direction of the light flux, divide the candlepower of the light source by the *square* of the distance in feet from the light source to the surface illuminated.

Example.—A light has an intensity of 25 candlepower downward in a vertical direction. What is the illumination in foot-candles on a horizontal table 4 ft. below this light. 183. Law of Inverse Squares.—Figure 378 shows that portion of the light emitted by a certain source which is included within a given solid angle. Let Ai be a perfectly transparent surface at a distance Dl from the source. Let $A2$ be a similar surface at a distance $D?$ from the source. By geometry, the areas A and $A2$ are proportional to the *squares* of their distances from the apex of the cone or pyramid. That is Fig. 378.—Variation of light intensity with distance from source.

The light flux passing through A is equal to the light flux passing through Az, as none of the light flux passes out through the sides of the solid angle. If the light flux passing through Ai and $A2$ is the same, then the density of the light flux or the lumens per square foot must be inversely as the areas. *Therefore, the intensity of illumination from a point source varies inversely as the square of the distance from the source.*

Let Ei be the illumination on surface Ai and Ez the illumination on surface $A2$. Then Ei _ZV $E2$ IV

The above law of inverse squares is strictly true only when the light source is a point. It is impossible to obtain a point source in practice. With the usual light sources, no great error is introduced in assuming a point source, unless the illuminated. areas under consideration are very close to the source. With mercury tube lamps, Moore tubes,

and lamps having certain types of reflectors, the law of inverse squares must be applied with great caution.

Example.—A drawing board directly under an incandescent lamp and 4 ft. distant has an average illumination of three foot-candles. What is the illumination on the drawing board when the lamp is raised 2 ft.? $E2 = E2 = (4)2$ 16 Ei 3 $(6)2$ 36 JE« "It "1-333 foot-candles. *Ans.* DO 184. Absorption; Brightness.—When light falls on a surface or object, a certain amount of the light is either absorbed or transmitted and the rest is reflected. No substance reflects all the light that it receives, although highly polished surfaces reflect a very large proportion, as is shown in the table which follows. A white surface reflects a high percentage of the light falling on it and reflects all colors equally well. A pure black surface should reflect no light at all, but practically all black surfaces reflect a certain amount of light. When illuminated with white light, the color of an object as seen by reflected light is determined by its ability to absorb, transmit and reflect the various colors of the spectrum. For example, a green object as seen by reflected light has the property of reflecting green and of absorbing or transmitting practically all other colors. Hence the object appears green by reflected light. The color of an object as seen by *transmitted* light is frequently different from its color as seen by *reflected* light. For example, the color of thin gold leaf as seen by reflected light is yellow, whereas its color as seen by transmitted light is greenish.

The unit of *brightness* in the metric system is the *lambert,* which is one lumen per square centimeter. Brightness may also be measured in candles per square inch. These units are used when the brightness of a luminous source, such as an incandescent filament, is under consideration.

The brightness of a surface is the number of lumens per unit area which the surface *emits* in the direction of the normal. Let it be expressed by E'. E' is always less than the illumination E, as the surface absorbs some light. The ratio E'/E is called the *coefficient of reflection* It is the ratio of the light emitted to the light received and is aways less than unity. Below are given the coefficients of reflection for various well-known surfaces.

Coefficient Of
Per Cent. Reflection
New aluminum bronze (unprotected) 54
Polished brass 60
Baked white enamel (paint) 72
Matt surface, porcelain enamel 79
Silvered mirror 83
White matt surface paper (smooth) 57
Light buff surface paper (smooth) 45
Embossed gilt paper 43
Light-blue paper 12 185. Light Distribution.—Light, from sources such as incandescent lamps, arc lamps, etc., varies in intensity in different direc

Fig. 379.—Distribution of light-intensity about a 16-candlepower, carbon-filament lamp.

tions. The distribution depends not only on the light source itself but also on the reflectors, refractors and fixtures which are used with the source. In most light sources the light giving element is so designed that the horizontal intensity is nearly the same in all directions. This is particularly true of incandescent lamps and of arc lamps. 1 "Standard Handbook," Section 14.

The intensity in vertical planes passing through the axis of the lamp varies considerably in different directions. This is illustrated in Fig. 379, which shows the distribution in one vertical plane of the light from an incandescent carbon-filament lamp without a reflector. The intensity of the light in the vertically upward direction (180) is small owing to the presence of the base and socket and to the disposition of the filament. The intensity of the light in a vertically downward direction is also small.

These distribution curves are particularly useful in determining the suitability of a lamp for any particular purpose. The distribution may be modified by shades, reflectors, etc. (See Fig. 400, page 444.) As will be shown later, the

area of these distribution curves is not proportional to the total light flux emitted by the lamp.

186. Light Sources; Incandescence; Luminescence.—Light is emitted by sources under two conditions, incandescence and luminescence. Incandescence is produced by heating a substance to a high temperature, as in the incandescent lamp, the carbon arc, and the Welsbach gas mantle. The amount of light emitted by a substance increases very rapidly with its temperaature. In fact, the light emitted increases as the fourth power of its absolute temperature. Hence,-a small increase in the temperature of an illuminant results in a very large increase in the light which it emits. The light becomes more nearly white as the temperature rises. This principle should be kept in mind because it is the reason for the different efficiencies which are obtained with different types of lamps.

Light emitted by incandescent bodies is accompanied by high temperature and a corresponding dissipation of considerable energy as heat..On the other hand, a luminescent substance may give out light at moderate temperatures. Examples of luminescent light sources are the vacuum tube, the mercury arc, the flaming and luminous arcs, all of which will be discussed later. The firefly is an excellent example of a luminescent source.

187. Carbon-filament Lamp.—The requirements of an incandescent filament are that it shall be highly refractory, that is, the filament shall be able to withstand high temperatures without rapid deterioration, and the filament shall be mechanically strong at these high temperatures. Also, the electrical resistivity of the filament material should be so high that the length of the filament for a given resistance shall not be excessive. For many years the carbon filament was the only one that proved satisfactory. It consists of a filament of carbonized cellulose, bent into horseshoe form and operated in a vacuum. A later development was the "G.E.M." lamp in which the carbon is "metallized," that is, it is flashed in a gas rich in hydrocarbons. This gives the carbon filament a metal-

lic appearance and a positive temperature coefficient such as metals have, and permits the filament to operate at a higher temperature.

The purpose of a vacuum in a lamp is two-fold. A vacuum prevents chemical action on the filament and is an excellent heat insulator.

The treated carbon-filament lamp has an efficiency of about 3.0 watts per mean horizontal candlepower and the metallized filament an efficiency of about 2.5 watts per mean horizontal candlepower.

The objection to the carbon-filament lamp is its low efficiency. Remembering that the amount of light emitted by an incandescent source increases as the fourth power of its absolute temperature, carbon-filament lamps can be made to have very high efficiency by operating them at high temperatures. However, this increased efficiency is accompanied by rapid evaporation of the filament, resulting in a very short life. Carbon-filament lamps are therefore operated at such a temperature that their life is from 700 to 1,000 hours. Their candlepower may become so reduced after Iong use that it is more economical to discard the lamps than to continue their use at this low candlepower.

188. The Tantalum Lamp.—The metal tantalum was next used as a lamp filament. Because of the high temperature at which tantalum can safely operate, tantalum lamps are able to develop an efficiency of 2.0 watts per mean horizontal candlepower and at the same time have an average life of about 600 hours. One peculiarity of the lamp is that when alternating current is used the filament disintegrates rapidly. Owing to the high efficiency reached by the tungsten lamp, the tantalum lamp has been absolutely superseded. 189. The Tungsten Type-A and Type-B Mazda Lamps.—Tungsten is a metal of high resistivity and is capable of withstanding very high temperatures. It is therefore well adapted for use as the filament in incandescent lamps. In the early days, it was not possible to draw it through a die as ordinary wire is drawn. Powdered tung-

sten was mixed with an organic binder, such as cellulose, and forced through a die. The binder was then driven out by heating, leaving a porous and pitted tungsten filament which was very fragile. This type of filament was used in the Mazda-A lamp. In 1911, the process of drawing tungsten wire was perfected, so that the present-day filaments are rugged. Lamps using this drawn filament in a vacuum are called Mazda-B lamps. .03.04.05.06.07.08 Time in Second Fig. 380.—Current variation in tungsten lamp when switched in circuit.

Type-B tungsten lamps have an efficiency of about 1.2 watts per mean horizontal candlepower. In the large sizes, 150 watts and above, the efficiency may be as high as 0.90 watt per mean horizontal candlepower. The reason for this high efficiency is the very high temperature (about 3,000C.) at which it is possible to operate the tungsten without too rapid deterioration. The life of tungsten lamps is longer than that of most other lamps, the guaranteed life being about 1,000 hours.

Tungsten has a positive temperature coefficient so that its resistance at operating temperatures is several times that when cold. This results in a high initial current when the lamp is first switched in circuit. This is called "overshooting. " The relation of the current to time is shown in Fig. 380. Fuses in tungsten-lamp circuits have been known to blow as a result of this first rush of current, although the fuses were of ample capacity to take care of the steady operating current of the lamp.

190. Gas-filled Lamps; Type C.—Remembering that the light emitted by an incandescent body increases as the fourth power of its absolute temperature, it is obvious that there is still opportunity for increasing the efficiency of tungsten lamps if the temperature of the filament can be safely increased. The factor which limits the operating temperature of the Type-A and Type-B filaments is the volatilization or evaporation of the filament itself. This volatilization has two effects. The evaporation removes useful material from the filament, resulting in a reduction of

efficiency and in an ultimate burning out of the lamp. The tungsten vapor condenses on the bulb, blackening it and so cutting down the useful light by absorption. This last factor has in part been remedied by injecting a chemical into the bulb which tends to keep the tungsten deposit semi-transparent.

The gas-filled or Type-C Mazda lamp.is based on the principle of vapor pressure. The higher the pressure, the higher the temperature at which water and other substances evaporate. Water will boil at a very low temperature in a rarefied atmosphere, whereas under pressures greater than atmospheric its boiling temperature may become quite high. The same rule applies to tungsten. In a vacuum, the tungsten evaporates quite readily, which results in a more rapid deterioration of the filament. In the Mazda-C lamps, the bulb is filled with an inert gas such as nitrogen. This gas does not-enter into chemical reactions with the filament and yet it causes sufficient pressure to increase materially the evaporation temperature of the filament. Fl-381.—Gas-filled lamp (300 -11-1 f watt) and section of filament.

Due to the higher temperature of evaporation, the filament may be operated at a higher temperature without rapid deterioration. This gives not only increased efficiency but also a whiter light, more nearly resembling sunlight. This inert gas serves another purpose. It will be noted from Fig. 381 that this type of lamp has a long narrow neck. The gas coming in contact with the filament becomes heated and rises. If the lamp is in an inverted position, as it should be in commercial use, (Fig. 381), this heated gas passes up into the neck. It carries with it the tungsten vapor, which is thus deposited in the neck of the lamp where it cuts off no appreciable amount of light.

This gas sweeping through the filament has one very undesirable effect which is not present in a vacuum lamp. The gas carries heat away from the filament very rapidly by convection. This cooling of the filament tends to decrease the lamp efficiency. To minimize this effect, the filament, instead of being a

straight wire, criss-crossed back and forth on supports as it is in the Mazda-B lamp, is wound in the form of a very fine helix with the turns very close together, as shown in Fig. 381(6). This keeps the filament in very compact form and so reduces the convection losses.

Below is given a table of efficiencies for gas-filled lamps. It will be noted that the larger sizes have the higher efficiencies, due to their having larger filaments and hence less proportionate convection losses.

Rating—Gas-filled Multiple Lamps Watts Per Lumens

Watts M.H.CP. M.H.CP. Peh Watt 75 88 0.85 11.5 100 120 0.83 12.6 200 267 0.75 14.0 500 714 0.70 16.1 750 1,154 0.65 17.1 1,000 1,667 0.60 18.0

By using a bulb having a blue tint, the gas-filled lamp gives a light closely resembling daylight.

191. Effect of Voltage Variation on Incandescent Lamps.— As the voltage on commercial lighting systems will vary more or less from time to time, it is important to understand the effect of this voltage variation on the operation of incandescent lamps. An increased voltage results in a higher operating temperature of the filament and hence a higher efficiency. This is accompanied however by a decreased life of the lamp.

With an untreated carbon filament the effect of increasing the voltage is to *decrease* the resistance, as carbon has a negative temperature coefficient. Because of the decreased resistance of the filament, the current increases more rapidly than the voltage. Hence the power taken by the filament increases even more rapidly than the voltage squared. This makes the carbon lamp more sensitive to voltage changes than it otherwise would be. On the other hand, the metallized carbon filament and the tantalum and tungsten filaments all have positive temperature coefficients. An increased voltage is accompanied by increased resistance which tends to prevent the lamp taking more power.

95 100 106

Per Cent Rated Voltage

Fig. 382.—Effect of voltage change

on candlepower and watts per candle.

Therefore, these last three types of lamps are less sensitive to voltage changes than is the carbon lamp. Figure 382 shows the variation of candlepower and efficiency with per cent. of normal voltage for carbon and for tungsten lamps.

192. Arc Lamps.—The arc lamp was the first successful electric lighting unit. Its principle is the heating of the tips of carbons, or other electrodes, to incandescence by means of an electric current. This is illustrated by Fig. 383. Two carbon rods are connected in series across the lighting mains, a resistance R being in series with the rods.. If the carbon tips are first touched together, the heat developed at the point of contact produces a hot vapor which immediately becomes conducting. If the carbons now be drawn apart, more vapor will form and this hot vapor becomes a conductor of electric current. A large amount of heat energy is developed in a very small space, resulting in a very high temperature. This heats the carbon tips to incandescence and results in their emitting a very white light. Because of its very high temperature, the arc is a very efficient illuminant.

The resistance of the arc itself varies nearly inversely as its cross-section. If the carbons were connected directly across the line, a slight increase of current would result in a greater crosssection of arc which would reduce its resistance. The arc would then take more current resulting in a still less resistance, etc.

Fig. 383.—Direct-current arc and candlepower distribution curve.

which would ultimately produce such an extremely low value of resistance as to be practically a short-circuit. To prevent this instability in multiple lamps, a series resistance R, called the *ballast,* is necessary. The power loss in the ballast, reduces the over-all efficiency of the lamp. In a 110-volt lamp, the drop across the ballast is about 50 volts.

Figure 383 shows the early type of open, direct-current arc. A crater, formed in the positive carbon, becomes filled with molten carbon. This electrode is at a higher temperature than the negative one. Most of the light comes, not from the arc itself, but from this *hot* incandescent positive crater, so that the upper carbon should always be the positive one. Figure 383 indicates the light-distribution curve of this type of lamp. It will be noted that most of the light is thrown downward. The positive carbon is consumed more rapidly than the negative and so requires more frequent renewals.

If alternating current be substituted for direct current, it will be found that both carbons consume equally and have the same shape of crater, as is indicated in Fig. 384. Hence, as much light is directed upward as downward, as shown by the lightdistribution curve. Such a light needs, therefore, a reflector to intercept this upwardly directed light and reflect it downwards where it can be effectively used. The alternating-current multiple arc has one advantage over the direct-current arc in

A.C. Supply (HJOOOOOOOlT1

Inductive Ballast Fio. 384.—Alternating-current arc and candlepower distribution curve.

that it has an inductive ballast which consumes no appreciable power, although it does lower the power-factor.

193. The Enclosed Arc.—The next improvement in arc lighting was to enclose the arc in a small opalescent globe. This enclosing globe prevents free access of the oxygen of the air to the arc and so reduces the carbon consumption and therefore the number of trims. The opalescent globe cuts off considerable light and so reduces the lamp efficiency, but the total cost of illumination is reduced because of the lesser labor charge incident to trimming. The enclosed arc will burn 100 hours per trim where the open arc burns but 8 hours. The enclosed arc is also steadier than the open arc, gives a more diffused light and so reduces the objectionable glare of the open arc.

The open arc, especially the direct-current type, is used extensively for projection purposes, as for stereopticans, movingpicture machines, search lights, etc. The high intensity of the arc, combined with its being concentrated, makes it very desirable for these purposes. The foregoing types of arc lamps are practically obsolete as far as general illumination is concerned, although they are used for street lighting in isolated instances.

194. Arc Regulation.—Arc lamps must be automatically regulated both in starting and in operation. In starting, the carbons must be in contact and then be drawn apart as the arc is formed. The carbons must also be fed together as they are consumed.

Arc lamps are divided into two general classes, the series lamp and the parallel or multiple lamp.

In the series lamp, the *current* is maintained constant by the generating apparatus and this type of lamp requires no ballast. As the current in the series lamp is maintained constant under all conditions, the arc regulation is accomplished primarily by maintaining the proper voltage across the arc. The general scheme is shown in Fig. 385. The coil S is a series coil of low resistance connected in series with the arc and Sz is a shunt coil of high resistance connected across the arc. Si and $2 operate differentially on the lever L which actuates the clutch. Si tends to lift the clutch and pull the carbons apart and 82 tends to lower the clutch and allow the carbons to come together. Should the arc become too long, the voltage across it rises and strengthens the shunt solenoid S4. This tends to feed down the upper carbon.

A cut-out switch is necessary to keep the circuit closed should the lamp become open-circuited. When this occurs a high it-out Switch

Fio. 385.—Mechanism and connections of series arc lamp.

lalliist voltage will exist across S2 and will cause it to pull up its end of the lever and so close the cut-out switch. The circuit is then closed through the resistance R. Upon starting, the cut-out switch is normally closed. More current passes through the low resistance path formed by Si and the carbons than through R. Hence Si immediately pulls the carbons apart. This action ceases when the voltage across *S2* becomes sufficient to counteract the effect of *S*. The feeding mechanism is connected to a dashpot which prevents any sudden movement of the plungers.

The mechanism of the multiple lamp, Fig. 386, is very simple. The current passes through the ballast in which about 30 to 50 per cent. of the power is lost. The current then passes through two solenoids connected in series, each of which operates on one of the legs of a U-shaped plunger. Should the current become too great, the plunger raises the upper carbon by means of the clutch and thereby increases the resistance of the arc. This causes the current to decrease to its normal value. When the power is turned off, the upper carbon drops and closes the circuit so that the lamp is ready for operation.

196. Flame Arcs. — Approximately eighty-five per cent. of the light in the direct-current open arc comes from the incandescent crater in the positive carbon. In the enclosed arc the percentage is lower than this. In either case, probably not more than 10 to 20 per cent. of the light comes from the arc flame itself. In flame arcs almost all the light comes from the flame. It is found that by impregnating the carbons with certain salts, the arc itself becomes luminescent due to the vaporization of these salts and the luminescence they attain at the high arc temperatures. The color of the arc can be readily controlled by the kind of salt used. For example, calcium salts produce a yellow color, strontium salts a reddish tint and barium and titanium salts a brilliant white light.

Fig. 386.—Mechanism and connections of multiple arc lamp.

Deposit

Fio. 387.—Shortburning flame arc.

As the electrode products are utilized in the arc, rapid consumption of the electrodes results. This is accentuated by the fact that air circulation is necessary to remove the white powder deposited by the arc vapor. In the open type of lamp this ventilation means a continuous supply of oxygen, resulting in rapid consumption of the electrodes. The open type of lamp, therefore, re-

quires very long electrodes in order to obtain a reasonable number of burning hours per trim. These electrodes are so long that it is not practicable to place them one above the other, so they are both fed down from above and are placed at an angle, as shown in Fig. 387. A small magnet *M* blows the arc down. To increase the conductance of the electrodes, a metallic wire is usually run down their centers. This type of lamp is very efficient, requiring only 0.43 watt per mean spherical candlepower. Owing to the short life of the electrodes and the consequent high cost of trimming, it cannot be profitably used in this country for general illumination purposes. An unpleasant odor makes its use undesirable for interior illumination. In this country its use is confined to outside display lighting. Because of its high efficiency, this type is used extensively abroad as the higher energy costs and the lower labor costs existing there make its use economically desirable.

To eliminate the short-burning feature of the flame arc lamp and also the objection of the unpleasant fumes which accompany its use, a long-burning type of lamp has been developed. The arc burns in a chamber, the air supply to which is restricted. There is a large condensing chamber just above the arc chamber. The hot gases rise and pass into this condensing chamber, where the white powder is condensed and deposited against the comparatively cool surfaces of the chamber. The gases are then allowed to re-enter the arc chamber. This re-utilization of the same air prevents the access of a continuous supply of oxygen and gives the lamp about 100 burning hours per trim. Figure 388 shows the multiple lamp, together with the diagram of connections. The watts per mean spherical candlepower run about 0.74. Comparing this figure with 0.43 watt per mean spherical candlepower already given for the open flame arc, it will be seen that long life is obtained at the expense of efficiency. This type of lamp has one decided advantage over the magnetite lamp in that it can be used with alternating current.

Fio. 388.—Mechanism and connec-

tions of Westinghouse long-burning flame arc, D. C. multiple type.

196. The Metallic Electrode or Magnetite Lamp.—The metallic electrode arc lamp or luminous arc or magnetite lamp, as it is often called, differs from other arc lamps in that it employs metallic electrodes and the light is derived from the arc itself, being due to the luminescence of the vapor which comes from the cathode or negative electrode. The positive electrode is a large cylinder of solid copper. The copper being an excellent conductor of heat tends to keep moderately cool. The negative electrode is made of an iron oxide containing titanium to give a white light. Other ingredients are added to give the electrode desirable burning characteristics.

A diagram of the lamp is shown in Fig. 389. The arc stream consists of negatively-charged luminescent gas particles which originate at the negative electrode. The iron oxides and the titanium of the negative electrode give a white light. Being negatively charged, the gas particles of the arc stream are repelled by the negative electrode and are attracted by the positive electrode. Therefore, the arc stream moves from negative to positive. The copper must always be positive. If it becomes negative, the arc will then consist of luminescent copper-vapor and be green in color. In connecting a lamp in circuit, care must be taken to insure correct polarity.

As the electrodes are comparatively cool under operating conditions, there is not sufficient heat to maintain the arc if the electric power is interrupted even for an instant. Hence, out

Fig. 389.—Mechanism and connections of luminous arc lamp, series type.

side the question of the greenish arc resulting from the copper operating as cathode, this type of lamp cannot be used with alternating current.

The feed mechanism is slightly different in principle from that of other types of arc lamp. Were the ordinary type of feed used, the hot metal of the cathode or negative electrode would weld to the copper and "freeze" when the current was turned off. Therefore, the mechanism is so designed that the feed is intermittent, the arc being maintained by re-striking. When the lamp is out of circuit its electrodes are separated by a gap. When current flows, the starting magnet, Fig. 389, brings the lower electrode into contact with the upper electrode, striking a sharp blow. This operation allows the current to flow through the series magnet, thereby opening the circuit of the starting magnet at the cut-out contacts, allowing the lower electrode to fall, starting the arc. The lamp is then burning with the lower electrode holder resting on its stop and with the series magnet holding the cut-out contacts open. As the lower electrode is consumed, the voltage across the arc rises, the shunt magnet (which is connected in series with the starting magnet and starting resistance across the arc) lifts its armature, and closes the cut-out contacts when the arc voltage has reached a predetermined value. The closing of these contacts puts the starting magnet again in circuit, thereby causing the lower electrode to strike the upper electrode a sharp blow, bringing about a shorter arc once more.

The copper anode, which is comparatively cool, is consumed very slowly and needs to be replaced only at long intervals. The magnetite electrodes, however, need to be replaced at frequent intervals. The 4.4-amp. lamp will burn from 160 to 200 hours per trim. The 6.6-amp. lamp burns about 125 hours per trim. Magnetite lamps have a high efficiency. The 4.4-amp. lamps have an efficiency of approximately 0.70 watt per mean spherical candlepower. The 6.6-amp. lamps have an efficiency of 0.45 watt per mean spherical candlepower. The intense white light of this type of lamp makes it very attractive, particularly when the lamp is mounted on ornamental poles. In addition to general street lighting, it is used to a considerable extent for "white way" and boulevard lighting. (See Fig. 397, Page 443.) 197. The Mercury-arc Lamp; the Moore Tube; the Nernst Lamp.—The mercury-arc lamp consists of a long tube containing metallic mercury. The pressure in the tube is very low. When an arc is formed in the tube the mercury is vaporized and gives a greenish-blue light. This light is due almost wholly to the luminescence of the mercury vapor, the temperature of the vapor being only 250 to 300C. The lamp owes its high efficiency to the low temperature at which it is able to give its light. Most of the light emitted comes from the blue-violet end of the spectrum, there being little red. This absence of red rays makes people appear ghastly when viewed under this light, an objectionable feature in its use. Red light is tiring to the eye and contributes little to the visibility of objects. Therefore, from the physiological standpoint, the mercury-vapor lamp is excellent for drafting rooms and for other places where fine detail is concerned. Very satisfactory results are obtained if tungsten lamps are used in conjunction with the mercury-vapor lamps, thus doing away with the ghastly appearance of the workers.

The mercury-vapor lamp has an efficiency of about 0.75 watt per mean spherical candlepower. It is fundamentally a direct-current lamp, but with a compensator it can be used on alternating-current circuits. The connections when used with direct current are shown in Fig. 390.

A higher pressure mercury-vapor lamp has been developed operating at a higher temperature. The lowmelting point of glass necessitates the use of quartz tubes with this type of lamp.

In the Moore tube, the light is obtained from a luminescent gas at a comparatively low temperature. This is produced by high-voltage electric discharge in vacuum tubes. These tubes are sometimes

They have an

Ballast

Shifter

Fio. 390.—Connections of mercury-vapor lamp.

luminescence taking place 200 ft. long and are bulky per candlepower. advantage in that the color can be controlled by means of the gas used in the tube.

If carbon dioxide be used, for example, a light closely approaching daylight is obtained. This characteristic, together with its uniform distribution, makes it an excellent lamp for color matching, as with textiles. Its efficiency is about 2.5 watts per candlepower. The power-factor is low.

The Nernst lamp operates on the principle that porcelain, at a red heat, is a conductor of electricity. This lamp has a heating coil which makes a porcelain filament conducting. The lamp gives a substantially white light and gives off no odor or dirt. It does not require trimming. Its maintenance is high. Tungsten lamps have driven it from the lighting field. Its efficiency is about 1.3 watts per mean hemispherical candlepower.

As a maximum, the foregoing lamps emit as light energy from 2 to 5 per cent. of the electrical energy supplied to them. The remainder of the energy is liberated as heat. On the other hand, the firefly has a light-giving efficiency of 96 per cent. It emits very little heat energy. Although great improvements have been made in the field of illuminants during the past few years, there is still a tremendous opportunity for further advances.

PHOTOMETRY 198. Photometry.—Photometry is the measurement of light. Light measurements are nearly all made by direct comparison. The chief source of inaccuracy in making such comparisons is the question of color. Unless the lamps under comparison are of almost exactly the same color, only an approximate photometric balance can be obtained. This balance varies with different persons owing to the effect of different colors upon the eye. 199. The Bunsen Photometer.—The Bunsen photometer is the simplest type of photometric measuring device and is illustrated in Fig. 391. Assume that it is desired to measure the candlepower of the incandescent lamp L, using the candle C as a standard for comparison. The two lights are placed some 10 to 20 ft. apart and a movable screen S is placed between them. The screen S consists of a piece of paper or parchment with a grease spot in the center. The grease spot on the screen is translucent and so allows light to pass. If viewed from the side which is illuminated, the spot will appear darker than the rest of the screen, owing to the fact that light passes through the translucent grease spot more readily than through the surrounding part of.the screen.

On the other hand, if this same screen be viewed from the non-illuminated side, the grease spot will appear brighter than the rest of the screen, since it is more translucent than the rest of screen. Now if both sides of the screen receive equal illumination, the grease spot will look the same in comparison with the surrounding portion of the disc, when viewed from either side. When this occurs, a photometrical balance is obtained.

In order that the observer may view both sides of the screen simultaneously, two mirrors M, Fig. 391, are set at an angle and reflect the light in the manner shown by the dotted lines.

The screen S is moved until the screen looks the same on both sides, and the distances I and Zi are read. Let E be the candlepower of the candle or standard and Ei the candlepower of the test lamp. Remembering that light intensity varies inversely as the square of the distance.

Fio. 391. — Bunsen photometer.

The candlepower of the test lamp

If a standard candle is used, E equals 1.0.

If the two lights have different color, the two sides of the screen will never appear alike and only an approximate balance can be obtained. The position of balance is to a considerable extent determined by the personal equation of the observer.

Because of the unreliability of candles, standard incandescent lamps are used. These may be obtained from the Bureau of Standards. Such lamps, when used at the voltage at which they are calibrated, are very accurate standards. An arrow on the lamp indicates the position in which it was calibrated. It is customary to use the standard lamp only to calibrate a working standard so that the candlepower of the ultimate standard will not change due to its being in too constant use.

Lino

Fio. 392.—Connections for photometric test.

As the candlepower of lamps is very sensitive to changes in voltage, the connections are often made as shown in Fig. 392. Both lights are fed through an adjustable resistance #2. An adjustable resistance R is in series with the standard lamp and another adjustable resistance Ri is in series with the test lamp. A voltmeter is in parallel with each lamp. Both lamps are brought to the desired values of voltage by adjusting roughly Rz, and then by separate adjustments of R and R. Any fluctuations of line voltage can be taken care of by the resistance $R2$, which affects both lamps. Further, any unnoticed change of line voltage affects both lamps approximately to the same degree if the lamps have similar voltage-candlepower characteristics.

200. The Lummer-Brodhun Photometer.—The LummerBrodhun screen presents two clearly denned elliptical areas, as

Fio. 393.—Lummer-Brodhun photometer.

shown in Fig. 393. When a photometric balance is obtained, the two ellipses merge into one if the lights have the same color. The operation of this screen is as follows: Si, Sz is a white, opaque screen. The light coming from one source falls on the side Si and that coming from the other source falls on the side S2. The brightness of Si and of S2 depends on the intensity of the source which illuminates each. The light from Si and S2 is reflected by the plane mirrors Mi and M2 to the total reflecting prisms Pi and P2. The hypotenuses of the two prisms are in contact over a circular area only. The light striking this area of contact can pass through; other light is totally reflected. That is, only the central beam, shown dotted, from the mirror Mi, passes through to the eye piece. The remainder of the light is turned away. On the other hand, the central beam shown by a solid line, from Mz passes through the center circle and is absorbed by the walls of the box. The remaining light is reflected to the eye piece. The observer sees two distinct ellipses if the photometer is out of balance. (The circle of contact of the two prisms appears as an ellipse to the observer because he is viewing the circle at an angle.) When the transmitted light from Mi (center dotted line) is equal in intensity to the reflected light from M2 (two outside solid lines) the two ellipses have the same appearance. That is, when the two ellipses blend, the same illumination is coming from each source and the photometer is in balance. To eliminate any errors due to differences in the two sides of the screen Si and S«, the photometer screen should be reversed. PORTABLE PHOTOMETERS 201. The Sharp-Millar Photometer.—A portable photometer is necessary for making such tometers involve the same principles as laboratory photometers. The difference lies in the compactness and in the ease of manipulation.

The Sharp-Millar Photometer is typical of the portable type. A plan of the instrument is shown in Fig. 394. T is a tube which can swing through 180. At the elbow R is either a mirror or a white diffusing surface. The light entering the tube is reflected at right angles by R and is directed towards P, a LummerBrodhun Screen. This light is balanced against the brightness of a screen W, illuminated by a 6-volt tungsten lamp S, which is standardized. The viewing aperture is at 0. The screens, Si, \$2, 83 prevent stray light from the lamp S falling on the window W. The balance is obtained by moving the lamp S, which varies the illumination on the window W. By reading the position of the lamp on a scale when balance is obtained, the candlepower can be determined. If illumination is being measured, a white translucent glass, called a test plate, is placed at T and the mirror used at R. The brightness of T is a measure of the total illumination falling upon it.

On the other hand, if the candlepower of a lamp is being measured, extraneous light must be excluded. Therefore, the test plate at T is removed and the mirror at R is reversed, which substitutes for R the white diffusing surface. The only light which enters the tube is that coming from the source towards which the tube is directed, the extraneous light being cut off by the sides of the tube.

If the light to be measured is too bright for obtaining a ,,,, 0 Flo. 395.— Connections for adjusting balance, either screen A or A i standard lamp of Sharp-Millar photomay be interposed between R meter, and the photometer head, thus reducing the light in a known ratio. On the other hand, if the measured light is found to be too dim, these screens may be turned so as to lie between W and the photometer head, thus reducing the light from the standard, and making a balance possible.

The standard lamp S is supplied by a 6-volt storage battery. Its candlepower is controlled by a rheostat. The correct adjustment may be determined by an ammeter connected in series with the lamp. A later method, which depends upon the wheatstone bridge principle, has been developed and is illustrated by Fig. 395. The lamp is one arm of a wheatstone bridge. A, C and D, made of materials with zero temperature coefficient, form the other three arms. The resistance of a tungsten filament is very sensitive to changes of temperature and therefore to changes of current. For this reason the bridge will balance only for one value of total current, which is controlled by the resistance R. The slider S allows small adjustments to be made. Instead of the galvanometer, G, a telephone receiver is often used, the balance being determined by interrupting the telephone circuit. If the system is out of balance a click is heard in the receiver. This equipment eliminates the weight of either an ammeter or a galvanometer.

202. The Rousseau Diagram.—The mean horizontal candlepower of an incandescent lamp may be determined by rotating

Fig. 396.—Rousseau diagram.

the lamp and at the same time measuring its candlepower in a horizontal direction. The speed of rotation must be sufficiently high so that flicker will not prevent the obtaining of an accurate balance. Too high a speed distorts the filament by centrifugal force. The rotator is usually so designed that the lamp can be turned through any vertical angle between 0 and 180. If the candlepower is measured at various angles between 0 and 180 and the corresponding values of candlepower are laid off on the radii of polar coordinate paper, a curve similar to that shown in Fig. 396 is obtained. As would be expected, an incandescent lamp in the ordinary position emits but little light in the upward vertical direction, owing to the presence of the socket and base. But little light is thrown vertically downward because of the small length of filament exposed in that direction.

The mean spherical candlepower of the lamp is obtained by the Rousseau diagram as follows:

A circle mnm' of some convenient diameter is drawn about the polar candlepower diagram, as shown in Fig. 396. The ends of the various radii are projected on a straight line $a'b$,. Perpendiculars are dropped from these projection points to ab which is parallel to $a'b'$. The distances od, of, etc., are laid off from ab on these perpendiculars, as shown at o,d', o,f. The *average* height of the curve $gf'e'$ 'to scale gives the mean spherical candlepower of the lamp. This average height may be obtained by measuring the area $abge,a$ and dividing by the base ab, the proper scale being used, or by taking the average of several *equally-spaced* perpendiculars.

The output of the lamp in lumens is obtained by multiplying the mean spherical candlepower by 4ir (12.57).

The *spherical reduction factor* of a lamp is the ratio of the mean spherical candlepower to the mean horizontal candlepower. It is usually less than 1. 0. This factor depends on the shape and disposition of the filament and also on the reflector, if one be used. For a given type of lamp, the spherical reduction factor is nearly a constant quantity. Knowing this factor for a given type of lamp, the mean spherical candlepower and the luminous output of a lamp can be readily calculated after making one measurement of the mean horizontal candlepower. Below are the spherical reduction factors of a few standard lamps.

Spherical Reduction Factors

Treated carbon, oval filament 0.82

Metallized carbon (GEM) oval filament 0.82

Tantalum 0.97

Mazda, 60 watts 0.78

Gas-filled 0.80 to 0.90 1 "Standard Handbook," Section 14.

Example. — In the polar diagram of Fig. 396, 1 in. radial distance equals 20 candlepower. The diameter of the circle and the line ab have a length of 4 in. The area of $abge,a$ is 8sq. in. The lamp takes 60 watts and the length of the line oh is 2.5 in. Determine: (a) The mean spherical candlepower. (6) The output of the lamp in lumens, (c) The mean horizontal candlepower. *(d)* The efficiency of the lamp in watts per mean horizontal and per mean spherical candlepower. *(e)* The spherical reduction factor of the lamp. ,. 8 X 20 (a) m.s.cp. = — j — = 40 cp. *Ans.* (b) 40 X 12.57 = 502.8 lumens. *Ans.* (c) The mean horizontal candlepower = the length of the line oh X 20 = 2.5 X 20 = 50 cp. *Ans.* 60 (d) gg = 1.20 watts per m.h.cp. *Ans.* 60 Jq = 1.5 watts per m.s.cp. *Ans.* 40 (c) The spherical reduction factor =? = 0. 80. *Ans.* 203. Reflectors. —-It is impossible, except in a general way, to design a lamp so that it will distribute its illumination in a desired manner. The distribution of light from light sources is readily controlled, however, by reflectors, the type of reflector being determined by the existing conditions. In "white way" illumination, for example, as used in the business districts, it is desirable that considerable light be delivered upward so as to illuminate the front of the buildings. Ornamental fixtures similar to that shown in Fig. 397 can be used. On the other hand, light sent upwards in the residential districts is either lost or is undesirable from the point of view of the residents, because such light comes in at their windows. Hence, some type of fixture with a reflector similar to the pendant type shown in Fig. 398 should be selected for this type of illumination. When it is desirable to throw the light into the lower hemisphere, a prismatic refractor is very efficient. This consists of a glass cylinder in which prisms are cut. A cross-section of such a refractor is shown in Fig. 399 (vertical section). The prisms bend the light beams downward by refraction, as shown. Oftentimes a second refractor is used in which the prisms are cut vertically, as shown in Fig. 399 (horizontal section). This breaks up the light in the horizontal planes and eliminates glare. Figure 404 shows a roadway illuminated by the use of prismatic refractors. Figure 400 shows the effect on the light-distribution curve of placing a reflector on a lamp used for interior illumination. The upward light is re-directed downward where it is more useful.

The type of reflector to be used is determined almost entirely by the conditions and the illumination desired. Such reflectors may absorb from 20 to 25 per cent, of the light emitted by the lamp, but they increase the light in the direction in which it is utilized.

15J 15' 30

Fig. 400.—Candlepower distribution curves of lamp with and without reflector.

204. Interior Illumination.—Interior illumination is a very big subject and requires considerable space for a comprehensive treatment. It is more or less complicated by the number of lamps that it may be necessary to use, by the reflection from ceilings, walls, etc. The more involved problems require the services of an illuminating engineer who has had considerable experience in the art of illumination. A few fundamental principles, however, underly the general problem and these will be treated briefly.

Interior illuminants are confined practically to the vacuum Mazda lamp and the gas-filled lamp, although carbon lamps are unfortunately still used in some instances. The purposes to be accomplished by interior illumination are to provide sufficient light for working, reading, writing, etc., and to distribute this light properly without glare. A lamp with a clear globe may give

Fio. 401.—Single lamp in room.

sufficient light and distribute it properly, but the glare of the bare filament may be objectionable. Further, the artistic aspect of illumination should be kept in mind. The fixtures should be pleasing, the proper light thrown on pictures, decorations, etc.

Reading and writing require an illumination of from 3 to 4 foot-candles, whereas drafting and detail work may require as many as 8 foot-candles. This is the illumination necessary on a plane about 30 in. above the floor. In an office, or drafting room, it is desirable that this light be provided entirely by overhead sources. Where a single desk requires a higher intensity of illumination than the rest of the room, desk lamps may be provided.

When lighting an interior, a single ceiling light in the center may suffice if the room is not too large. In this case the reflector should distribute the light in a manner similar to that indicated in Fig. 401 so that the walls will receive some of the light. If a large room is to be illuminated, a single center light may be insufficient. Such a room should be divided approximately (6) Reflector distribution in large room Fig. 402.—Distribution of lighting units and lamps in large room.

in squares, as shown in Fig. 402 (a), and a light placed at the center of each square. Under these conditions, the reflector should distribute the light in the manner indicated in Fig. 402 (6). This secures a more uniform lighting of the room.

Fig. 403.—Reflection and illumination from an indirect lighting source.

Indirect lighting is used to a considerable extent. An opaque inverted bowl is suspended from the ceiling, as shown in Fig. 403. The light is first directed to the ceiling and is then diffused throughout the room. This method of lighting conceals the source and insures a fairly uniform distribution. The method is inefficient and the complete hiding of the source of light is considered objectionable by many persons.

An improvement over the indirect is the semi-indirect system. The opaque bowl is replaced by a translucent bowl, allowing a considerable portion of the light to be transmitted through the fixture. As in the indirect system, a large amount of light is directed to the ceiling and then reflected through the room. This system is more efficient than the indirect and is more pleasing because the source of light is not entirely concealed. Both systems require clean, light-colored ceilings for their most effective use.

Factory and shop lighting is a field in itself. Individual machines, where fine work is being done, may require individual lamps. These lamps should be provided with reflectors which concentrate the light on the work. Overhead belting militates against good illumination and this fact constitutes a good argument for individual and group drives. Cranes often necessitate lamps being placed much higher than is efficient from the standpoint of illumination. Where units are placed high above the floor, large units are more efficient than small ones. The cost of lighting a factory is a very small percentage of the total cost of operation. Moreover, good illumination results in increased production, more accurate work, and fewer accidents. Particular care should be taken to provide good illumination in factories, mills, etc.

206. Street Illumination.—Street illumination differs materially from interior illumination. Interior illumination must be such that small details can be clearly seen. This requires intensities of from 1.25 to 8.0 foot-candles. When this high intensity is obtained, there is no difficulty whatever in seeing and recognizing objects. On the other hand, the purpose of street lighting is not to show details but to enable one to see and recognize objects and persons. Obviously this must be accomplished with intensities very much less than those used in interior illumination. In a room, the ceilings, walls, paper, etc. are nearly always light in color and considerable light is obtained by reflection from

See *Electrical World* Vol. 70, 1917. Various articles on Illumination by Prof. C. E. Clewell.

them. In street lighting, on the other hand, the objects illuminated are the street and sidewalk surfaces, trees, etc. , all of which are dark in tone, and any light not falling directly on them is lost. The light reflected from buildings is so small that it need not be seriously considered. The light which goes upwards and into trees is lost.

Formerly, the idea of good street lighting was to imitate daylight as closely as possible by securing uniform illumination. Those responsible for the design and placing of street lighting units directed their efforts towards obtaining this result and the greater the uniformity the more successful the installation

Fig. 404.—Road objects recognized by silhouette.

was considered. This method of illumination has one fault which was not appreciated at first. With the dim illumination which necessarily accompanies street lighting, objects are seen mostly by *shadow* and *silhouette.* This is illustrated by Fig. 404, in which the automobile is recognized almost entirely by its silhouette. With uniform lighting, obtained by using a large number of small units placed close together, shadows and silhouette are almost entirely eliminated. Depressions and objects in the road become much more difficult to distinguish. This is particularly true when the object and the road have the same general color.

Uniform lighting is considered good engineering in the great white ways which have become popular in recent years. The conditions here are, however, quite different from those of average street lighting. The intensity of illumination is very high. Such white ways are almost always situated in the down-town business streets where the light from the lamps is supplemented by illumination from shop windows, illuminated signs, and display lighting. Objects and persons are recognizable by very slight differences in color or in outline, because of the high in

Fig. 405.—Night illumination, Nahant Road, Mass., with General Electric luminous arc lamps equipped with prismatic refractors. (Illustrates specular reflection.) tensity of illumination which exists. This is analogous to interior illumination.

Another feature of road illumination

has only recently become recognized. Automobiles give oiled, tarred roads a glazed surface. When illuminated at night, light is reflected from the roadway to the observer as from a mirror. It is the same phenomenon as the moon shining on the water. Figures 404 and 405 show examples of this phenomenon, which is called "specular reflection." Just as in the case of a floating object coming into the reflected moonlight on the water, so an object on the road is clearly silhouetted against this "specular reflection" from the road. This is well illustrated by Figs. 404 and 405, where even the very slight road ripples can be clearly seen. This reflection is better obtained by using a few large units, spaced some distance apart, than by using a large number of small units placed close together, as in attempting to obtain uniform illumination. The use of automobiles has added to the problems of street illumination. Owing to the high speed of this type of vehicle, im

Fig. 406.—Illumination of Public Library, Lynn, Mass., with flood-lighting projectors.

proper illumination may lead to many accidents. A very common cause of such accidents is the improper location of the lighting units at curves, etc. Care should be taken to so locate the units as to eliminate the glare in the drivers' eyes and at the same time to make clearly visible any object approaching in the opposite direction. 206. Flood Lighting. —Many spectacular and pleasing effects have resulted from flood lighting. This type of illumination is obtained by projecting the light on the building or object to be illuminated by means of properly located projectors. Flood lighting is commonly applied to the illumination of public buildings, statues, etc. (the Statue of Liberty in particular). It was extensively used at the Pan-American Exposition in San Francisco in 1915. In flood lighting it is necessary not only to illuminate the object but the projectors must be so placed that architectural details are brought out by relief and shadow. This is well illustrated by Fig. 406 showing the Public Library at Lynn, Mass.

During the past few years, street lighting has been studied not only from the photometric standpoint, but from the architectural point of view as well. In many instances crude poles, mast arms etc. have been replaced by ornamental fixtures of the type shown in Fig. 397.

APPENDIX A

Circular Measure—The Radian

The *radian* is circular angle subtended by an arc equal in length to the radius of its circle as shown in the figure. The circle has a radius of *r* units and the radian is subtended by an arc whose length is *r* units.

As the circumference of a circle is 2-*irr* units, there must be 2i r or 6.283 radians in 360. Therefore 1 radian equals 360/2ir = 57.30. It follows that 180 = Tt radians.

Angular velocity is often expressed in radians per second, and the accepted symbol is u (omega). In every revolution a rotating quantity completes 2ir radians. If the rotating quantity makes *n* revolutions per second, its angular velocity o = 2irn radians per second.

Ratio of sides in a right isosceles triangle

Ratio of sides in a 30-60 degree right triangle 15. sin B =-= cos *A* = cos (90-*B),* since *A* = 90-*B* C 16. cos *B* =-= sin *A* = sin (90 -*B)* C 0 90 180 270 360 Graphic Representation of Trigonometric Functions.

Law of Cosines.—In any triangle the square of any side is equal to the sum of the squares of the other two sides minus twice the product of these two sides into the cosine of their included angle.

That is: $V + c2-26c$ cos A

$V + c2-o2$

$26c$

cos A = — 46. cos B = 47. cos C 2ca - c' 2o6 APPENDIX E MATHEMATICAL TABLES

Natural Sines and Cosines

Note.—For cosines use right-hand column of degrees and lower line of tneths APPENDIX F

Natural Tangents and Cotangents

Note.—For cotangents use right-hand column of degrees and lower line of tenths APPENDIX H

Resistance of Copper Wire, Ohms per

Mile 26C. (77F.)

For more detailed tables, see Vol. I, pages 45 and 46.

APPENDIX I

Inductive Reactance per Single Conductor, Ohms per Mile1

Stranded

APPENDIX J

Charging Current per Single Wire, Amperes per Mile per 100,000 Volts from Phase Wire to Neutral1

Stranded QUESTIONS ON CHAPTER I 1. State briefly some of the advantages of direct over alternating current for industrial purposes. Which type of power supply is best adapted to traction purposes? Why is direct current necessary for electrolytic work?

2. In spite of the many advantages of direct current why is a large percentage of power at the present time generated as alternating current? Name some secondary reasons for generating power as alternating current. 3. How does the weight of transmission conductor vary with the transmission voltage? Give reasons why it may be more economical to generate power in large quantities and transmit it over expensive transmission systems rather than to generate it at the point of use. 4. What is meant by a sine wave? Do commercial alternators as a rule generate sine waves? Why are sine waves assumed in making alternatingcurrent calculations? 6. Describe a graphical method of producing a sine wave. Show how such a wave may be plotted by the use of sine tables. 6. Through how many space-degrees must a coil, rotating in a bipolar field, turn before one cycle is completed. Under these conditions what is the relation of the space-degree to the electrical degree? What is meant by an alternation? 7. In a four-pole machine, through how many space-degrees must a coil turn before a cycle is completed? Why? In this case what is the relation between electrical and space-degrees? How fast in r. p.s. must such a coil rotate in order to produce a frequency of 60 cycles per second? In r.p.m? 8. What are two advantages of the higher frequencies for commercial generation and utilization? Name two distinct disadvantages of the

higher frequencies. 9. Why is 60 cycles per second usually chosen as the system frequency when a power company supplies both lighting and power loads? Under what conditions is 25 cycles used? What is the advantage of this frequency over 60 cycles? 10. What is the average value of an alternating-current wave over one complete cycle? Upon what is the value of an alternating-current ampere based? Define an alternating-current ampere. 11. How does the heating effect of a current vary with the current? How does the squared current wave compare with the original current wave as regards frequency, maximum value, and its axis of symmetry? What is the ratio of the maximum to the effective value of a sine wave? What is the ratio of effective to average for a half cycle and what is this ratio called? 12. Compare 1 ohm resistance for alternating current with 1 ohm resistance for direct current. How is an alternating-current volt defined? 13. Define a scalar quantity; a vector quantity. How are vectors represented? How are they added? What is meant by the parallelogram of forces? The triangle of forces? How are vectors subtracted from one another? 14. What is meant by a current and a voltage being in phase with each other? In what terms is phase difference expressed? A certain wave crosses the zero axis in a positive direction to the right of another wave. Is the first wave lagging or leading the other? Explain. 15. If two current waves are plotted, how can the sum of the currents be determined? If two currents are in phase how is their sum found? 16. Is the sum of two current waves necessarily equal to their algebraic sum? Explain. How may this be proved? 17. Explain how a sine wave is produced by a rotating vector. How is the value of the wave determined at any instant? What is the relation of the speed of the rotating vector to the circuit frequency? 18. If two current waves differ in phase by a certain angle, what is the relation existing between the vectors which produce these waves? Illustrate with sketches. 19. What is the fundamental method of adding two currents? How is the resul-

tant current determined? 20. What relation exists between the resultant wave and the vector sum of the rotating vectors? Does this suggest a ready method of adding alternating currents or voltages? Why may vectors representing effective values be used as well as vectors representing maximum values? PROBLEMS ON CHAPTER I 1. An alternating current has an effective value of 28.3 amp., making its maximum value 40 amp. Draw this wave to scale by the method of Fig. 2, page 4 and also construct this wave from a table of sines. (See page 5.) Indicate the effective and average values of this wave. 2. Find the instantaneous values of the current in problem 1 for angles of 30, 60, 270, 290, using a table of sines. If the frequency of this wave is 25 cycles per second, to what values of time do the above angles correspond assuming that the time is zero when the wave crosses the axis in a positive direction? 3. An alternating voltage, following a sine law, has a maximum value of 155 volts and a frequency of 60 cycles per second. What is the value of this voltage 0.001, 0.004 and 0.01 sec. after crossing the zero axis in a positive direction? 4. A 60-cycle water-wheel generator has a speed of 120 r.p.m. How many poles has this generator? How many electrical space-degrees correspond to one actual space-degree? 6. What is the speed in r. p.m. of a 60-cycle, 10-pole, turbo-driven alternator? How many electrical space-degrees correspond to one actual space-degree? 6. What is the frequency of a 750 r.p.m, four-pole alternator? Of 'a 600 r.p.m. four-pole alternator? 7. What is the equation of a 110-volt, 60-cycle voltage? What is the instantaneous value of this voltage when the time is 0.002 sec.? 0.004 sec.? 8. Plot the squared values of the current wave of problem 1, using a much smaller scale for the squared values. (a) What is the frequency of the squared wave? (6) What distance in amperes is its axis above the original axis? (c) What is its average value? (d) What is the square root of this average value? (e) What relation exists between (d) and the maximum value? 9. A certain current has an effective value of 20 amp. What is its maximum value? What direct current will produce the same heating in a given ohmic resistance. 10. A certain direct-current series-arc system is not guaranteed to operate safely at any voltage in excess of 10,000 volts between wires. This system is later supplied with alternating current by a constant-current transformer. What is the effective value of the highest alternating electromotive force that can be safely used? 11. Sketch a current and a voltage wave having effective values of 12 amp. and 120 volts respectively and in which the current lags the voltage by 45. If the voltage wave is passing through zero in a positive direction at 1:00 o'clock at what time will the current wave be going through its corresponding phase? The frequency is 60 cycles per second. 12. Plot two waves corresponding to currents having effective values of 20 and 30 amp. respectively, these two currents differing in phase by 90, the 30 amp. current lagging. Add the ordinates of these two waves point by point and plot the resulting wave, (a) What is its maximum value? (6) What is its effective value? (c) What result is obtained by adding two current vectors of 20 and 30 amp. laid off at right angles? *(d)* What is the angle between the resultant wave and the 20-amp. wave? 13. Repeat problem 12 when the two waves differ in phase by 60. 14. Two generator coils generate 200 volts and 400 volts respectively. These two voltages differ in phase by 120, the 200 volts leading. If connected in series, what is the voltage across their open ends? 16. If the 400-volt coil is reversed the two voltages now differ in phase by 60. Find the resultant voltage. 16. If both coils, problem 14, generate 200 volts, find the resultant voltage. 17. Repeat problem 16 except that one of the voltage coils is reversed, making the two voltages differ in phase by 60. QUESTIONS ON CHAPTER II 1. How may the power in an alternating-current circuit be determined at any instant? What is the general character of the power curve when the voltage and the current are in phase? (Illustrate by sketch.) What is the maximum val-

ue of this curve in watts? The average value? 2. In what way does the power curve for a circuit in which the current lags 90 behind the voltage differ from that in which the current is in phase with the voltage? What is the *average* power in such a circuit? *3.* What is the *average* power of a circuit in which the current *leads* the voltage by 90? 4. When the current and the voltage differ in phase by an angle which is less than 90 but greater than zero, what is the general character of the power curve? Make a sketch. What is meant by *power-factort* What is power-factor equal to, numerically? 5. When an alternating voltage is impressed across a resistance, what phase relation exists between this voltage and the resulting current? How may the value of the current be calculated, knowing the voltage and the resistance? 6. What is the effect of inductance on the building up of a current in a circuit across which a steady direct-current voltage is impressed? What occurs when the current attemps to die out in an inductive circuit? 7. State how the current may be prevented from reaching its Ohm's law value in an inductive circuit. 8. What effect does inductance in an alternating-current circuit have upon; (a) the phase angle between the current and the voltage; (6) the magnitude of the current? What is the effect of frequency upon the magnitude of the current? What is "inductive reactance?" 9. What in general is the effect of capacitance upon the flow of current in any electric circuit? How does capacitance affect (a) the phase angle between the current and the voltage in an alternating-current circuit; (6) the magnitude of the current? 10. What is the effect of frequency upon the magnitude of the current in a condensive circuit? What is condensive reactance? What is the value of the average power taken by a perfect condenser? 11. What is the phase relation existing between the current and the voltage across the resistance in a circuit containing resistance and inductance in series? Between the current and the voltage across the inductance? How is the line voltage obtained? 12. What is meant by impedance? How may

the angle between the line voltage and the current be determined? What relation does this angle bear to the power in the circuit? 13. What is the phase relation existing between the current and the voltage across the resistance in a circuit containing resistance and capacitance in series? Between the current and the voltage across the capacitance? What phase relation exists between the line voltage and the current? 14. Sketch the vector diagram of a circuit containing resistance, inductance and capacitance all in series. How may the circuit phase-angle be found? 16. What is meant by "resonance" in an alternating-current circuit? What phase relation exists between the line voltage and the line current? What is the numerical relation existing between the inductive voltage and the condensive voltage? Show that with both inductance and capacitance in a series circuit, the voltage across each can be several times the line voltage. 16. In practice, why is parallel grouping of resistances, inductances, etc., more common than series grouping. How may the resultant of several currents be found? In what way does parallel resonance differ from series resonance especially with regard to the value of the current? In what way are the two similar? 17. Explain why the alternating-current resistance of an iron-cored impedance coil differs from the direct-current resistance. How may this resistance be taken into consideration when the impedance coil is connected in circuit with resistance, etc? 18. How may the phase relations existing in a series circuit having two component voltages be determined when the voltages across the various parts of the circuit are known? Make a sketch. How may the power and the power-factor of all parts of the circuit be determined? 19. Explain why a circuit in which the line voltage and three component voltages are all known in magnitude only, cannot be determined unless one more factor be known. What additional information makes the circuit relations determinable. 20. In what way is the polygon of currents similar to the polygon of voltages? In what way do the two polygons differ? 21. What is meant by energy current? What relation does it bear to the power? What is quadrature current and what relation does it bear to the power? Why is quadrature current usually undesirable? PROBLEMS ON CHAPTER II 18. A certain lamp load consists of 50 60-watt lamps, each lamp having a resistance of 220 ohms. Compare the power taken by this load when connected across 113 volts direct current and across 113 volts alternating current. 19. A certain transformer takes 30 kw. at 2,200 volts and the current is 15 amp. What is the power-factor of this load? 20. A single-phase motor takes 4.68 kv-a. (kilovolt-amperes) at 220 volts and at a power-factor of 0.80. How much power and how much current does it take from the line? 21. A non-inductive resistance of 12 ohms is connected across 120-volt alternating-current mains. What current and what power does it take? 22. Determine the current taken by a coil having 0.3 henry inductance, when connected across 115-volt 60-cycle mains; 115-volt 25-cycle mains. 23. A coil whose resistance is negligible but which has an inductance of 0.2 henry is connected across 110-volt 25-cycle mains. (a) What current flows? (b) For what value of frequency will the current be 2 amp.? 24. A condenser whose capacitance is 12 m.f. is connected across 220volt 60-cycle mains. (a) What current does it take? (6) What would be the current if the frequency were 133 cycles per second? 26. It is desired to obtain 16 amp. leading current by the use of a condenser connected across 220-volt 60-cycle mains. (a) How many m. f. are necessary? (6) What is the kv-a. capacity of this condenser? (c) What would the kv-a.. capacity of the condenser, found in(a), be at 440 volts? (d) How does the kv-a. capacity of a condenser vary with the voltage? 26. A certain condenser when connected across 115-volt 25-cycle alternating-current mains takes 8 amp. What current does it take if the frequency and capacitance are both doubled? 27. Repeat problem 26 for an inductance taking 8 amp. under the same conditions. 28. In a circuit having an inductive reactance of 8 ohms and an ohmic resistance of 6 ohms, 12 amp. flow, (a) What is the voltage across the resistance? (6) Across the reactance? (c) What is the circuit voltage? (d) How much power does the circuit take and what is its power-factor? (e) What is the phase angle between the circuit voltage and the current? 29. What is the power-factor of a circuit which has a resistance of 3 ohms and a reactance of 4 ohms? What current flows if the voltage is 60 volts and how much power does the circuit consume? 30. Repeat problem 29 making the inductive reactance twice the value given. 31. What current flows in a circuit consisting of a 40 m.f. condenser and 30 ohms resistance in series if connected across 120-volt 60-cycle mains? What is the voltage across the resistance? The condenser? What power does the circuit take and what is its power-factor? 32. Repeat problem 31 substituting a 50 m. f. condenser for the 40 m.f. condenser. 33. A series circuit consisting of 10 ohms resistance, 15 ohms inductive reactance, and 20 ohms condensive reactance is connected across 220-volt 60-cycle mains. Find (a) the current. (6) The voltage across each circuit member, (c) The power taken by the circuit. (d) The power-factor and power-factor angle of the circuit. Draw a vector diagram of the circuit to scale. 34. A potential difference of 220 volts at 60 cycles is impressed across a circuit having 50 ohms resistance, 15 m.f. capacitance and 0.2 henry inductance all in series. (a) What current flows? (6) What is the voltage across the resistance, the inductance and the capacitance? (c) What power does the circuit consume? (d) What is its power-factor and power-factor angle? Draw a vector diagram of this circuit to scale. 36. A resistance of 10 ohms, an inductance of 0.1 henry and a 40 m.f. condenser are connected in series across 220-volt 60-cycle mains. (a) What current flows? (6) What power is taken from the line? (c) What is the voltage across the resistance, the inductance and the capacitance? (d) What is the circuit power-factor and power-factor angle? Draw a vector diagram. 36. If the inductance of problem 35 were ad-justable, for what value would the current be a maximum? Find the current, the power, the power-factor and the voltages under these conditions. 37. A circuit contains a resistance of 10 ohms and an inductance of 0.352 henry and a variable condenser all in series. If the frequency is 60 cycles, for what value of the capacitance will the current be a maximum? If the line voltage is 20 volts, what are the current, power, power-factor and the various voltages? 38. A resistance of 10 ohms, a 50 m.f. condenser and an inductance of 0.05 henry are connected in parallel across 100-volt 60-cycle mains. What current does each take and what is the total line current? 39. An inductance having negligible resistance and a condenser are connected in parallel across 25-cycle mains. Each takes 2 amp. (a) How much current does the line supply? (6) If the frequency is doubled, how much current does the line supply? 40. A non-inductive resistance and an impedance coil are connected in series across 120-volt 60-cycle mains. The voltage across the resistance is found to be 90 volts and that across the impedance 60 volts. If the current is 5 amp., determine: (a) the circuit power; (6) the circuit power-factor; (c) the impedance power-factor; (d) the inductance of the impedance coil. 41. To measure the power consumed by a 115-volt single-phase induction motor, it is connected in series with a non-inductive resistance across 220volt mains. When the motor takes 20 amp. the resistance is so adjusted that the voltage across the motor terminals is 115 volts and that across the resistance is 127 volts. What power is the motor taking and what is the power-factor of the motor and of the system? 42. A series alternating-current circuit consisting of a resistance, an impedance coil and a condenser having negligible leakage takes 420 watts at 115 volts, 60 cycles and 5 amp. lagging current. The voltage across the resistance is 75 volts and that across the capacitance is 90 volts. Find (a) the power consumed in the resistance, (6) The power consumed in the impedance coil. (c) The voltage across the impedance coil. (d) The capacitance of the condenser in m.f. *(e)* The inductance of the impedance coil in henrys. 43. A series circuit consisting of a resistance of 10 ohms, an inductance coil (resistance negligible) and an unknown condenser all in series is connected across 220-volt 60-cycle mains. When 10 amp. leading current flow in the circuit the combined voltage across the capacitance and inductance when added numerically is 250 volts. What is the value of the unknown capacitance? 44. A non-inductive resistance and an impedance coil are connected in parallel across 110-volt 60-cycle mains. When the impedance and resistance each take 20 amp. the line current is 32 amp. What power is the impedance taking? The circuit? What is the power-factor of each? 46. A non-inductive resistance, an impedance and a condenser having negligible power loss are all connected in parallel across 110-volt 60-cycle mains. The resistance takes 12 amp., the condenser 10 amp., the impedance 13 amp. and the line supplies 15 amp. to these three. Find (a) the circuit power; (6) the circuit power-factor; (c) the power taken by the impedance; *(d)* the impedance power-factor. 46. A certain load takes 2 kw. at 220 volts 60 cycles and at a power-factor of 0.707, lagging. How much energy and how much quadrature current does this load take? If this load is supplied over a line each wire of which has one ohm resistance, what is (a) The total line loss? (6) The line loss due to the energy current? (c) The line loss due to the quadrature current? 47. In problem 46 how large a condenser would be necessary in order to make the power-factor of the load unity? What is the line loss under these conditions? 48. A certain transmission line delivers 30 kw. at 2,300 volts and the power-factor of the load is 0.80. The line loss is 5 kw. (a) What current flows over the line? (6) How much is quadrature current? (c) How much is energy current? (d) What is the line loss due to energy current and what is that due to quadrature current? QUESTIONS ON CHAPTER HI 1. Describe the principle of the Siemens Electro-dynamometer. How are its coils connected and what is the relation exist-

ing between the turning moment and the current? What are the disadvantages of this type of instrument? 2. In what way is the indicating electro-dynamometer similar to the Siemens dynamometer? In what ways do the two instruments differ? 3. Explain how the electro-dynamometer principle may be applied to a voltmeter. What is the general character of the scale divisions? Compare its current with that taken by a direct-current instrument of the same range. Discuss the accuracy of such an instrument when used with direct current. 4. Describe the inclined-coil voltmeter, giving the principle upon which it operates. 6. What difficulty arises when attempt is made to apply the dynamometer principle to the alternating-current ammeter? 6. Describe the construction of a wattmeter and give the principle of its operation. Show how it is connected to a circuit. Give the best method of connecting the potential-circuit, especially when the instrument is used in connection with considerable voltage. 7. Show two possible methods of connecting a wattmeter in circuit. Discuss the corrections that should be made in each case, if the exact value of the power is desired. What compensation for these errors can be made in the construction of the instrument? 8. What precautions should be taken against over-loading a wattmeter? How are wattmeters rated and why? 9. Give the advantages of a polyphase wattmeter over single-phase instruments. How is the polyphase wattmeter constructed? 10. How are wattmeters calibrated? Give a diagram of connections. 11. Describe how the Weston iron-vane type of voltmeter utilizes the principle of magnetized iron. Upon what fundamental electrical principle does this instrument operate? How is the instrument damped? 12. Show how the iron-vane principle has been adapted to an inclinedcoil instrument. What two methods are used to damp this instrument? 13. What change should be made in the construction of the above two types of iron-vane voltmeters in order that they may be used as ammeters? What are the limitations of iron-vane instruments for direct-cur-

rent measurements? 14. Discuss the use of direct-current watthour meters upon alternatingcurrent circuits. 15. Describe the construction of the induction watthour meter. What should be the phase relation existing between the potential-coil flux and the circuit voltage? How is this phase relation obtained? 16. How is friction compensation effected? Discuss this principle very carefully. 17. Show by simple sketches how the driving torque is developed. Why does the disc tend to rotate in the direction of the sliding field? 18. How is the induction watthour meter calibrated? What adjustments, not used for the direct-current type, are necessary? What are the advantages of this type of meter over the direct-current type? 19. Describe one common type of frequency meter. Upon what principle does it operate? Why are the vibrating reeds kept polarized? 20. Describe the Tuma phase meter. How is this instrument adapted for power-factor measurements? What control is exerted on the moving system? Why are the coils of the moving system not exactly 90 apart? What modifications of the instrument are necessary in order that it may be used on three-phase circuits? 21. For what purposes are synchroscopes used? Describe the construction of some one type. In what way is it related to the phase meter? 22. What are the commercial uses of the oscillograph. What is its principle of operation? In what way does the moving element differ from that of the ordinary galvanometer? How are the time abscissas obtained? Why is it desirable to immerse the moving element in oil? 23. Sketch the general arrangements of the laboratory type giving the relative positions of the lamp, prisms, vibrators, lenses, rotating mirrors, film drum, etc. 24. Make a diagram of connections showing how the voltage vibrator and the current vibrator are connected in circuit. QUESTIONS ON CHAPTER IV 1. Give three reasons why three-phase power supply is superior to singlephase supply. 2. Why is it desirable at times to use symbolic notation in the solution of problems? Why is this system particularly applicable to polyphase systems?

State briefly the principles upon which one such system is based.

3. Describe an elementary three-phase generator. What relations exist among the three voltages of such a generator? How may three independent single-phase systems be obtained from such a generator? 4. What is meant by Y-connection? How many wires are necessary? What is the numerical relation and the phase relation of the line voltage to the coil voltage in this system? The line current to the coil current? What relation exists among the three-coil currents if there is no neutral wire? 6. At unity power-factor what is the total power generated in a Y-connected generator in terms of coil volts and coil current? If the power-factor is other than unity? What is the line power in terms of line current, line voltage and power-factor? 6. Why is the line power-factor the cosine of the *coil* power-factor angle? What significance has power-factor in an unbalanced system? 7. Why is the delta-connection not a short-circuit for the three coil voltages? What is the numerical relation and the phase relation which exists between the coil current and the line current? 8. What is the total power produced by a delta-connection equal to in terms of coil voltage, coil current and coil power-factor? In terms of line voltage, line current and coil power-factor? 9. Sketch the connection of the three-wattmeter method of measuring power, (a) When the neutral of the system is accessible. (6) When the neutral is not accessible. To what is the total power equal in terms of the wattmeter readings? 10. Illustrate the principle of the Y-box and state the conditions under which it can be used. 11. Sketch the connections of the two-wattmeter method of measuring power. Under what conditions do the wattmeters read the same? Different? 12. Under what conditions does one wattmeter reverse? Give two methods of obtaining power-factor from the two-wattmeter readings alone. Under what conditions can the two-wattmeter method not be used to measure power in a three-phase system? 13. What phase relations exist between the voltages of a two-phase system? Show the connections of four different types of two-phase system. What relations exist among all voltages in each of these systems? 14. Sketch two methods of connecting the coils of a two-phase generator. What relation exists between the coil voltage and the line voltage in each of the two systems? Between coil current and line current? PROBLEMS ON CHAPTER IV 49. A certain three-phase Y-connected alternator is rated at 500 kv-a. 2,300 volts, at unity power-factor. What is its current rating per terminal? What is the rated coil current and coil voltage? 60. What is the current and voltage rating of the machine in problem 49 if it is changed over to a delta-connection? 61. A Y-connected alternator delivers a balanced three-phase load of 50 amp. at 230 volts and 0.8 power-factor to a delta-connected motor. Find the current and voltage per coil in both the generator and the motor. What power is involved? 52. Three resistances of 70 ohms each are connected in Y across a 3-phase 230-volt supply. How much current does each take and what is the total power taken by the three? 53. If the three resistances of problem 52 are connected in delta, how much current does each take and what is the line current? Determine the total power and compare it with that obtained in problem 52. 64. Three condensers each have a capacitance of 40 mf. Compare the kv-a. that they take when connected in Y and then in delta across the 230volt, three-phase, 60-cycle mains. 55. A certain three-phase induction motor takes 30 kw. at 550 volts. The line current is 40 amp. What is the power-factor of this motor? What is the angle between the coil current and the coil voltage? 56. The electromotive forces generated in two alternator coils differ 120 in phase. When the end of one coil is connected to an end of the other, the voltage across their open ends is 190 volts. If the voltage of each coil is 190 volts, what will be the voltage across their open ends if one coil is reversed? 67. In the test of a 230-volt, 3-phase induction motor the two-wattmeter method of measuring power is employed. One wattmeter reads + 4,350 watts and the other reads + 2,200 watts, (a) What is the motor powerfactor at this load? (6) What is the line current? *(c)* What is the coil phase angle? 68. At light load on the motor of problem 57, the first wattmeter reads 1,820 watts and the second —450 watts, (a) What is the motor powerfactor and the line current at this load? (6) What is the coil phase-angle? 69. Each of the two coils of an alternator generates 2,300 volts and these voltages differ in phase by 90. If these two coils are connected together at their center points and this connection brought out with the others, indicate all the voltages that can be obtained. 60. (a) If the two coils of problem 59 are connected together at one end, what is the voltage across the open ends? (6) If the current per coil is 70 amp., what will be the current in each of the three wires leading from the machine? (c) How much power does the generator deliver, assuming unity power-factor? 61. A two-phase generator rated at 1,500 kv-a., 2,300 volts, 60 cycles, has two coils. What is the current rating per coil? If this machine supplies a five-wire system show how the wattmeters would be connected in order to measure its output. 62. Each of the coils of the alternator of problem 61 consists of two separate sections connected in series. The machine may be then connected in mesh. Determine its voltage and current per terminal under these conditions. 63. A two-phase four-wire system has a potential difference of 115 volts between adjacent wires giving 163 volts, across diametrically opposite wires. Four resistances of 10 ohms each are connected between adjacent wires, as shown in Fig. 63(A). Determine the total power and the current flowing in each line. 64. How much current and how much power is supplied by the two alternator coils to the load in problem 63? QUESTIONS ON CHAPTER V 1. Why can a rotating field and a stationary armature be used for alternators where they cannot be conveniently used for direct-current machines? Give two reasons why it is advantageous for alternators to be of the rotating-field type. 2. What two conditions must a coil of

an alternator armature winding fulfill? Compare the wave and the lap winding of alternators with these same types of winding in direct-current machines. 3. Illustrate by a simple sketch the difference between a half-coil and a whole-coil winding. What is the difference between a single-and a twolayer winding? What are the objections to using one slot per pole? 4. In what way does the spiral winding differ from the lap and wave windings (barrel type) as regards mechanical disposition of the coils? Why is the interior coil usually omitted? What is meant by "single range"? 6. Show that a two-phase winding is an extended application of the singlephase winding. How may the chain winding be adapted to two-phase? Why is the two-range feature necessary? State the advantages and the disadvantages of this type of winding. 6. State the advantages of the lap winding. In the full-pitch lap winding what relation exists between the coil sides in any one slot? 7. Show that a three-phase winding consists of three single-phase windings properly spaced. 8 Under what conditions are coils of special shape required in threephase, two-range, chain windings? 9. State the advantages and disadvantages of fractional-pitch windings. 10. Compare the types of stampings required for machines of large and machines of small diameter. Why are there perforations back of the slots in the stampings used for the larger machines? 11. How are the armature laminations held in position in engine-driven generators? Why is the frame usually of the hollow-box type? 12. Why is it necessary to brace very strongly the coil end? of large capacity? 13. Sketch the shape of two common types of alternate; advantages and disadvantages of each type. 14. Show that with large units, even when operating at very high etu the amount of energy to be dissipated per minute is very large. Whu,. method is used to carry the resulting heat away from the machine? 15. Into how many classes are the rotating-field structures of alternators divided? Give reasons why different designs of field structure are necessary. What is the general construction of the field cores in all types of salient-pole alternators? How are these field cores held in position? 16. Describe the field spiders for (a) very slow-speed machines of large capacity; (6) moderately high-speed water-wheel generators. Why is it not possible to use salient poles in high-speed turbo-alternators? 17. Describe the construction of (a) the parallel-slot type of rotor; (6) the radial-slot type of rotor. Why must the end-flanges be of nonmagnetic material? Under what conditions are these types of rotor used? 18. Describe the method of conducting the field current into the field winding. What different methods are used for supplying excitation? What precautions are often taken to insure continuity of excitation? 19. Derive from a fundamental relation the equation of the induced electromotive force in an alternator. What is meant by "breadth factor" and "pitch factor"? 20. What relation exists between the flux distribution and the shape of the electromotive force wave per conductor? How may the shape of the actual electromotive force wave of a generator be made nearly sinusoidal even though the electromotive force wave in the individual conductors differs considerably from a sine wave? 21. Sketch the flux distribution curve for a distributed field winding. Explain why such machines usually have a better wave shape than machines of the salient-pole type. 22. State the procedure in phasing out the coils of a three-phase alternator so that they may be Y-connected. 23. Repeat for a machine which is to be delta-connected. 24. Upon what factors does the rating of an alternator depend? Why is a kilovolt-ampere rating more rational than a kilowatt rating? Upon which rating does the rating of the prime mover depend? PROBLEMS ON CHAPTER V 66. Draw a single-phase, full-pitch, four-pole lap winding in which there are four slots per pole and the winding and slots occupy only 60 per cent. of the armature periphery. 66. Repeat problem 65 for a wave winding. 67. Draw a single-phase, single-range, spiral winding for an armature having eight slots per pole, the winding occupying but six of these slots. 68. Draw a two-phase, chain winding for a six-pole alternator having 12 slots per pole, utilizing all the slots. 69. Repeat problem 68 for a-three-phase winding. Show tho windings of the three phases each in a different color. 70. An eight-pole alternator has 80 slots. Draw a two-phase, %-pitch, lap winding for this machine. 71. A four-pole alternator has 48 slots. Design a three-phase, fullpitch, lap winding, showing the connections of one phase only. 72. Repeat problem 71 for a J-pitch winding. 73. Repeat problem 71 for a full-pitch, wave winding. 74. A four-pole, 60-cycle, single-phase alternator has a concentrated winding similar to that shown in Fig. 102, page 102. There are eight conductors per slot and 2,800,000 magnetic lines per pole. Assuming that the flux distribution under the pole is practically sinusoidal, determine the electromotive force of this generator. 76. A six-pole, 25-cycle, two-phase alternator has one slot per pole per phase and 12 series-connected conductors per slot. Determine the induced volts per phase if there are 3,000,000 lines per pole. Assume that the flux distribution is practically sinusoidal. 76. A single-phase, four-pole, 1,800 r.p.m. alternator has eight slots per pole. Only half of these slots are occupied by the winding so that the breadth factor, 0.907, is the same as that of a two-phase winding having four slots per pole per phase. There are four series-connected conductors in each slot and the winding is full pitch. There are 3,000,000 lines per pole and the flux may be assumed to be distributed sinusoidally. Determine the electromotive force of this alternator. 77. A three-phase, 12-pole, 600 r.p.m. alternator has 12 slots per pole and a full-pitch winding having six series-connected conductors per slot. There are 3,500,000 lines per pole. What is the open-circuit terminal electromotive force, if the machine is Y-connected? 78. Repeat problem 77 for a %-pitch winding. 79. An alternator is rated at 6,000 kw. at 70 per cent. power-factor. (a) What is its kilovolt-ampere rating. (6) How many kilowatts can it safely deliver at unity power-factor? (c) If it has an efficiency of 95 per cent. at 70 per cent. power-factor, what should be the rating of its prime mover in horsepower? *(d)* If the prime mover speed is 1,200 r.p.m. what torque does it develop? 80. A three-phase, 60-cycle, 13,200-volt alternator is rated at 20,000 kw. at 70 per cent. power-factor, *(a)* What is the current per terminal? *(b)* What would be the safe current per terminal if the power-factor were unity? (c) What would be its kilowatt rating at unity power-factor? *(d)* At 70 per cent. power-factor, what should be the approximate rating of its prime mover in horsepower? QUESTIONS ON CHAPTER VI 1. Why is the question of the regulation of alternators more important than the regulation of direct-current generators? What factor other than the magnitude of the current determines the regulation of alternators? Why is it usually not desirable to determine the regulation of alternators by actual loading? 2. Show by a simple sketch that the inductors of an alternator armature have inductance. Compare the relative inductances, other conditions being equal, of (a) a smooth-core armature; (6) an iron-clad armature in which the slots are deep and narrow; (c) an iron-clad armature of the same number of slots but in which the slots are shallow and broad; *(d)* an ironclad armature having semi-closed slots. 3. What is the effect of the number of conductors per slot upon the armature inductance? How does the reactance of a 25-cycle armature compare with that of a 60-cycle armature, other conditions being equal? 4. Give two reasons why the resistance of an alternator armature to alternating current is greater than its resistance to direct current. What is the order of magnitude of this increased resistance? How may this effective resistance be determined? 6. What is the effect of the current in an alternator coil upon the main field. (a) When the current is in phase with the induced electromotive force? (6) When the armature current lags the induced electromotive force by 90? (c) When the current leads the induced electromotive force by 90? *(d)* When the current

lags and leads the induced electromotive force by an angle *91* 6. Compare the effects of (5) with corresponding effects in direct-current machines. Under what conditions does the armature magnetomotive force for a given armature current have its greatest effect upon the main field of salient-pole alternators? 7. Show by a vector diagram how the induced electromotive force in an alternator armature may be calculated knowing the terminal voltage, the armature resistance drop and the armature reactance drop, (a) When the power-factor of the load is unity, (6) When the current lags the terminal voltage by an angle 6. *(c)* When the current leads the terminal voltage by an angle 6. Give the trigonometric solution of the diagram in every case. 8. Why is the induced electromotive force in an alternator armature, when loaded, not equal to the no-load voltage? Why are open-circuit and short-circuit tests used in obtaining data for calculating alternator regulation rather than actually loading the machine? 9. How is the armature *reaction* taken into consideration in the synchronous impedance method of determining regulation? 10. Show that when a coil has moved 90 electrical space-degrees from the point where the flux linking it is a maximum the induced electromotive force becomes a maximum. Distinguish between a *space* diagram and a *time* diagram. When can the two be combined? 11. Why is it rational to combine the space magnetomotive force diagram of an alternator with the time-voltage and time-current diagrams of the same machine? What is the phase relation existing between; (a) the armature current and the armature magnetomotive force; (6) the resultant field and the induced electromotive force; (c) the impressed field and the no-load electromotive force? 12. Show that a fictitious voltage of the proper value leading the current by 90, and therefore in phase with the voltage which balances the armature reactance drop, can be substituted for the effect of armature reaction, and the no-load electromotive force therefore be determined. What armature constant may be increased to include

this fictitious voltage, and what assumption is made in doing this? 13. What is meant by synchronous reactance? Synchronous impedance? Describe carefully the method usually employed to determine these quantities. 14. What error occurs in the value of the synchronous reactance when it is determined under short-circuit conditions? How does this affect the regulation determined by using this value of synchronous reactance? 16. Why does the synchronous impedance method of determining regulation give unsatisfactory results with single-phase machines? Why are results obtained with polyphase machines more in accord with the actual performance of the machine? 16. Describe the open-circuit test, giving the connections used. Repeat for the short-circuit test, giving two methods of connecting the ammeters. Compare the ammeter readings in each case with the line current and the coil current of a delta-connected machine. 17. How is the regulation of a Y-connected machine calculated? Of a delta-connected machine? How do the respective coil resistances and reactances compare in the two cases for the same machine? What care should be taken when either method is used? 18. In what fundamental way does the magnetomotive-force method differ from the synchronous impedance method? Show by a vector diagram the various voltages and the magnetomotive forces which are substituted for voltages in this method. How is the resultant field obtained? The no-load electromotive force? 19. How do results obtained by the synchronous impedance method compare with those obtained by the magnetomotive force method? Why do the two methods give different results? Which should be used? 20. Fundamentally, how does the A. I. E. E. method of determining regulation differ from the synchronous impedance and the magnetomotive force methods? What difficulty is encountered in this method which is not encountered in the other two methods? How do the results which it gives compare with the actual performance of the machine? 21. Compare the construction of the Tirrill regulator as applied to alternators with the regulator as used on direct-current machines. (See Vol. I, page 307, Fig. 280.) What is the function of the "main contacts" and how are they operated? What is the function of the relay contacts and how are they operated? Explain the operation of the entire regulator from the time that the exciter voltage commences building up until the machine has reached rated voltage, and load then applied. 22. Explain why the prime mover characteristics alone determine the kilowatt division of load between alternators. Why is this true of alternators and not true of direct-current generators? Why is it undesirable that the prime movers have flat speed-load characteristics? 23. What effect has a temporary change of speed of one prime mover on the phase relation of the electromotive force induced in the two machines which they drive? What effect does the resultant voltage produce? Why is the resultant current called the synchronizing current? Show that the action of this current is such as make the parallel operation of alternators a condition of stable equilibrium. 24. If the field of one of two alternators operating in parallel is strengthened, in what two ways is its internal electromotive force affected? Its current? Why? How is the electromotive force and the current of the other machine affected at this same time? Show that the reactions resulting from changing these field excitations cannot change the kilowatt division of load between the machines. What is the objection to having two alternators in parallel, one operating with a leading current and the other with a lagging current? 25. Sketch the connections of a simple method which may be used to show the proper time for switching alternators in parallel. How should the voltage rating of the synchronizing lamps compare with that of the system? How do such lamps indicate the relative phase relation of the incoming machine and the bus-bars? When should the line switch be closed? 26. State two disadvantages of the "three-dark" method of synchronizing. How may these disadvantages be in part eliminated by a different grouping of the lamps? Why is

the use of a synchronism indicator superior to the foregoing methods, especially with certain types of alternators? 27. What types of prime mover have pulsating torques? How may the effect of these torque pulsations be magnified several times by direct-connected alternators? Why is it undesirable that pulsations of frequency be communicated to the system? State the general remedies which may be used to reduce "hunting" and the reason for the use of each of these. PROBLEMS ON CHAPTER VI 81. A 550-volt, 50-kv-a. single-phase alternator has an effective armature resistance of 0.18 ohm and an armature *reactance* of 0.75 ohm. What is the induced electromotive force of this armature when the machine delivers its rated current at rated voltage and at unity power-factor? 82. Repeat problem 81 for 0.8 power-factor, lagging current; leading current. 83. The core loss, friction and windage of the alternator of problem 81 is 1,700 watts. The field takes 15 amp. at 115 volts. What is the efficiency of this alternator at its rated load and unity power-factor? 84. The synchronous reactance of the alternator of problem 81 is equal to twice the armature leakage reactance. If it is carrying full non-inductive load at rated voltage and at unity power-factor, find the no-load voltage of the machine when this load is thrown off. 85. Find the no-load voltages in problem 82, when these loads are thrown off the alternator. 86. What is the regulation of the machine under each of the conditions given in problems 84 and 85? 87. A 25-cycle, 25-kv-a., 550-volt, single-phase alternator has an effectivearmature resistance of 0.20 ohm and a synchronous impedance of 1. 8 ohms. What is its synchronous reactance and what is the regulation of this alternator at unity power-factor and at 0. 7 power-factor, lagging current? 88. A 2,000-kv-a., 3-phase, 2,300-volt, 60-cycle alternator has an effective armature resistance of 0.0354 ohm per phase and a synchronous reactance of 0.62 ohm per phase. The machine is Y-connected. Determine its regulation at unity power-factor and at 0.8 power-factor, lagging current. 89. Following are the constants

of a 50-kv-a., 220-volt, three-phase alternator:

Average ohmic resistance between terminals, 0.0233 ohm.

Field current adjusted to 6.2 amp., open circuit terminal volts = 125 volts.

Field current adjusted to 6.2 amp., generator short-circuited, three line currents each equal to 164 amp.

Ratio of effective to ohmic resistance = 1.5.

Calculate the regulation of the machine when the power-factor is 0.8 lagging and 0.8 leading current. Assume that the machine is Y-connected.
90. Repeat problem 89, assuming that the machine is delta-connected. 91. Below are given open-circuit and short-circuit data of a 200-kv-a. 500-volt, 60-cycle, three-phase alternator. The machine is Y-connected and has an ohmic resistance of 0.012 ohm per coil. The ratio of effective to ohmic resistance is 1.5. Find the regulation of the machine by the synchronous impedance method at 0.7 power-factor for both leading and lagging current.

Open-circuit Short-circuit

Field Current Terminal Armatuf.b

Volts Current 10 245 300 15 365 450 20 460 25 525 35 623 40 665 50 725 92. Determine the regulation of the alternator of problem 91 by the magnetomotive force method, at unity and 0.7 power-factor for both leading and lagging current.

93. A 1,000 kv-a., three-phase, 2,200-volt, 60-cycle alternator is tested for its regulation by the A. I. E. E. method. A load of practically zeropower-factor lagging current is applied, this load being adjusted until the rated current of the machine is flowing. The field current is then adjusted until the machine terminal voltage is 2,200 volts. Under these conditions the field current is 250 amp. When the field current is 250 amp. on open circuit the terminal voltage is 2,700 volts. The machine is delta-connected and has an effective resistance of 0.3 ohm per coil. Find the synchronous reactance of this machine and its regulation at unity and at 0.8 powerfactor, lagging current. 94. Two similar 1,000 kw. alternators operate in parallel. The

speedload characteristic of the first alternator is such that its frequency drops from 63 to 60 cycles from no load to full kilowatt load. The frequency of the second alternator drops from 64 to 58 cycles under the same conditions. When the combined load on the two alternators is 1,500 kw. how much load does each alternator supply? 96. The tension in the governor spring of the second alternator of problem 94 is so adjusted that both alternators have the same frequency when the load on each is 1,000 kw. This change in the speed-load characteristic of the second machine raises its speed-load characteristic two cycles at every point. How much power does each alternator deliver when there is no load on the system? 96. Two alternators are operating in parallel supplying single-phase power at 2,300 volts to a load of 400 kw. whose power-factor is unity. No. 1 alternator supplies 109 amp. at 0.8 power-factor, lagging current. What power and what current does alternator No. 2 supply? 97. Two three-phase alternators are operating in parallel to supply a 2,000 kw. load at 6,600 volts, this load having unity power-factor. The current delivered by alternator No. 1 is 83 amp. at 0.85 power-factor, leading current. What power and what current is alternator No. 2 delivering? What is its power-factor? QUESTIONS ON CHAPTER VII 1. Define a transformer. What distinct advantages do transformers possess over most other types of electrical machinery? 2. By what means is energy transferred from one circuit to the other? Which winding is called the primary? The secondary? 3. Show that the induced electromotive force of a transformer winding is proportional to the number of turns. To what three factors is the induced electromotive force in any transformer winding proportional? 4. What current flows into a transformer primary when the secondary is open? What is its order of magnitude? What is the relation of the direction of primary current to the direction of flux in the core? Of the secondary current? Explain. 5. Explain the sequence of reactions which cause the primary to take more power from the line when load is applied to the secondary. 6. Why is the mutual flux in a transformer nearly constant from no-load to rated load? What is the magnitude of the variation of the magnetizing current under these conditions? 7. What relation exists between primary ampere-turns and secondary ampere-turns? What relation exists between primary current and secondary current? 8. Distinguish between primary leakage flux, secondary leakage flux and mutual flux. Which of the foregoing depend upon the voltage and which depend upon the current? 9. What effect have the two leakage fluxes upon the operation of the transformer? 10. In the complete vector diagram of the transformer, why are the primary and the secondary induced voltages shown equal in magnitude and in phase with each other? Why is a voltage equal and opposite to the primary induced voltage necessary in order to find the voltage at the terminals of the primary? 11. Show that the total primary current is not equal and opposite to the secondary current even when both windings have the same number of turns. Resolve the primary current into two components explaining why one of these components varies with the load on the transformer secondary. 12. What is meant by transformer regulation and what assumptions are usually made in obtaining it. 13. Explain what approximation is made in obtaining the simplified transformer diagram. What advantages result in so making the diagram? 14. What is the relation ordinarily existing between the primary and secondary resistances in a transformer? What is meant by "equivalent resistance referred to the primary" and how is this quantity used? 16. Discuss the relation usually existing between the primary and the secondary reactance, giving reasons for the exisance of this relation. What is meant by "equivalent reactance referred to the primary." How is this quantity used in determining the transformer characteristics? 16. What relation exists between the equivalent reactance referred to the primary and that referred to the secondary? 17. Show that if one side of a transformer is open and the other side is connected to the line, practically the entire input goes to supply the core loss. How does this core loss vary with the voltage? Why? 18. Why do both the magnetizing current and the core loss increase very rapidly after the rated voltage of the transformer has been reached? Why is it practically impossible to operate transformers at voltages very much in excess of those for which they are rated? How is the true magnetizing current found? 19. When one side of a transformer is short-circuited and the other side is connected to an alternating supply, show that the input goes almost entirely to supplying the copper losses of the primary and secondary coils. How is the equivalent impedance and the equivalent reactance referred to either side determined from the short-circuit test? 20. What losses exist in a transformer operating under load? How may these losses be computed for different loads? Indicate the method of calculating the efficiency over the working range of the transformer? What are the advantages of this method over direct measurements of output and input? 21. In what way does the core type of transformer differ in construction from the shell type? Compare the dimensions of the electrical and magnetic circuits in the two cases. Which type is better adapted for high voltage and why? How is the leakage flux reduced to a minimum in the two cases? 22. What advantage is claimed for the Type-H transformer of the General Electric Co? What provisions are made for keeping this type of transformer cool? 23. Describe one method of keeping transformers cool. Outside of its cooling properties, what other advantage is obtained by using oil? 24. Describe two other methods used to dissipate the heat from selfcooled transformers when the surface of the case itself becomes inadequate? 26. What are the advantages and disadvantages of air-cooled transformers? How is the circulation of air maintained? 26. Describe the method ordinarily used for artificially cooling oil-filled transformers. What care chould be taken when this method is used? 27. Explain the principle upon which three-phase transformers operate. What are the advantages and the disadvantages of this type of transformer? From the operating standpoint, in what ways do the shell type and the core type of three-phase transformer differ? 28. In what way does the auto-transformer differ from a resistance drop wire? From an ordinary transformer? 29. Under what conditions is it advantageous to use an auto-transformer? Under what conditions should an auto-transformer not be used? How may an ordinary transformer be connected to operate as an autotransformer? 30. Indicate the different connections that can be used for three-phase transformer banks. State the conditions for which each connection is best adapted. 31. What is meant by a "floating neutral" and by what connection is it produced? How may it be eliminated? 32. Under what conditions cannot three-phase transformer banks be operated in parallel, even although the ratios between line voltages are alike for the several banks? 33. Give the reasons why, at no load, the three three-phase voltages existing across delta-delta-connected transformer secondaries are not in any way disturbed by the removal of one of the transformers, if the voltages are balanced? 34. What is the ratio of the kv-a. capacity of the delta-connected bank to the V-connected bank? Under what conditions is the V-connection used? 35. Make a diagram of the T-connection when used for transforming three-phase to three-phase. How does the total three-phase kv-a. capacity of the T-bank compare with the sum of the kv-a. capacities of the individual transformers? 36. Show how the T-connection may be used for obtaining a two-phase, three-wire system. What connection is necessary if the three-wire system is to have equal voltages on both legs? 37. How may a two-phase (or four-phase) four-or five-wire system be obtained from the T-connection. Where is the neutral of the "T"? 38. How does the construction of a constant-current transformer differ from that of a constant-potential transformer? Assuming a change of load, analyze the reactions which cause the transformer to maintain

the current constant. Why is the power-factor of this type of transformer usually low? 39. Sketch the connections of a constant-current transformer as used with a mercury-arc tube to obtain unidirectional current for magnetite arcs. Why is reactance necessary? 40. For what reasons is it necessary to use instrument transformers for measuring power on high-voltage, alternating-current circuits? 41. Describe potential transformers. What is the usual voltage rating of their secondaries? Why should the secondaries always be well-grounded at one point? 42. Describe the construction of a current transformer. What prevents it from giving a ratio of transformation that is exactly proportional to the ratio of secondary to primary turns? Why should the secondary always be kept closed? In what ways does a current transformer differ from a constant-potential transformer? PROBLEMS ON CHAPTER VII 98. A transformer, rated at 2,200/220 volts, has 2,400 turns on the primary or high-side winding. How many secondary turns are there if the secondary no-load voltage must be increased 4 per cent. to allow for the 4 per cent. voltage drop in the transformer when under load? 99. A 2,000/110-volt transformer has 80 turns in the low-side winding How many turns are there in the high-side winding if compensation is made for a 4 per cent, voltage drop in the transformer when load is applied to the low side? 100. What voltage is induced in a transformer winding of 1,200 turns if the frequency is 60 cycles per second and the maximum value of the flux is 1,880,000 lines? (Assume sine wave.) 101. A certain 60-cycle, 13,200/660-volt transformer is to have a primary winding of 4,000 turns and a secondary winding of 200 turns. What must be the cross-section of the core if the maximum flux density in the iron is 00,000 lines per square inch? The ratio of net iron to the.total volume of the core is 0.9. 102. Repeat problem 101 for a 25-cycle transformer operating at 75,000 lines per square inch. Which transformer has the more iron and the more copper? Why? Explain why the iron in a 25-cycle transformer can be

operated at a higher flux density than it can be in a 60-cycle transformer. 103. A 2,200/550-volt transformer is rated at 20 kw. at unity powerfactor. What is the current rating of each winding? What is the current rating of each winding at 0.8 power-factor, lagging current? 104. A certain transformer at no load takes 12 amp. at 220 volts and the power-factor is 0.16. What is the energy component and what is the magnetizing component of this no-load exciting current? If a load is applied to the transformer so that an additional 30 amp., substantially in phase with the line voltage, is taken from the line, what does the total primary current become? What is the power-factor of the primary? If the transformer steps up the voltage one to ten, what is the approximate value of the secondary voltage? 106. A 50-kv-a., 2,200/600-volt transformer has the following constants: High-side resistance, 0.7 ohm; low-side resistance, 0.052 ohm; equivalent reactance referred to the high side, 1.7 ohms. Determine the regulation of the transformer at unity power-factor and at 0.7 power-factor, lagging and leading current. 106. Determine the efficiency of the transformer of problem 105 at %, y±, %, rated, and %-kv-a. load at unity power-factor. Plot a curve with current as abscissas. The no-load core loss is 500 watts. 107. Repeat problem 106 for 0.8 power-factor, lagging current. 108. The following are the data taken from open-and short-circuit tests of a 10-kv-a., 2,080—208-volt, 60-cycle transformer:

High side open, instruments on low-voltage side: $E = 208 / = 2.8$ amp. $P = 124$ watts

Low side short-circuited, instruments on high side: $E = 110 / = 6.3$ $P = 370$ watts

Ohmic Resistance: High side, 4.85 ohms. Low side, 0.0381 ohm.

(a) Determine the regulation of this transformer at unity power-factor and at 0.8 power factor, lagging and leading current. (&) Determine the transformer efficiency under each of the above conditions. (c) What is the ratio of effective to ohmic resistance in this transformer? 109. A 20-kv-a., 1,100/110-volttransformeris connected as shown in Fig. 109*A* so that it acts as a booster to raise the line voltage from 1,100 volts to 1,210 volts. Neglecting the voltage drops in the transformer, determine: (a) the power received by the system; (6) the power delivered by the system; *(c)* the power transformed; *(d)* the power which flows through without transformation; *(e)* if the efficiency of the transformer is 97 per cent., what is the efficiency of the entire system? In the above problem the transformer currents must not exceed the values given by the transformer rating. 110. Repeat problem 109 with the secondary reversed so that the ultimate voltage is 990 volts. Which coil is the primary and which is the secondary under these conditions? 111. Figure 111 A shows a compensator used for reducing the voltage from 150 volts to 120 volts for a 10 kw. non-inductive load. Indicate all the currents and all the voltages existing in the system. How much power is transformed and how much passes through without transformation? Neglect losses. 112. Power generated in a station by a three-phase, 2,500-kv-a. alternator at 6,600 volts, is transformed for transmission to 33,000 volts by a deltaY-connected transformer bank. The voltage is then stepped down by a delta-delta bank to 13,200 volts for distribution. What is the current, voltage and kv-a. rating of each transformer? Give the current and the voltage rating of the primary and secondary of each transformer. Neglect losses. 113. A certain sub-station receives power at 6,900 volts and this power is stepped down to 2,300 volts by two200-kv-a., V-connected transformers. If the power-factor of the secondary load is unity, what is the maximum power which the transformers can deliver without exceeding their ratings? What is the current and the voltage of each winding? 114. Repeat problem 113 for three 200-kv-a. transformers connected delta-delta. Compare the per cent. increase of capacity with the per cent. increase of investment. 116. Power is received at a certain factory at 2,300 volts, three-phase. It is desired to transform it to 230 volts, two-phase, four-wire. If the to-

tal power is 50 kw. at 0.8 power-factor, what should be the ratings of the transformers if the Scott connection is used? Make a sketch showing the method of connecting these transformers. 116. If the transformers of problem 115 were used to obtain a 230-volt, two-phase, three-wire system, indicate the currents and the voltages in each part of the system. Make a sketch. 117. A single-phase line delivers 60 kw. at 1,100 volts. An ammeter, a voltmeter, an indicating wattmeter, and a watthour meter are all necessary in operating this circuit. Sketch the connections of the instrument transformers and the instruments. Give the ratios of the transformers and the factor by which each instrument reading must be multiplied in order to obtain the corresponding value of the current, voltage, power, etc., existing in'the high-voltage circuit. The ammeter and the wattmeter shouH have 5-ampere ratings. Allow for fifty per cent. overload. QUESTIONS ON CHAPTER VIII 1. Describe a simple experiment illustrating the underlying principle of induction-motor action. Show that thetendency of the rotor to follow the inducing magnetic field is another illustration of Lenz's law. Why cannot the rotor attain the speed of the inducing magnetic field? 2. Make a sketch of a two-phase gramme-ring winding and sketch the position of the magnetic field for three or four different values of the currents. Repeat for a three-phase drum winding. What is the relation between the space-advance of the magnetic field and the time-change of the currents? 3. What is meant by revolutions slip? Per cent. slip? Show how the rotor frequency is related to the slip. 4. Upon what three factors does the torque developed by an alternatingcurrent motor depend? Plot a sine of wave current and a sine wave of flux about 45 out of phase and then plot the resulting torque curve. 6. Describe the construction of a squirrel-cage rotor, indicating the various methods of making the end-ring connections. 6. Why does the rotor slip increase with increased load? What is the order of magnitude of the slip in commercial motors? 7. What direct-current motor

characteristics do the squirrel-cage motor characteristics resemble? Why do the power-factor and the efficiency of the induction motor increase rapidly with load? 8. State one very serious objection to the squirrel-cage motor for certain types of service. Analyze carefully the reasons why this type of motor develops but little torque at starting although it takes an unusually large current. Under what conditions is the torque a maximum when the current and flux are fixed in magnitude? 9. Sketch a typical slip-torque curve of an induction motor. What is meant by the breakdown torque? Upon what three factors does it depend? . 10. Name several industrial applications to which the squirrel-cage motor is particularly well adapted. 11. Sketch the special connections used for starting squirrel-cage motors when no starting-compensator is necessary. Why are starting compensators necessary? Make a diagram of two different types. 12. What is the effect upon the slip of an induction motor of introducing resistance into the rotor circuit? Explain. What is the distinct disadvantage of controlling the speed of the motor by inserting resistance into the rotor circuit? 13. Why are wound-rotor induction motors often necessary? Compare their starting characteristics, their operating characteristics and their cost with those of squirrel-cage motors. State a few of the industrial applications of the wound-rotor motor. 14. What is the effect upon the operation of the induction motor; (a) of increasing the length of the air-gap; (6) of using open slots; (c) of using semi-closed stator slots; *(d)* of using semi-closed rotor slots? Discuss the mechanical construction of the motor with special reference to air-gap requirements. 15. What three quantities determine the speed of the induction motor? State briefly the underlying principle of a method of speed control which does away with some of the disadvantages of introducing resistance into the rotor circuit. Make a diagram of connections and discuss the efficiency of this method. Where would such methods of speed control be used? 16. Give an example where speed may be controlled by change of frequency, stating the limits of such speed control. 17. State how the number of poles of an induction motor may be changed in order to give different speeds. What are the limitations of this method? 18. What is meant by concatenation? Discuss this method of speed control, giving a diagram of connections when two similar motors are used. To what direct-current method of speed control does this correspond? 19. Under what conditions will an induction machine develop electrical energy? State: (a) the rotor reactions which cause the reversal of electrical energy in the rotor; (6) the effect of these reactions upon the stator. (c) How is the load of the induction generator controlled.? (d) From where does the machine obtain its exciting current? (e) What determines its frequency and voltage? 20. State the advantages and the disadvantages of the induction generator as compared with the synchronous generator. Why is the machine power-factor determined by the machine itself and not by the load? Illustrate with a vector diagram. To what type of work is the induction generator particularly well adapted? 21. What measurements are necessary in order to obtain data for the construction of the circle diagram? Why is reduced voltage used in the blocked run? How is the diameter of the semi-circle determined? What construction is necessary in order to separate the primary and secondary copper losses? 22. How are the following factors determined for any given value of primary current: (a) Secondary current; (6) power input; (c) core and friction losses; (d) primary copper loss; (e) secondary copper loss; (/) output; (g) efficiency; (h) torque; (i) slip; (f) power-factor. 23. Why is it inaccurate to determine the slip by measuring the rotor and synchronous speeds and then subtracting? Describe the principle of the stroboscope method. How may slip be measured mechanically? 24. What types of common alternating-current machinery does the induction regulator resemble? What windings has the regulator and where are they placed? Why is a tertiary winding necessary in the single-

phase regulator and where is it placed? 26. How is the regulator operated? How is it connected to the circuit? Compare the three-phase regulator with the single-phase regulator. PROBLEMS ON CHAPTER VIII 118. A three-phase, 60-cycle induction motor has 10 poles. Through how many space-degrees will the rotating field advance during one cycle? What is the speed of the rotating field in revolutions per second? In revolutions per minute? 119. Repeat problem 118 for a 25-cycle motor of the same number of poles. 120. It is desired to obtain an induction motor which shall have a speed in the neighborhood of 400 r.p.m. How many poles should this motor have if 60-cycle power is available? 25-cycle power? 121. The rotor of a six-pole, three-phase, induction motor rotates at 1,160 r.p.m. when the motor is connected across 60-cycle mains. What is the slip of the motor in per cent? What is the frequency of the currents in the rotor when it is running at this speed? 122. A certain squirrel-cage induction motor develops a starting torque of 40 lb.-ft., when it is connected to the 40 per cent. taps of a three-phase starting compensator, and the line current is 80 amp. What is the approximate starting torque and the line current when it is connected to the 60 per cent. taps? 123. What is the ratio of the break-down torques of two similar induction motors, one of 25-cycles and the other of 60-cycles if the rotor inductance, the flux and the currents are the same in each? 124. A 10 hp., 230-volt, three-phase, 60-cycle induction motor, when connected directly across 230-volt, three-phase mains, takes 125 amp. the instant that the circuit is closed. What current will the motor take and what will be the line current if a compensator similar to the one in Fig. 238, page 251, is used and the motor is connected to the 50 per cent. taps? 125. Repeat problem 124 for the 60 per cent. taps. What will be the ratio of starting torques in the three cases, of full voltage, 50 per cent. and 60 per cent. taps? 126. A six-pole, 60-cycle, 3-phase wound-rotor, induction motor is taking 10,000 watts from the line. The core loss plus friction losses

is 700 watts and the stator $I2R$ loss is 300 watts. The rotor $I2R$ loss is 400 watts. What is the motor efficiency under these conditions? At what speed does it run and what torque does it develop? (Hint: The ratio of the rotor PR loss to the rotor input is proportional to the slip.) 127. Find the motor efficiency in problem 126 if resistance is introduced in the rotor circuit so that the motor runs at (a) 900 r.p.m. (6) 600 r.p.m. 128. A 10 hp., 230-volt, six-pole, 60-cycle, 1,140 r.p.m., induction motor is tested by means of a prony brake. The data are as follows:

The brake arm is 2 ft. long and its dead weight is 1.85 Ib. The 2-wattmeter method is used. From the above data compute the following: torque; per cent. slip; speed; horsepower output; efficiency; power-factor. Plot the above data with horsepower as abscissas. Why does the efficiency increase to a maximum and then decrease? Why does the power-factor increase with the load? From the two wattmeter readings determine the power-factor, using either equation 34, page 91, or Fig. 90, page 92. Compare these values with those obtained from dividing the total power by the volt-amperes, QUESTIONS ON CHAPTER K 1. What suggests that both the shunt and the series direct-current motors might possibly be used with alternating current? Why is it not possible to use the shunt motor effectively with alternating current? What characteristic of the series motor makes it possible for this type of motor to operate effectively with alternating current?

2. In what way does the field structure of an alternating-current series motor differ from that of the direct-current series motor? How does the number of series turns of the alternating-current motor compare with the number ordinarily used with the direct-current motor of corresponding rating? Why are the poles short and of comparatively large cross-section? Why is the air-gap short? Why is low frequency necessary? 3. Why does the alternating-current series motor have a large number of armature turns? Give two reasons why armature reaction must be compensated. Show

two methods of compensating. 4. What commutating difficulty exists in the alternating-current motor, which is not present in the direct-current motor? How is this difficulty met? Why is there a large number of commutator segments? 5. Sketch a vector diagram of the motor? How is the speed controlled? Where is this type of motor used? 6. What is the nature of the induced electromotive forces in a gramme-ring armature having a commutator, when the armature is placed in a singlephase alternating-current field? What occurs when the brushes are shortcircuited and placed in the geometrical neutral? When these brushes are placed parallel to the pole axis? Why is no torque developed in either case? 7. Why is torque developed when the brush axis makes some angle greater than zero and less than 90 with the pole axis? How is the direction of rotation controlled? How may the field structure be wound so that the brushes may be left in the geometrical neutral? 8. Why are repulsion motors made with uniform air-gaps rather than with salient poles? What is the nature of the speed and torque curves of the repulsion motor? 9. Show that a single-phase alternating-current field can be replaced by two fields rotating around the air-gap in opposite directions. Sketch the slip-torque curve due to each of these two fields. How may the fact that the single-phase induction motor has no starting torque be explained by these curves? How do they explain the fact of the motor accelerating in the direction in which it is started? 10. By means of a sketch show the position of the rotor ampere-turns of an induction motor when the transformer currents alone are considered. Show the direction of the magnetic field which these ampere-turns produce. 11. Show that a speed electromotive force in time-phase with the singlephase flux is produced by the rotation of the armature conductors. What flux is due to the current produced by this speed electromotive force and what is its space position? Why do the combined effects of this field and of the speed field produce a rotating magnetic field? How does this

in part explain the operation of the single-phase induction motor? 12. What is the approximate ratio of weights of single-phase to polyphase induction motors of the same ratings? How may a three-phase induction motor be operated single-phase? What is often the cause of a polyphase motor over-heating when carrying its normal load? 13. Describe the manner in which the initial starting torque of a singlephase motor may be obtained by splitting the phase. What is the order of magnitude of this starting torque? Show how the phase may be split by the use of resistance and inductance; by the use of resistance and capacitance. 14. Discuss the operation of the "shaded-pole" as a method of starting single-phase motors. How is the repulsion motor principle utilized in starting the Wagner Single-Phase Induction Motor? What operation converts the motor from a repulsion motor to an induction motor? 16. Upon what principle does the phase converter operate? What advantage is derived by its use on railway locomotives? Make a diagram of connections showing how the single-phase power received at high voltage from the trolley is converted into low voltage three-phase power for use in the motors of the railway locomotive. 16. Make a diagram of connections showing the relation of field windings to armature, etc., in the General-Electric Repulsion-Induction Motor. How is this motor reversed? 17. What unique feature is embodied in the armature of the Wagner, type BK motor? What excellent operating characteristics are claimed for this motor? QUESTIONS ON CHAPTER X 1. Compare the design of the alternator with that of the synchronous motor. 2. Show that at standstill the average torque of the single-phase synchronous motor is zero and that in order to develop a continuous torque either the moving conductor or the moving pole must cover a distance equal to one pole pitch every half cycle. What is the relation between speed, frequency and number of poles? 3. What reaction occurs in the direct-current shunt motor which enables it to take more current when additional load is applied? Show that the re-

action in the synchronous motor under similar conditions cannot be exactly the same as that of the shunt motor. 4. What is the first reaction which occurs when load is applied to any motor? What resulting reaction follows in the case of the synchronous motor? Show that the current taken by a synchronous motor when the angular position of the rotor is slightly retarded, is mostly *energy* current. 6. What are the reactions which follow an increase of the excitation of a direct-current shunt motor? Why cannot these reactions occur in a synchronous motor? 6. What two reactions permit the synchronous motor to operate when its field current i? increased above the normal value? Show that the induced armature voltage can be numerically greater than the terminal voltage. When the synchronous motor is over-excited what must be the phase relation between its current and its terminal voltage? Illustrate by a vector diagram. 7. What effect is noted when the field of a direct-current shunt motor is weakened? Why cannot these reactions occur when the field of the synchronous motor is weakened? 8. What is the effect of a lagging current upon the field of a synchronous motor? Upon the relation of the induced to terminal voltage? Make a vector diagram for the motor when operating under-excited. 9. Why is a synchronous motor able to operate, even without directcurrent excitation? Whence does it obtain its excitation under these conditions? 10. Sketch a synchronous motor V-curve. Show the point of unity powerfactor, the region of lagging current, and the region of leading current. Sketch another V-curve in which the power is twice that of the original curve. How is the position of this curve determined? What is meant by "normal" excitation and how does this vary with the motor load? 11. Give two reasons for building squirrel-cage or "amortisseur" windings around the poles of a synchronous motor. Analyze the reactions by which an amortisseur winding stabilizes the operation of the synchronous motor. 12. Describe the method of starting a synchronous motor by means of an auxiliary motor. What types of motor are used for this purpose? What are the objections to their use? 13. What is the sequence of operations in starting a synchronous motor by means of its direct-current generator? What objection is there to starting a motor in this way? 14. By what process may the synchronous motor start of itself? Why is a compensator used? Of what order of magnitude is the starting torque? When should the direct-current field be closed? 15. Analyze closely the method by which the synchronous motor, when starting as an induction motor, is able to pull into synchronism even without direct-current excitation. 16. What happens at the time of closing the field switch if the directcurrent excitation opposes the field built up by armature reaction? What should be the position of the starting device when the field switch is closed, and why? 17. How may correct polarity of the field poles be insured so that little or no disturbance results when the field switch is closed? 18. Why is it necessary to insulate the field coils of a synchronous motor for a voltage several times the normal operating voltage? How may the electromotive force induced in the field be reduced? 19. Why are synchronous motors, running without load, often installed at various points of power systems? What is the motor called when operating under these conditions? 20. What is the distinct advantage of using a synchronous motor drive in certain instances? 21. Show by a vector diagram how it is possible to control the voltage at some point on a system by means of a synchronous motor or synchronous condenser. What condition is necessary in order that the voltage at the motor be raised to a value higher than that of the rest of the system? What degree of excitation is necessary in order that the voltage may be raised7 Sketch the connections of a motor, together with the necessary instruments for making tests when the motor is used as a voltage-controlling device. 22. Why are single-phase synchronous motors not common? 23. What are the advantages of the polyphase synchronous motor over the polyphase induction motor? What

are its disadvantages? Under what conditions should it be used? 24. Describe two simple types of synchronous motor which are of very small size. Upon what property of the magnetic circuit do they depend for their operation? From where do they obtain their field excitation? For what purposes are such motors used? PROBLEMS ON CHAPTER X 129. A 100-kv-a., 600-volt, Y-connected, three-phase synchronous motor has the following constants.

Armature resistance per coil = 0.08 ohm.

Synchronous reactance per coil = 1.0 ohm.

Determine the back electromotive force of the motor for a current of 100 amp.

(a) When the power-factor of the motor is unity.

(6) When the current leads the terminal volts by approximately 90.

(c) When the current lags the terminal volts by approximately 90.

(Note: In (6) and (c) the resistance drop may be neglected.)

130. The motor of problem 129 requires a field current of 30 amp. at 110 volts and the core and friction losses are 3,000 watts when the motor is operating at unity power-factor at its rated load. What is its output in hp. and what is its efficiency at this load? 131. Figure 131 (A) shows four V-curves for a 200-kv-a. , 600-volt, synchronous motor. (a) Indicate the kilowatt input at which each curve was obtained. (6) Draw lines through points of unity power-factor and of 0.75 powerfactor, leading and lagging current. 132. The synchronous motor of problem 131 is connected in parallel with a load of 150 kilowatts at 600 volts and 0.6 power-factor, lagging current. The motor runs light without load. To what value must its excitation be adjusted in order to bring the system power-factor up to unity? (Use lowest curve, Fig. 131A.) 133. Repeat problem 132 except that the power-factor is brought up to 0.9. How many kv-a. must the motor take in each case? 134. What is the maximum power-factor which the system can have in problem 132 if the motor carries a load of 150

kw.? It must not be overloaded. What is the value of the field current? 136. A 500-kw., 2,300-volt, three-phase load has an average power-factor of 0.5, lagging current. How many leading quadrature kv-a. are necessary to raise its power-factor to 0.6? 0.8? 0.9? 1.0? Plot a curve with powerfactor as abscissas and kv-a. as ordinates. 136. What size synchronous motor would be necessary to raise the powerfactor of problem 135 to unity and at the same time take 500 kw. from the line in order to carry its mechanical load? 137. Repeat problem 136 except that the power-factor is raised only to 0.9. 210 10 20 30 40 60 60 70 80 90 100 Field Current

Fio. 131A.

1S8. A 48-pole, 60-cycle, synchronous motor shows a tendency to hunt. How many space-degrees either side of its true position may its rotor swing if its back electromotive force may have a phase displacement of not more than 10 time-degrees? QUESTIONS ON CHAPTER XI 1. Give several uses of electrical power in which it is impossible to employ alternating current. 2. Make a sketch showing the method of operation of the rectifying commutator. What are the disadvantages of this type of rectifier? For how large capacities is it possible to operate this type of rectifier? 3. Upon what principle does the mercury-arc rectifier operate? Why are two anodes usually employed? Why is it necessary to have reactance in circuit? Sketch the diagram of connections which would be used for charging a low-voltage storage battery and trace the current flow, explaining carefully the operation of the auto-transformer. Why is a starting anode desirable? 4. What is the underlying principle of the tungar rectifier? Why are the electrons repelled from the incandescent filament during one half cycle and attracted during the other half-cycle? What are the approximate efficiency and maximum capacity of this type of device? 6. What property has aluminum, when immersed in certain salt solutions, which makes it possible to utilize it in the rectification of alternating currents? Sketch a wiring diagram of such a rectifier which rectifies every half-cycle. What are the disadvantages and the limits of capacity for this type of rectifier? 6. What common methods may be used to convert alternating to direct current on a large scale? Name the disadvantages of each type of apparatus. 7. Name the machines whose principles are embodied in the synchronous converter. Just how is the converter armature connected? How is power supplied to the ordinary converter armature? What power is taken from the armature? Name the different types of familiar machines for which the converter may be used. 8. Under what operating conditions is the synchronous converter called "direct"? "Inverted"? 9. Indicate the points at which the slip-ring taps connect to the winding in a four-phase, two-pole converter. Four-phase, four-pole converter. 10. Repeat (9) for a three-phase, two-pole converter and a three-phase, four-pole converter. How many taps will an eight-pole, six-phase converter have? What special restriction, not necessary with the ordinary direct current winding, is imposed on the converter winding? Why? 11. Compare the number of active conductors between brushes with the number between slip-ring taps in the single-phase converter. How is the resulting voltage between direct-current brushes obtained? Between slip-ring taps? What is the relation between the two? 12. Show by a circle and an inscribed polygon how the individual inductor voltages of a converter add. Indicate how; (a) the single-phase voltage is obtained; (6) the three-phase voltage; (c) the four-phase voltage; *(d)* the six-phase voltage. 13. Knowing the voltage relations in a converter armature, derive the ratio of the direct current to the alternating current per terminal in (a) the single-phase converter, (6) the three-phase converter, (c) the four-phase converter; *(d)* the six-phase converter. Show the effect of efficiency and power-factor on these ratios. 14. Sketch the variation of the direct current in a single conductor midway between slip-ring taps, as it takes successive positions in its rotation. Sketch the alternating current in this same conductor for corresponding positions

when the current is in phase with the induced emf. Find the resultant current. 16. Repeat (14) for a conductor at one of the slip-ring taps. 16. Repeat (15) for a power-factor considerably less than unity. 17. What is the effect on the resultant current curve of increasing the number of phases? 18. Why does increasing the number of phases materially increase the rating of a converter? Why does the efficiency of a converter decrease more rapidly with a decrease in power-factor than it does in most other types of apparatus? 19. Compare commutation in a converter when operating as such and when operating as a direct-current generator carrying the same load. Why does the very materially increased armature current resulting from low power-factors have little distorting effect on the main fields? What is its effect on commutation? 20. Why are commutating poles desirable in synchronous converters even although the main field is not distorted to any considerable extent by armature reaction? 21. Why are the voltage ratios in a converter almost constant under operating conditions? Why is it possible to modify the ratio of the direct to the alternating voltage a small amount by changing the excitation? 22. Explain how a series reactance may be used to control the directcurrent voltage. When may a separate reactance be omitted? State the disadvantages of this method of voltage control. 23. Explain tho use of the induction regulator as a means of controlling the direct-current voltage. What is the objection to the use of the regulator? 24. Explain the operation of the series booster. What are its advantages and its disadvantages? 25. Why is it impracticable to control the direct-current voltage by changing to different transformer taps when the converter is in operation? 26. Explain the underlying principle of the split-pole method of voltage control. 27. Sketch a diagram of connections, including all instruments, which would be used in determining the various characteristics of the converter. What characteristics is it instructive to determine? How should they be plotted? 28. Why are transformers almost always necessary with syn-

chronous converters? Sketch the connections of the double-Y, six-phase secondary connection, showing the primaries in either Y or delta. Indicate the voltage at each point, assuming 220 volts between the three-phase lines on the primary side. What is the advantage of this system? 29. Repeat (28) for the double-delta connection of secondaries. 30. How does the rating of a synchronous converter, when operating inverted, compare with its rating when operating direct? Why? How do the speed relations in the two cases compare? Show by careful analysis the sequence of reactions which may cause an inverted converter to race. What means are used to prevent racing? 31. By what reactions does a synchronous converter armature start rotating when polyphase currents are supplied to its slip-rings? What is a sectionalizing switch, and what should be its position when starting the converter from the alternating-current side? Why is it necessary to open the series-field shunt, etc.? When starting, why does sparking take place under the brushes even with no direct-current load? Why are brush-lifting devices necessary? 32. How does the armature pull into synchronism? What effects occur if the shunt-field current opposes the field built up by armature reaction? How may the continual "slipping of a pole" be stopped? 33. Give two methods by which the direct-current polarity may be reversed, if necessary. 34. Describe how the speed of rotation in space of the field produced by the armature currents becomes less and less, during starting, as the armature speed approaches synchronism. How does this affect commutation? Describe the behavior of a direct-current voltmeter connected across the brushes during the starting period. When should the field switch be closed? 35. How may the armature be induced to build up the field poles to the right polarity and so insure the correct direct-current polarity at the brushes? 36. Describe briefly the procedure of starting a synchronous converter by means of an auxiliary machine. 37. Give the connections of both the shunt-and the series-field circuits of a synchronous converter when it is started from the direct-current side. Why should the switch between the transformer secondaries and the slip-rings be opened during the starting period? What difficulty is encountered in synchronizing? 38. Discuss the operation of synchronous converters in parallel. How many equalizers may be required? How are the loads between machines adjusted? Why is it preferable that each converter have its own transformer bank? 39. Why may synchronous converters operating in parallel show a tendency to run away under some circumstances? Describe methods which are used to prevent synchronous converters from thus running away. 40. What is the principle by which a neutral is obtained in the three-wire generator? Why is it undesirable to use three single secondaries connected in Y when obtaining a neutral? How may a Y-connection be used and at the same time direct-current magnetization of the core be prevented? 41. Sketch the complete connections of a three-wire, six-phase synchronous converter having two series fields, where the transformer secondaries are connected 6-phase star. PROBLEMS ON CHAPTER XI 139. It is desired to secure a 200-kw. synchronous converter and its transformers for changing three-phase, 6,600-volt power into 115 volts direct current. The converter has three slip-rings and the transformers are connected, primaries in delta and secondaries in Y. The converter has an efficiency of 90 per cent. and the transformer bank an efficiency of 97.8 per cent. When the converter operates at 95 per cent. power-factor determine: (a) The direct-current rating of the converter. (6) The alternating current per slip-ring. (c) The rating in kilovolt-amperes, amperes, and volts of the transformer secondaries. (d) The power input, the current, and the voltage of each transformer primary when the converter is delivering rated output. 140. A converter similar to that of problem 139 has six slip-rings and is operated six-phase. This increases its efficiency to 92 per cent. The transformers are connected, the primaries in delta and the secondaries six-

phase diametrical. When the converter is delivering 200 kw. and operating at 95 per cent. power-factor (see Fig. 140A) determine: (a) The voltages £i-2, Ex-,, etc. (6) The voltages -3, Ef-b, Es-. (c) The diametrical voltages Ei-t, Er-t,, etc. (d) The current in each transformer secondary.

What is the probable rating of the converter under these new conditions, and have the transformers now the proper rating?

Fig. 140A.

141. If three other transformers having the same kilovolt-ampere rating as those in problem 140 be connected primaries in Y and secondaries in double-Y, with connected neutrals, determine: (a) The rating of each primary, in kilovolt-amperes, volts, and amperes, assuming 200 kw. output of the converter. (6) All possible voltages obtainable from the transformer secondaries. 142. A 500-kw. synchronous converter is to be installed for supplying 230-volt, three-wire, direct-current service. The alternating-current supply is 13,800 volts, 60-cycles.

Make a complete diagram of connections such as would be necessary for obtaining the required direct-current service. Indicate the currents and voltages at each point. Obtain the efficiencies of the various parts of the system from data already given.

QUESTIONS ON CHAPTER XII 1. Why is alternating current particularly well adapted for transmitting power over considerable distances? What difficulties are encountered when direct current is similarly used? 2. State the advantages of polyphase transmission. Which of the polyphase systems is most commonly used and why? Under what conditions is single-phase occasionally used? 3. Why are 6,600-volt generators commonly used when the transmission voltage is high? What rough basis can be used for determining the transmission voltage? What economic considerations are involved in determining this voltage? 4. Give the principal links in a power system which distributes power to large and small consumers located at a considerable distance from the point

of generation of power. State the considerations which govern the selection of each of these links. 6. Why are the voltages ordinarily selected for power and for lighting purposes usually different? Why should the secondaries of lighting transformers be grounded? 6. Name the various types of apparatus which may be installed in a substation, giving the type of service which each supplies. 7. Make a sketch of the magnetic field existing between the two parallel conductors of a single-phase transmission line. What effect does this field have on the operation of the line? 8. On what two factors does the inductance of such a line depend? Distinguish between the inductance of the circuit loop and that of a single wire. 9. Sketch the magnetic field which may exist at some particular instant in the region between the three conductors of a three-phase transmission line, these conductors being symmetrically spaced. What is the general nature of the field existing in this region and what is its effect on the operation of the transmission system? 10. On what three factors does the reactance per conductor of a threephase system depend? 11. Sketch the electrostatic field which exists between the two conductors of a single-phase transmission system. On what two factors does the capacitance existing between two such wires depend? 12. Show that a thin fictitious plane may be inserted midway between two parallel wires and perpendicular to their plane without disturbing the electrostatic field between these conductors. With this as a basis, replace the capacitance between conductors by two series-connected condensers. What is the ratio of the capacitance of each of these condensers to the capacitance between the line conductors? 13. Replace the actual capacitance which exists between symmetricallyspaced three-phase lines by two different arrangements of condensers. Which of these two arrangements is ordinarily considered and why? 14. What close approximation as to wire spacing may be used when transmission conductors are not located at the corners of an equilateral triangle? 16. State some of the advantages of splitting a single-phase transmission line along a fictitious neutral and using the quantities to neutral when working out the line characteristics. 16. Why can the ground be considered as having no resistance and no inductance although such is actually not the case? 17. Given the line resistance and reactance, the load voltage, current and power-factor, show by means of a vector diagram the method of obtaining the voltage at the sending end of the line. 18. Show that a three-phase line may be split into three single lines, any one of which may be used for purposes of calculation. Why can the ground be considered as having zero resistance and zero reactance under these conditions? 19. How is the capacitance of a line actually distributed? For purposes of calculation how may this total capacitance be distributed? What effect does the current taken by each condenser have on the line behavior? The generator load? 20. What is the general nature of "corona"? Upon what factors does its appearance depend? Upon what parts of a conductor does it first appear? How may corona loss on transmission lines be minimized? 21. What factors may cause abnormal voltage rise in a power system? What is the purpose of a lightning arrester? What four properties should lighting arresters possess? 22. Describe the multigap arrester, discussing its operation. What is the weak point in this type of arrester? 23. On what principle does the horngap operate? Why is it necessary to use resistance or reactance in series with such a gap? What are the disadvantages of the horngap arrester? 24. What is the underlying principle of the aluminum cell arrester? Why can an alternating current flow into such an arrester when a direct current cannot, even though the film is intact in both cases? 25. Describe the characteristics which make such a device an excellent lightning arrester. What is the general arrangement of the individual cells in the actual arrester? Why is oil used? 26. Why is there a short horngap in series with each arrester? Why is it necessary to "charge" such arresters at intervals? Sketch the connections of an arrester designed to protect a three-phase ungrounded system. 27. Sketch the connections showing the position of the arrester, the choke coil with relation to the incoming (or outgoing) line, and of the apparatus which it is designed to protect. 28. State the advantages of pin-type insulators for low and moderate voltages. What are their limitations at the higher voltages? What materials are used for these insulators, and what are their relative advantages and disadvantages? Why are the larger units made up in sections? 29. In what manner does the suspension-type of insulator support the line conductors? What are the advantages of this type of insulator over the pin-type? 30. Under what conditions are wooden poles employed as line supports? Steel poles? Steel towers? Compare steel towers and steel poles. 31. What is meant by "flexible tower" construction? Under what conditions are flexible towers used and what are then-advantages? 32. What is the function of the sub-station? Sketch roughly the connections of a transformer sub-station. 33. By what types of apparatus is direct current obtained from alternating-current supply? Compare the advantages and disadvantages of these different types. 34. Why is it difficult to break a high-voltage power arc in air? Discuss briefly the construction of an oil switch for high voltages. Why is it necessary that the case be designed to withstand high pressures? 36. What special care must be used in carrying high-voltage lines into stations? What care should be taken in the location of the various types of apparatus? 36. What are the economic necessities which have developed the outdoor substation? In what way does the apparatus for such a station differ from that of an indoor station? PROBLEMS ON CHAPTER XII 143. Calculate the inductance per mile of a two-conductor distributing line of 2/0 solid conductors spaced 2 ft. apart. What is the inductance if the distance between the wires is 4 ft.? 144. What is the reactance per conductor at 60 cycles of 4 miles of the lines of problem 143? What is the maximum allowable current if it is necessary that the reactive voltage drop shall not exceed 250 volts in the two wires? What is the total *impedance* drop per wire? 146. A three-phase distribution line consists of three 4/0 solid conductors symmetrically spaced, the distance between conductors being 30 in. (a) What is the reactance per wire per mile of this line when the frequency is 25 cycles per second? 60-cycles per second? (6) What is the impedance under these conditions? 146. What is the capacitance to neutral per mile of the line in problem 143? What is the charging current when there is 6,900 volts between conductors and the frequency is 60 cycles per second? 147. A 30-mile transmission line consists of two 4/0 conductors spaced 3 ft. apart. (a) What is the capacitance to neutral of this system? (6) What is the 60-cycle charging current if the voltage is 26,400 volts between conductors? 148. Compute the capacitance to neutral per conductor and the charging current per conductor in problem 147, if the system is changed to a threephase system by the addition of a third conductor similar to the other two and equally spaced from them. The voltage between wires is the same in both cases. 149. The voltage at the receiving end of the line with the 2-foot spacing in problem 144 is 2,300 volts, when the load is 100 kw. and the load powerfactor is unity, (a) What is the voltage at the generating end? (6) What is the line regulation? (c) What is the power-factor at the generating end? Neglect the charging current. 160. Repeat problem 149 for a power-factor of 0.8, lagging current. 161. Repeat problem 149 for a power-factor of 0.8, leading current. 162. It is desired to transmit 2,000 kw., single-phase, a distance of 15 miles with 10 per cent. line loss. The load voltage is 15,000 volts, the load powerfactor is 0.7, lagging current, the frequency is 25 cycles per second, and the conductors are spaced 30 in. on centers. Find '(a) the voltage drop per wire; (6) the generator voltage; (c) the line regulation; and (d) the generator powerfactor. 153. Given transmission distance 20 miles; load, 5,000 kw.; load voltage, 33,000 volts between conductors; system, three-phase; frequency, 60 cycles per second; load power-factor of 0.85, lagging current; spacing of conductors, 48 in.; permissible line loss, 10 per cent. of receiver power. Find (a) line regulation; (6) generator power-factor. Neglect the charging current. 154. Repeat problem 153 for a power-factor of 0. 85 leading current. 155. A hydroelectric station transmits 50,000 kw., three-phase, 60 cycles, a distance of 120 miles over a line consisting of three 0000 copper conductors spaced 13 ft. apart. The voltage between line wires at the substation is 140,000 volts; the power-factor of the load is 0.9, lagging current. Find *(a)* line regulation; (6) efficiency of transmission. The line charging current should be taken into consideration. 166. Repeat problem 155 for a power-factor of 0.9, leading current. QUESTIONS ON CHAPTER XIII 1. How may light be described? What is illumination? Photometry? 2. What is luminous intensity? In what units is it measured? What is the objection to the use of the candle as a photometric standard? What photometric standards are used at the present time? 3. Define a unit solid angle or "steradian." What are its geometrical properties? How many solid angles exist about a point? 4. In what way may light flux be considered? In what way is it comparable to magnetic flux? What is a lumen? Why is a given cone of light flux confined to the solid angle? How many lumens would a standard candle emit if its intensity in all directions were the same as its horizontal intensity? 6. Why were carbon lamps rated in mean horizontal candlepower? What is the objection to this method of rating? Why is it highly desirable at the present time to rate lamps in lumens? What is the relation between lumens and mean spherical candlepower? 6. What is illumination? What is the unit? To what does it correspond in magnetism? 7. What is the law of in verse squares? How is this law proved geometrically? What assumption may introduce error into the application of · this law? What is the magnitude of this error and how may it be made negligible? 8. What is meant by "absorption?" What types of surfaces reflect the largest proportion of incident light? What types reflect the least?

What determines the color of a surface? What is meant by the "coefficient of absorption"? 9. Why does the intensity of light emanating from light sources usually vary in different directions? How does light intensity in general vary in horizontal planes or zones? In what way may this distribution be represented? In what regions is the light from an incandescent lamp of the greatest intensity? Of the least intensity? Of what commercial use are these distribution curves? 10. What is meant by incandescence? How does the light emitted by an incandescent substance vary with its temperature? Does this bear any relation to the efficiencies of various illuminants? 11. In what way does luminescence differ from incandescence? Give examples of luminescent sources of light. 12. Name two essential requirements for an incandescent lamp filament. Why was a carbon filament practically the only satisfactory filament for a number of years? How does a "G. E. M." lamp filament differ from the ordinary carbon filament? What are its advantages? What are the efficiencies of these two filaments? 13. What are the objections to a carbon filament? Is it possible to increase the efficiency of a carbon filament lamp? What must be sacrificed in order to do this? What is the approximate life of a carbon filament lamp? When does it become more economical to throw away a lamp rather than continue its use? 14. Why does a tantalum lamp develop a higher efficiency than a carbon lamp? About what efficiency does the tantalum lamp develop? What is the average life of such a lamp? What is its peculiarity when used with alternating current? 16. Why is tungsten particularly well-adapted to being used for lamp filaments? What is the difference between a drawn and a pressed filament? Compare the two. What are the approximate efficiencies of "Mazda" lamps? What is the guaranteed life of such a lamp? Why does the tungsten lamp take an excessive current when it is first switched on? 16. What factor limits the operating temperature of the vacuum tungsten lamp? What two effects are produced? How may one of

these factors be minimized? 17. What is the basic principle of the gas-filled lamp? Why is an inert gas necessary? Why does this type of lamp have a long narrow neck? Why does the filament differ in shape from that of the vacuum lamp? Why are the efficiencies higher in the larger units? How may daylight be approximated with these lamps? 18. What is the basic principle of the arc light? Why is the arc a highefficiency illuminant? What is the reason that the arc itself cannot be connected directly across the line? What is the "ballast" and approximately what percentage of the power is lost in the ballast? 19. What is the principal source of light in the direct-current are? What determines the relative positions of the two electrodes? Which is consumed the faster? 20. Compare the alternating-current arc with the direct-current arc. Why is a reflector very desirable in the alternating-current lamp? What advantage has the alternating-current multiple arc over the direct-current multiple arc? 21. What is the effect of enclosing the arc upon its efficiency and upon the cost of operation? What are the advantages of the enclosed arc over the open arc? Where are open direct-current arcs now commonly used? 22. Distinguish between a series arc and a multiple arc. Upon what does the series arc depend for its regulation? What is the object of the series coil? The cut-out switch? Describe the operation of the lamp on starting. 23. Upon what factor does the regulation of a multiple arc depend? Why are the carbons touching when the lamp is not in operation? Why are dashpots used in lamp regulating mechanisms? 24. In what way does the flame arc differ from the ordinary direct-current arc as regards the light source? How may the color of the light be controlled? Why are the electrodes consumed very rapidly in this type of lamp? 26. Give one reason why both electrodes feed from above in the open flame arc. How is the arc flame kept burning in the proper position? Why is this type of lamp used abroad more than in this country? How does the efficiency of this lamp compare with that of other illuminants? 26.

How has the electrode life of the flame arc been increased? How does the efficiency of this type of lamp compare with that of the previous type? Name one good feature of the flame arc. 27. Of what materials do the positive and negative electrodes in the magnetite arc consist? Why must the copper always be the anode?

In general how does the regulation of this type of lamp differ from that of other arcs? Why is it different? What is the number of hours that the lamp burns per trim? What is the efficiency of the 4.4-amp. arc? Of the 6.6-amp. arc?

28. What is the principle of the mercury arc? To what is the light due? What is the color effect produced by this type of illuminant? Is it injurious? What is its physiological effect? How may the illumination be made more pleasing when these lights are used? What is the efficiency of this type of lamp? 29. What is the source of light in the Moore tube? What are its advantages? Why are these tubes not in more common use? 30. What is the principle of the Nernst lamp? What can be said of its light and its efficiency? Why is it not in common use? 31. Why are there excellent opportunities for itill further increasing the efficiencies of illuminants? 32. What is photometry? In what way are the measurements generally made? Name a very marked source of error in photometric measurements. 33. Upon what principle is the Bunsen photometer screen based? How is the position of balance determined? How is the candlepower calculated after a balance has been obtained? 34. Why is a candle itself seldom used at the present time as a working photometric standard? What are used as standards? Sketch the connections which facilitate voltage adjustment when a photometric measurement is being made. 35. How is the position of balance determined when a Lummer-Brodhun photometer screen is used? How may errors due to differences in the white screens be eliminated? 36. Where is the use of a portable photometer often necessary? In what way do portable photometers resemble the laboratory type? How do they differ? 37.

What is the standard in the Sharp-Millar photometer? In what way is the standard accurately adjusted? How is brightness or illumination measured? How is the candlepower of a lamp measured? How is a balance obtained and read? How may the range of the instrument be greatly increased? 38. How is the mean horizontal candlepower of a lamp determined with one photometer setting? How may the candlepower in other zones be measured in a similar manner? How does the candlepower in a vertical direction of an incandescent lamp compare with that in the horizontal direction? 39. Sketch the Rousseau diagram, showing how the mean spherical candlepower may be determined from the polar distribution diagram. What is meant by "spherical reduction factor?" Of what use is a knowledge of its value? 40. Why are reflectors useful even although they absorb a considerable amount of light? When may it be desirable to throw light upward? Downward? How is this accomplished? 41. What are the illuminants most generally used in interior illumination? What important factors must be considered when illuminating an interior? 42. What illumination is required for reading, writing, etc.? When should such illumination be furnished entirely by overhead fixtures? By individual desk lamps? 43. When a room is to be illuminated by a single unit how should the reflector differ from those required when a number of units are used? 44. What is indirect lighting? What is its chief advantage? What are its disadvantages? How are these disadvantages overcome by the use of semi-indirect fixtures? 46. Where are individual lamps required in factory illumination? How do overhead belts and traveling cranes affect the placing of the lighting units? 46. In what way does street illumination differ from interior illumination? How do the lighting intensities compare in the two cases? What can be said of reflected light in comparing the two? 47. Why may uniform street illumination be objectionable? Where is street lighting of uniform intensity used? Why is its use considered good engineering under

these conditions? 48. How does the road surface assist in illumination? What is meant by "specular reflection?" How does automobile traffic affect street illumination problems? How may the improper placing of lights on curves, etc. be the cause of accidents? 49. What is meant by "flood lighting" and where is this type of illumination used? PROBLEMS ON CHAPTER XIII 167. A sphere has a radius of 2 ft. (a) How many steradians or unit solid angles will an area of 4 sq. ft. on its surface subtend? (6) 5 sq. ft.? 168. Repeat problem 157 for a sphere whose radius is 3 ft. 169. A light source has a mean spherical intensity of 40 candles. How many lumens does it emit? 160. A light source has a mean spherical intensity of 40 candles, (a) How many lumens will fall on the inside surface of a sphere of 2 ft. radius having the light source at its center, assuming that the light radiates with equal intensity in all directions. (6) How many lumens will fall on an are of 4 sq. ft. on this surface? 161. Determine the illumination in foot-candles in (a) and (6), problem 160. 162. A certain lamp is equipped with a reflector which sends all the light into the lower hemisphere. This light has a mean lower hemispherical candlepower of 200. How much light flux (in lumens) does this lamp emit? 163. The candlepower of the lamp in problem 162 at a 45 angle is 250. What is the illumination in foot-candles on a surface 10 ft. from this lamp when the surface is perpendicular to the direction of the light? 164. A certain lamp placed 25 ft. above the street has an intensity of 400 candlepower in a vertically downward direction. What is the illumination on the street directly below the lamp? 166. What does the illumination in problem 164 become if the lamp is raised 5 ft.? 166. A nitrogen-filled lamp has a rating of llOm.s.cp. If it takes 0.9 amp. at 110 volts, what is its efficiency in watts per mean spherical candlepower? In lumens per watt? 167. A tungsten vacuum lamp takes 0.52 amp. at 115 volts and is rated at 50 m.h.cp. What is the efficiency in watts per mean horizontal candlepower? In lumens per watt if the spherical reduction factor is 0.75?

168. One-hundred watt tungsten vacuum lamps are furnished by a lighting company at 20 cts. each. Seventy-five watt gas-filled lamps are supplied at 50 cts. each. Assuming that the candlepower of each is practically the same and that energy costs 8 cts. per kw.-hr., how many hours must the lamps burn before the total cost of the two is the same? 169. Assuming that each lamp of problem 168 has a life of 1,000 hr., at what cost of energy per kilowatt hour will the total cost of the lamps be equal? 170. In a series direct-current open-arc lamp the arc itself takes 450 watts and the current is 9.6 amp. The mean spherical candlepower is 410 and the maximum candlepower in an obliquely downwards direction (see Fig. 383, page 426) is 1,200 cp. (a) What is the efficiency in watts per mean spherical candlepower? (6) In lumens per watt? 171. Repeat problem 170, giving (a) and (6) when the arc lamp is supplied from 115-volt multiple mains. 172. A multiple, direct-current, 115-volt, enclosed arc takes 550 watts at 115 volts and has a mean spherical candlepower of 165. (a) What is the efficiency in watts per mean spherical candlepower? (6) In lumens per watt? 173. A series, direct-current, 6.6-amp. , enclosed arc lamp requires 495 watts and has a mean spherical candlepower of 270. (a) What is its rating in watts per mean spherical candlepower? (6) In lumens per watt? Compare this efficiency with that of the arc lamp of problem 172, which is similar. Account for the difference. 174. A series, alternating-current, 6.6-amp., enclosed-carbon, arc lamp takes 430 watts at a power-factor of 0.85, and gives a mean spherical candlepower of 135. The arc itself takes 420 watts, and the arc current and the arc voltage are substantially in phase with each other, (a) What is the efficiency of the entire lamp in watts per mean spherical candlepower? (6) In lumens per watt? (c) What is the voltage across the arc? (d) Draw to scale a vector diagram. (e) Why is the power-factor of this lamp less than unity? 175. A multiple alternating-current arc which takes 0.54 amp. at 115 volts and a power-factor of 0.70, has a mean spherical candlepower of 145. (a) What is the efficiency of this lamp in watts per mean spherical candlepower? (6) In lumens per watt? (c) The voltage across the arc is 70 volts and is substantially in phase with the current. Draw a vector diagram of this arc. (d) How much power is consumed by the auxiliary devices? (e) Compare this power with similar power consumed in the direct-current, multiple arc lamp. (See problems 170, 171 and 172.) 176. A series magnetite lamp takes 6. 6 amp. at 75 volts and has a rating of 550 m.s.cp. What is the efficiency in watts per mean spherical candlepower? Lumens per watt? 177. Assume that a lamp similar to that of problem 176 is arranged to go in multiple across 110-volt direct-current mains, and that the arc itself takes the same volts and the same watts and has the same candlepower. What is the over-all light efficiency of this lamp in lumens per watt? 178. A lamp trimmer receiving $3.70 per day can trim 75-500-watt magnetite lamps in that time. Allowing 1.1 cts. per kilowatt hour for energy, 120 burning hours per trim, 10 cts. per electrode and $51.00 per year maintenance, compute the approximate operating cost per year of such a lamp assuming that a lamp is in service 4,000 hr. during that time. 179. In a photometric measurement the standard lamp has a candlepower of 15 and the photometric balance is obtained when the photometer screen is 40 in. from the standard lamp and 80 in. from the test lamp. What is the candlepower of the test lamp? 180. Compute the candlepower of another test lamp in problem 179 when the screen is 20 in. from the standard lamp. The distance between the standard and test lamps is the same in each case. 181. The test lamp in problem 179 has a spherical reduction factor of 0.82. What is its mean spherical candlepower and what is its output in lumens? 182. Find the mean spherical candlepower and the output in lumens of the lamp in problem 180 if it has a spherical reduction factor of 0. 82. 183. If the lamp of problem 182 is equipped with a reflector which absorbs 25 per cent. of the light emitted by the source, determine the mean spherical candlepower and the lamp output in lumens. 184. A lamp is tested for candlepower and the polar and Rousseau diagrams are plotted. The area of the Rousseau diagram is 14.2 sq. in. and the base is 6 in. The scale on the polar diagram is 20 candlepower = 1 in. What is the mean spherical candlepower and the output of the lamp in lumens? 186. If the 90 radius problem 184 (see Fig. 396, page 440) is 3 in. and represents the mean horizontal candlepower of the lamp, what is the spherical reduction factor of the lamp?

CPSIA information can be obtained at www.ICGtesting.com
Printed in the USA
BVOW07s1415080414

350076BV00011B/781/P